Green Chemistry and Sustainable Technology

Series Editors

Liang-Nian He
State Key Lab of Elemento-Organic Chemistry, Nankai University, Tianjin, China

Pietro Tundo
Department of Environmental Sciences, Informatics and Statistics, Ca' Foscari
University of Venice, Venice, Italy

Z. Conrad Zhang
Dalian Institute of Chemical Physics, Chinese Academy of Sciences, Dalian, China

Aims and Scope

The series Green Chemistry and Sustainable Technology aims to present cutting-edge research and important advances in green chemistry, green chemical engineering and sustainable industrial technology. The scope of coverage includes (but is not limited to):

– Environmentally benign chemical synthesis and processes (green catalysis, green solvents and reagents, atom-economy synthetic methods etc.)
– Green chemicals and energy produced from renewable resources (biomass, carbon dioxide etc.)
– Novel materials and technologies for energy production and storage (bio-fuels and bioenergies, hydrogen, fuel cells, solar cells, lithium-ion batteries etc.)
– Green chemical engineering processes (process integration, materials diversity, energy saving, waste minimization, efficient separation processes etc.)
– Green technologies for environmental sustainability (carbon dioxide capture, waste and harmful chemicals treatment, pollution prevention, environmental redemption etc.)

The series Green Chemistry and Sustainable Technology is intended to provide an accessible reference resource for postgraduate students, academic researchers and industrial professionals who are interested in green chemistry and technologies for sustainable development.

More information about this series at http://www.springer.com/series/11661

Mark B. Shiflett
Editor

Commercial Applications of Ionic Liquids

 Springer

Editor
Mark B. Shiflett
Department of Chemical and Petroleum
Engineering
University of Kansas
Lawrence, KS, USA

ISSN 2196-6982   ISSN 2196-6990   (electronic)
Green Chemistry and Sustainable Technology
ISBN 978-3-030-35247-9   ISBN 978-3-030-35245-5   (eBook)
https://doi.org/10.1007/978-3-030-35245-5

This Springer imprint is published by the registered company Springer Nature Switzerland AG
The registered company address is: Gewerbestrasse 11, 6330 Cham, Switzerland

Dedication to Prof. Kenneth "Ken" Richard Seddon

31 August 1950–21 January 2018

By Natalia V. Plechkova, The QUILL Research Centre, School of Chemistry, The Queen's University of Belfast, Belfast, Northern Ireland, UK, BT9 5AG

A man with encyclopaedia in his head

Ah, but a man's reach should exceed his grasp, Or what's a heaven for?

Robert Browning (1812–1889) Andrea del Sarto

Sometimes he spent hours together in the great libraries of Paris, those catacombs of departed authors, rummaging among their hoards of dusty and obsolete works in quest of food for his unhealthy appetite. He was, in a manner, a literary ghoul, feeding in the charnel-house of decayed literature.

Washington Irving (1783–1859) The Adventure of the German Student

Professor Kenneth Richard Seddon (Ken as he insisted to be called) was born on the 31st of August 1950 into a working-class family. Ken was the only child of Muriel (née Pope) and Richard Seddon, who was in the merchant navy before joining the engineering company Otis.

As Ken's cousin Christine recalls, "Ken was not the most energetic of children— I am reliably informed that he didn't walk until he was two—and he hated the walks our families would go on, by the seaside or in the country lanes opposite his family home. He would continually moan about walking, something he never changed his views on—he later told me that if people were meant to walk, taxis would not have been invented. Exercise of any sort was anathema to him, and at school he would take a book out onto the football or cricket pitch and return to the changing rooms pristine—something of a family trait, I must admit."

Ken attended Hillfoot Hey grammar school and loved chemistry from a very young age, and he said that it was due to his brilliant chemistry teacher.

Ken's mother Muriel recognised Ken's scientific capability at a young age, and she made sure that Ken could concentrate on his studies without distractions, so Ken mentioned that he never had to get any part-time jobs thanks to his mother.

Ken progressed through Liverpool University, finishing his Ph.D. at the age of 22.

In Liverpool, Ken got interested especially in inorganic chemistry due to the fact that periodic elements had very different properties yet some similarities in their periodicity (periodic table principle), and also due to the fact that inorganic chemistry is the chemistry of colours. Ken's Ph.D. supervisor was Prof. David Nicholls.

At the age of 24, Ken got a prestigious CEGB (Central Electricity Generating Board) research fellowship at St Catherine's and Corpus Christi College, which allowed him to join Oxford.

He married Elaine Eastwood (now Elaine Seddon, Professor of Physics at Manchester), whom he met at Liverpool University in 1978 and who was attending his tutorials; the marriage was dissolved in 1990.

As Elaine recalls, "Ken was a giant of a presence. A star. He filled whatever space in which he found himself. Precocious, articulate, hard-working, inspirational, a great conversationalist, and both loyal and supportive to those he loved. A natural leader, he had extraordinary clarity of thought (both in- and out-of-the-box) and strong courage of his convictions. He also had a strong sense of humour and of the absurd. Life was never mundane or dull. He valued people for what they were and not their status. His friends, chemistry and his group were his life, his family."

Ken and Elaine wrote a comprehensive book on ruthenium. The reason for choosing this specific element was because it has a large number of oxidation states, and Ken thought he could learn to write books by writing one for which a lot of information would need to be gathered and summarised.

The second book of Ken and Elaine, as well as two of their friends Christine and Tony Ryan, was on phosgene. The choice for the topic of this book was due to Ken's long-lasting interest in chemical weapons.

In 1982, he joined the University of Sussex, Brighton, UK, as Reader in Experimental Chemistry and remained there until 1993. In Sussex, Ken made friends with Harry Kroto (later winner of the Nobel Prize for his work on buckminsterfullerene). Harry Kroto was a great painter, as well as a chemist, and named his molecule after an American architect Richard Buckminster "Bucky" Fuller who popularised geodesic spheres that were structurally and mathematically similar to buckminsterfullerene. Ken had long conversations with Harry and was very supportive of his elegant concept.

Ken accidentally started working with ionic liquids. That is how he described it:

"I was working in Oxford, back in about 1980, on vanadium(IV) phosphine compounds. We sent in reports to the U.S. Navy, which had funded the work, and about six months later, we got a letter back asking if we could make a salt—potassium hexachloromolybdate(IV), and I said sure. So I went to the library to see how to make these things, and I discovered it was pretty well impossible. I'm sitting in the library with all these journals open, every known article on the subject, and I'm getting bored, thinking, 'Me and my big mouth,' and on the page opposite one of these articles was a paper by Bob Osteryoung on room-temperature molten salts, which is what ionic liquids were then called. And I thought, 'That's a jolly good idea—these would be an ideal environment in which to try and make these compounds.' So, I wrote a proposal to the U.S. Air Force, and three weeks later they flew me out to their laboratories in Colorado. What I didn't know then was the reason they wanted the compound I was supposedly making was to make batteries in room-temperature molten salts. It was a complex coincidence. John Wilkes was also there, and Chuck Hussey from Mississippi was visiting the lab at the same time. And they taught me everything they knew about these room-temperature molten salts: how to make them, purify them, work with them—they were extremely generous. And these were just extremely interesting materials. My one original thought was, 'I bet these would be pretty good solvents for doing chemistry with.' That was 1981. That's where everything started."

Ken did not discover ionic liquids. In fact, they were discovered a century before the 1980s, although they were not called ionic liquids straight away. What Ken did was spot their huge potential and popularised them. The main advantage of ionic liquids is that they have negligible vapour pressure, unlike molecular solvents, which are dangerous, flammable, and environmentally unfriendly. Ionic liquids therefore could potentially belong to the category of green solvents. Ionic liquids are not intrinsically green in their own right, but they can help to make the process greener by bringing their advantages and changing the philosophy of a conventional process performed using organic solvents.

At that time in the late 1980s, the Green Chemistry movement started raising concerns about solvent usage as a result of the U.S. Clean Air Legislation: solvents had high vapour pressures, and it is hard to imagine organic chemistry reactions and chemical processes without the use of organic solvents. Ken realised that these ionic liquids could be the new category of solvents that could change the way chemistry worked. Additionally, molecular solvents would dissolve compounds in a different manner than ionic liquids, as they have different bonds in them (prevalently, covalent). Ionic liquids are dominated by coulombic (or ionic) bonds, so they can dissolve compounds that cannot be easily dissolved using organic solvents, for instance, kerogen, cellulose, wood, bananas and their skin, and so forth. Hence, they gained the name of "super solvents".

Ken managed to secure funding to carry out research on ionic liquids:

"At that time, I was in Oxford. I moved to the University of Sussex to a Lectureship in Experimental Chemistry, as it was called. And it takes a year or two to get up and running. I then sent an application to the Engineering and Physical Sciences Research Council (EPSRC), our equivalent of your NSF. It was a proposal for ionic liquids and catalytic chemistry, and we got a gamma rating. Now, alpha means this is wonderful and if we have the money, we will fund you. Alpha-plus means we will definitely fund you. Beta means it has some merit, but it may not get funding. Gamma means never darken our doors again; we never want to hear from you, ever. Along with our gamma rating, they sent the referees report. Referee #1 said, 'this chemistry is so complicated it will never work'. Referee #2 said, 'this chemistry is so trivial, it's not worth doing'. Referee #3 said, 'why isn't he doing the neutron diffraction of vanadium bronzes?', which had no relationship to our proposal. He obviously had the wrong proposal in front of him.

So, we were basically rejected as a joke. And the EPSRC is supposed to fund speculative and interesting work. A year and a half later, we took the same proposal to British Petroleum, which had a special Venture Research Unit, headed up by Professor Don Braben, who was helped by Dr. David Ray; they looked at the proposal and said this is really exciting. Now, this is industry. You would think they would be hard-nosed, but this Venture Research Unit was also sometimes called the 'blue sky research unit'. They sent the proposal off to internal review at British Petroleum, and we got very strong backing from Professor Mike Green, who was at BP at the time. About a year later, we got a grant for over a quarter of a million pounds. In 1987, that was a massive amount of money. So big it hit the newspapers —The *Daily Telegraph*—not the science pages, but page five: 'BP funds super-solvent.' On that grant, I hired Tom Welton, who was completing his D.Phil. at the time, and that was the start of everything that happened. Without Don Braben's support, insight, and encouragement, the field on ionic liquids as we now know it would not have happened."

In 1993, Seddon was appointed Chair of Inorganic Chemistry at Queen's University Belfast.

As co-founder of the industry–university consortium centre QUILL (Queen's University Ionic Liquid Laboratories), with Co-director Jim Swindall in 1999, Ken built a multidisciplinary faculty of chemists and chemical engineers in Northern Ireland. Ken liked different cultures and was very open-minded, so he tended to hire people from all over the world. At one point, QUILL recruited researchers from 40 different countries. By 1999, there was sufficient industrial support from a dozen companies, such as Shell, Chevron, and Procter & Gamble.

QUILL invented new applications for ionic liquids in various areas of life: from triggered release of perfume, Brønsted and Lewis catalysts for various processes, treatment of nuclear waste, safer battery materials, and lubricants. Sectors as diverse as oil and gas refining, biofuels, pharmaceutical manufacturing, and novel antimicrobial (disinfectant) treatments have all benefited. Research by the team sponsored by the Malaysian company PETRONAS and in close cooperation with them led to the first-ever technology to scrub toxic mercury from natural gas production streams.

Now, there are many prominent centres for ionic liquids around the globe including the UK, USA, Australia, Denmark, Portugal, and others.

Professor Tom Welton OBE, Dean of the Faculty of Natural Sciences at Imperial College, London, says, "At Queen's, Ken became a key figure in changing ionic liquids' research from being a quiet academic backwater into a major area for modern chemistry, with thousands of papers published every year.

It is remarkable how many of today's established ionic liquids' researchers had their first paper in the field as the result of a collaboration with Ken."

Tom obtained his B.Sc. from Sussex in 1985 and completed his D.Phil. (supervised by Ken) in 1990.

Ken had a very special work rate. He could manage many very different projects at the same time. Luckily, Ken had a great memory. He could recall the contents of multiple papers and their bibliographic data, which would be very handy during writing. Additionally, Ken only needed 4 h of sleep after he turned 40 and could sit and work throughout the whole day without food, maybe with the occasional drink. Lunch was not needed, as Ken did not want to be interrupted and lose the flow.

Additional areas of Ken's interests included: the preservation of ancient manuscripts (in collaboration with the Oriental Institute in St. Petersburg and the Hermitage and British Library), crystal engineering, chemical weapons (when he died, he was writing a book on them), and ancient dyestuffs. In the last 5 years of his life, Ken took over a project on writing introductions of the works of famous scientists—Galileo, Newton, and Darwin—for the Delphi Complete Works Series. He covered not only scientific achievements of these revolutionary scientists, but also their other interests. So, Darwin is appreciated not only for his work on evolution, but also for his early use of photographs instead of drawings to illustrate his observations.

Ken did not necessarily think that ionic liquids had to be applied in every area of chemistry and technology. He explained his ideas, "If you are an industrialist and you have got a process that's giving you 100% yield that is working beautifully, then you don't need ionic liquids. If you have a problem; if the yield is too low; if the solvent you're using is going to be banned in two years because it's too toxic, then ionic liquids are an extremely attractive option for changing a known process or initiating a new one. Chemistry in ionic liquids is totally different than chemistry in the molecular environment of a normal solvent. The kinetics are different. The thermodynamics are different. The outcome is different. Everything is new."

From 1998, as the field grew at a faster than exponential rate with the founding of QUILL, Ken led an international community on ionic liquids research and was coined as the Father of Ionic Liquids, even though he did not invent them. Ken also had an early appreciation that ionic liquids by default could have impurities easily derived from their preparation. These impurities dramatically changed the properties of ionic liquids, so scientists had to be very careful and good-quality papers (especially physicochemical ones) would have to include at least three types of characterisations of ionic liquids, especially halide and water content by Karl Fischer titration.

Ken was very generous in assisting career development to his colleagues and friends. He would write very kind references, and even when he assisted in writing papers and gave good advice, he would allow his students to be corresponding authors if it could help their career.

Ken authored more than 400 papers and wrote and edited 14 books. Ken was named as inventor on more than 50 patents.

Ken worked hard to the end despite illness, systematically moving his students towards successful theses, so for instance, during his last meeting he was working on a neutron paper in collaboration with ISIS, Rutherford.

In 2006, QUILL was awarded the Queen's Anniversary Prize for Higher and Further Education.

Ken was very proud of meeting the Royal Family and talking to them.

In 2011, he came first in the UK in a global listing of the 100 Top Chemists of the Past Decade. He was ranked 46 overall in the world. Dr. (now Prof.) John D. Holbrey,

also from QUILL, took second in the UK and 59 in the world. They are two of only four UK chemists to be listed. The list is based on how often a chemist's work is cited or referred to during the last decade and the number of publications have to be at least 50.

In 2013, the Science Museum public poll voted his research into ionic liquids as "the most important British innovation of the 21st century".

As one of Britain's most influential chemists, Prof. Seddon was awarded an Officer of the Order of the British Empire OBE in 2015.

In fact, if you google "best UK chemists", Ken's name and picture are there. He was Professor Catedrático Visitante at the Instituto de Tecnologia Química e Biológica (ITQB), New University of Lisbon, Portugal, and Visiting Professor at the Chinese Academy of Sciences. Among other contributions, he served as Associate Editor of the *Australian Journal of Chemistry.*

Ken liked to develop short names for his Ph.D. students. They normally contained three letters. For instance, a Malaysian student Shieling Ng was called Bob, as the word *bob* was sometimes used for a monetary value of several shillings (old British currency). My name Natalia was shortened to Nat. Małgorzata Swadźba-Kwaśny was shortened to Mal. Any objections were not accepted. Luckily for Ken, Ken was a three-letter name.

Ken was not liked by everybody. He could be brutally honest if he found scientific mistakes. He lived with high professional standards, and he expected the same from others. Ken would forgive his staff and students if they did not know some scientific concept and would gladly explain it to them. Ken genuinely cared about the health of his colleagues. If he saw that someone was struggling because of sickness or tiredness, he would send them home in a caring manner.

He also could be very humorous and welcoming to the many students over the years. Ken was extremely generous, so we, his colleagues and friends, would wine and dine and watch shows and go to the theatre in the best restaurants around the globe. Ken loved international food, as long as it was no longer alive, but hated green food: salads. He particularly loved sweets and puddings.

During an interview for a scientific position, Ken would normally ask you about your favourite book, movie, or music band. If you applied for a clerical position, he would normally ask "the 50-pence" question: what this coin was made out of and why, why it had several edges (basically it was not just round), and then he would observe how stress would get hold of the candidate. He would be very happy to learn new information: for instance, one of the QUILL Ph.D. students was hired because her knowledge of Icelandic rock was better than Ken's.

Ken was also an enthusiastic appreciator of the latest technology and gadgets of the time: phones, cameras, e-books, Kindles, Alexa, and various software that could make his life easier and that could copy his life from one device to another within a minimum of fuss. Every time Ken received a message, his work computer announced, "Captain, you've got an incoming message", a reference to Star Trek of which he was a great fan.

Apart from chemistry, Ken loved literature, art, sculpture (he was especially fond of gargoyles since his time at Oxford), theatre, music, and photography. Ken's tastes were so broad that if you went to a concert with him, you could be rubbing shoulders with pink-haired students or pompous opera lovers.

He could not drink alcohol because of a violent allergic reaction. A couple of accidental encounters for instance, when the apple juice was off almost finished him, so he was always very careful and used his colleagues as experimental rats to check if in doubt.

Ken was very involved with his friends and colleagues. He would remember their birthdays and often would come up with original presents, chocolates from all over the world or cheeses from a quirky cheesery in Liverpool. He would keep in touch with many of his ex-students and staff. He would try to help you with anything and gave very good advice.

Some of Ken's distinctive characteristics included a gargantuan laugh, positivity, a large collection of artistic ties depicting chemistry and mythical creatures, side-locks à la Charles Dickens, and long nails inspired by Fu Manchu, a favourite of his childhood reading (he said they let him pick up dry ice, at -78 degrees).

Ken was brilliant teacher to multiple students in different universities and companies, although he personally did not regard teaching, as it was time-consuming and repetitive. Ken hated repetition. He mentioned that he would not be able to play musical instruments because it was like repetitive teaching and prevented him from concentrating on research and travelling in pursuit of new QUILL members.

Once I gave a lecture about Ken at a conference, and I was approached by a professor who told me that in the past he had cancer, but his country did not produce the drug he needed. Once Ken found out about this, he brought a full bag of the drug and this professor recovered.

Ken often had many Ph.D. students at the same time with their different projects and could handle them well.

Ken formed many friendships with world-renowned scientists, for instance, Chuck Hussey recalls, "Earlier in our careers we travelled a lot together to meetings, etc. I was visiting him at Sussex, we decided to go together by train to EuChem at the DTU in Lyngby. Things were going well until we got to Copenhagen. 'You stay on the train, Chuck, with our 8 bags and I'll go see it if I can get us a ticket to Lyngby.' A smiling worker came up and unhooked my car from the train which went on to Lyngby with Ken on board, and I sat there trying to figure out what to do. Fortunately, most of the Danes speak English…"

When Ken was very ill, he was in a hospital getting treatment, and he was chatting to doctors and nurses. One male nurse, who was helping Ken with his treatment, and with whom Ken shared various conversations, was very impressed with Ken's knowledge and coined Ken as "a man with encyclopaedia in his head".

Ken passed away on the 21 January 2018. Ken left a very big hole in many people's hearts, and his last wish was for his ashes to be shot into space; hence, we played "Rocket Man" by Elton John at his funeral.

Preface

The purpose of this first book on *Commercial Applications of Ionic Liquids* is to provide an overview of known processes and products using ionic liquids. There are 57 known commercial or pilot-scale applications using ionic liquids. Based on conversations with the authors, there are several applications that are maintained as trade secrets or are confidential and the actual number of applications is more likely greater than 100. A brief history of the field is included along with detailed discussions of several processes and products using ionic liquids.

The editor, Dr. Shiflett, was Technical Fellow in DuPont Central Research and Development (CR&D) located at the Experimental Station in Wilmington, Delaware. Dr. Shiflett retired from the DuPont Company in September of 2016 and joined the faculty in the Department of Chemical and Petroleum Engineering at the University of Kansas to continue his teaching and research in the field of ionic liquids.

Lawrence, KS, USA

Mark B. Shiflett
Foundation Distinguished Professor
mark.b.shiflett@ku.edu

Acknowledgements

I would like to thank first the authors for writing the excellent chapters contained in this book. I would also like to thank the following reviewers for their time to carefully read and provide constructive feedback on each chapter, in particular Dr. Reginald B. Shiflett and Dr. Dirk Tuma who reviewed all the chapters.

Dr. Reginald B. Shiflett, retired, Meredith College, Raleigh, North Carolina, USA
Dr. Dirk Tuma, Federal Institute for Materials Research and Testing, Berlin, Germany
Prof. Edward Maginn, University of Notre Dame, Notre Dame, Indiana, USA
Prof. Carlos Nieto de Castro, University of Lisbon, Lisbon, Portugal
Prof. Jason Bara, University of Alabama, Tuscaloosa, Alabama, USA
Prof. Rasmus Fehrmann, Technical University of Denmark, Lyngby, Denmark
Prof. João Coutinho, University of Aveiro, Aveiro, Portugal
Prof. Gary Baker, University of Missouri, Columbia, Missouri, USA
Prof. Aaron Scurto, University of Kansas, Lawrence, Kansas, USA
Dr. Joe Magee, National Institute of Standards and Technology, Boulder, Colorado, USA

I would also like to thank the team at Springer, in particular Thomas Spicer, Cindy Zitter, Loyola D'Silva, and Megana Dinesh, for their support. It is always a pleasure to work with such a professional group of people.

Finally, I would like to dedicate this book to my wife, Lori A. Shiflett, for her patience and understanding the amount of time required to put this book together.

Contents

Editor and Contributors

About the Editor

Prof. Mark B. Shiflett obtained his B.S. in Chemical Engineering at North Carolina State University (1989) and both his M.S. (1998) and Ph.D. (2001) in Chemical Engineering at the University of Delaware under the direction of Prof. Henry C. Foley. He is currently a Foundation Distinguished Professor of Engineering at the University of Kansas in the Department of Chemical and Petroleum Engineering and is associated with the Center for Environmentally Beneficial Catalysis. He is an inventor on 44 U.S. Patents and has published over 90 papers. He worked for the DuPont Company in Wilmington, Delaware, at the Experimental Station for 28 years on a variety of challenging projects including the development of non-ozone depleting refrigerants, a thermal plasma reactor, nanoporous carbon membranes, hydrogen storage materials, a next-generation TiO_2 process, new applications for ionic liquids, and novel gas separation processes. His research interests include ionic liquids, green chemistry and engineering, energy efficient chemical processes, product design, desalination, thermodynamics, kinetics, and reaction engineering.

Contributors

Aida R. Abouelela Department of Chemical Engineering, Imperial College London, London, UK

Daniel W. Armstrong Department of Chemistry and Biochemistry, University of Texas at Arlington, Arlington, TX, USA

Leslie Brown AECS-QuikPrep Ltd., London, UK

Michael R. Buchmeiser German Institutes of Textile and Fiber Research (DITF), Denkendorf, Germany;
Institute of Polymer Chemistry (IPOC), University of Stuttgart, Stuttgart, Germany

Elizabeth Carter Honeywell UOP, Des Plaines, IL, USA

Bong-Kyu Chang Chevron Energy Technology Company, Richmond, CA, USA

Matthew Cole Honeywell UOP, Des Plaines, IL, USA

David P. Durkin Department of Chemistry, United States Naval Academy, Annapolis, MD, USA

Martyn J. Earle School of Chemistry, The QUILL Research Centre, The Queen's University of Belfast, Belfast, Northern Ireland, UK

Manuela A. Gilea School of Chemistry, The QUILL Research Centre, The Queen's University of Belfast, Belfast, Northern Ireland, UK

Florence V. Gschwend Department of Chemical Engineering, Imperial College London, London, UK

Jason P. Hallett Department of Chemical Engineering, Imperial College London, London, UK

Marco Haumann Friedrich-Alexander-Universität Erlangen-Nürnberg (FAU), Lehrstuhl für Chemische Reaktionstechnik (CRT), Erlangen, Germany

Luke M. Haverhals Natural Fiber Welding, Inc., Peoria, IL, USA

Frank Hermanutz German Institutes of Textile and Fiber Research (DITF), Denkendorf, Germany

Roland S. Kalb proionic GmbH, Raaba-Grambach, Austria

Huping Luo Chevron Energy Technology Company, Richmond, CA, USA

Joe W. Magee National Institute of Standards and Technology, Boulder, CO, USA

Francisco Malaret Department of Chemical Engineering, Imperial College London, London, UK

Rahul A. Patil Department of Chemistry and Biochemistry, University of Texas at Arlington, Arlington, TX, USA

Natalia V. Plechkova School of Chemistry, The QUILL Research Centre, The Queen's University of Belfast, Belfast, Northern Ireland, UK

Robin D. Rogers 525 Solutions, Inc., Tuscaloosa, AL, USA

Thomas J. S. Schubert IOLITEC Ionic Liquids Technologies GmbH, Heilbronn, Germany

Kenneth R. Seddon School of Chemistry, The QUILL Research Centre, The Queen's University of Belfast, Belfast, Northern Ireland, UK

Julia L. Shamshina Mari Signum Mid Atlantic, LLC, Richmond, VA, USA

Mark B. Shiflett Department of Chemical and Petroleum Engineering, University of Kansas, Lawrence, KS, USA

Mohsen Talebi Department of Chemistry and Biochemistry, University of Texas at Arlington, Arlington, TX, USA

Hye Kyung Timken Chevron Energy Technology Company, Richmond, CA, USA

Paul C. Trulove Department of Chemistry, United States Naval Academy, Annapolis, MD, USA

Dirk Tuma Federal Institute for Materials Research and Testing (BAM), Berlin, Germany

Marc Philip Vocht German Institutes of Textile and Fiber Research (DITF), Denkendorf, Germany

Nomenclature

Ionic Liquids

APIL	Aprotic ionic liquid
CBIL	Carbonate-based ionic liquid
IL	Ionic liquid
PIL	Protic ionic liquid
poly(RTIL)	Polymerised room-temperature ionic liquid
RTIL	Room-temperature ionic liquid
TSIL	Task-specific ionic liquid

Cations

$[bm(2\text{-}Me)im]^+$	1-butyl-2,3-dimethylimidazolium
$[(bz)C_1C_5im]^+$	1-benzyl-2-methyl-3-pentylimidazolium
$[C_{10}btz]^+$	3-decylbenzothiazolium
$[C_{11}btz]^+$	3-undecylbenzothiazolium
$[C_{12}btz]^+$	3-dodecylbenzothiazolium
$[C_nim]^+$	1-alkylimidazolium
$[C_1im]^+$ or $[Hmim]^+$	1-methylimidazolium
$[C_2im]^+$	1-ethylimidazolium
$[C_4im]^+$	1-butylimidazolium
$[C_nC_1im]^+$	1-alkyl-3-methylimidazolium
$[C_1C_1im]^+$	1,3-dimethylimidazolium
$[C_2C_1im]^+$	1-ethyl-3-methylimidazolium
$[C_3C_1im]^+$	1-propyl-3-methylimidazolium
$[C_4C_1im]^+$	1-butyl-3-methylimidazolium
$[C_6C_1im]^+$	1-hexyl-3-methylimidazolium
$[C_8C_1im]^+$	1-octyl-3-methylimidazolium

$[C_nC_m\text{im}]^+$	1-alkyl-3-alkyl'imidazolium
$[C_1C_2\text{im}]^+$	1-methyl-3-ethylimidazolium
$[C_1C_4\text{im}]^+$	1-methyl-3-butylimidazolium
$[C_1C_6\text{im}]^+$	1-methyl-3-hexylimidazolium
$[C_1C_1C_1\text{im}]^+$	1,2,3-trimethylimidazolium
$[C_2C_1C_1\text{im}]^+$	1-ethyl-2,3-dimethylimidazolium
$[C_3C_1C_1\text{im}]^+$	1-propyl-2,3-dimethylimidazolium
$[C_4C_1C_1\text{im}]^+$	1-butyl-2,3-dimethylimidazolium
$[C_1C_1(2\text{-NO}_2)\text{im}]^+$	1,3-dimethyl-2-nitroimidazolium
$[C_1C_1(2\text{-Me-4-NO}_2)\text{im}]^+$	1,3-dimethyl-2-methyl-4-nitroimidazolium
$[C_1C_1(4\text{-NO}_2)\text{im}]^+$	1,3-dimethyl-4-nitroimidazolium
$[C_2C_1(4\text{-NO}_2)\text{im}]^+$	1-ethyl-3-methyl-4-nitroimidazolium
$[C_3C_3\text{im}]^+$	1,3-dipropylimidazolium
$[^iC_3C_1\text{im}]^+$	1-*iso*propyl-3-methylimidazolium
$[^iC_3{}^iC_3\text{im}]^+$	1,3-di*iso*propylimidazolium
$[(^iC_3)_2\text{im}]^+$	1,3-di*iso*propylimidazolium
$[^iC_4C_1\text{im}]^+$	1-*iso*butyl-3-methylimidazolium
$[^sC_4C_1\text{im}]^+$	1-*sec*butyl-3-methylimidazolium
$[^tC_4C_1\text{im}]^+$	1-*tert*butyl-3-methylimidazolium
$[(^iC_4)_2\text{im}]^+$	1,3-di*iso*butylimidazolium
$[C_4C_4\text{im}]^+$	1,3-dibutylimidazolium
$[^tC_4{}^tC_4\text{im}]^+$	1,3-di*tert*butylimidazolium
$[C_4C_1(4,5\text{-Br}_2)\text{im}]^+$	1-butyl-3-methyl-4,5-bromoimidazolium
$[C_5C_1\text{im}]^+$	1-pentyl-3-methylimidazolium
$[C_6C_6\text{im}]^+$	1,3-dihexylimidazolium
$[C_7C_1\text{im}]^+$	1-heptyl-3-methylimidazolium
$[C_9C_1\text{im}]^+$	1-nonyl-3-methylimidazolium
$[C_{10}C_1\text{im}]^+$	1-decyl-3-methylimidazolium
$[(C_{10})_2\text{im}]^+$	1,3-didecylimidazolium
$[C_{11}C_1\text{im}]^+$	1-undecyl-3-methylimidazolium
$[C_{12}C_1\text{im}]^+$	1-dodecyl-3-methylimidazolium
$[(C_{12})_2\text{im}]^+$	1,3-didodecylimidazolium
$[C_{13}C_1\text{im}]^+$	1-tridecyl-3-methylimidazolium
$[C_{14}C_1\text{im}]^+$	1-tetradecyl-3-methylimidazolium
$[C_{15}C_1\text{im}]^+$	1-pentadecyl-3-methylimidazolium
$[C_{16}C_1\text{im}]^+$	1-hexadecyl-3-methylimidazolium
$[C_{17}C_1\text{im}]^+$	1-heptadecyl-3-methylimidazolium
$[C_{18}C_1\text{im}]^+$	1-octadecyl-3-methylimidazolium
$[C_8C_3\text{im}]^+$	1-octyl-3-propylimidazolium
$[C_{12}C_{12}\text{im}]^+$	1,3-bis(dodecyl)imidazolium
$[C_{1O}C_2C_1\text{im}]^+$	1-(2-methoxyethyl)-3-methylimidazolium
$[C_6C_{7O1}\text{im}]^+$	1-hexyl-3-(heptyloxymethyl)imidazolium
$[C_2F_3C_1\text{im}]^+$	1-trifluoroethyl-3-methylimidazolium
$[C_4\text{vim}]^+$	3-butyl-1-vinylimidazolium

$[C_9(vim)_2]^{2+}$	1,9-di(3-vinylimidazolium)nonane
$[(^iC_3)_2(4,5\text{-}Me_2)im]^+$	1,3-di*iso*propyl-4,5-dimethylimidazolium
$[^iC_3C_1(4,5\text{-}Me_2)im]^+$	1-*iso*propyl-3,4,5-trimethylimidazolium
$[(^tC_4)_2(4\text{-}SiMe_3)im]^+$	1,3-di*tert*butyl-4-trimethylsilylimidazolium
$[(allyl)C_1im]^+$	1-allyl-3-methylimidazolium
$[P_nC_1im]^+$	polymerisable 1-methylimidazolium
$[C_2C_1mor]^+$	1-ethyl-1-methylmorpholinium
$[C_2py]^+$	1-ethylpyridinium
$[C_4py]^+$	1-butylpyridinium
$[C_6py]^+$	1-hexylpyridinium
$[C_8py]^+$	1-octylpyridinium
$[C_{14}py]^+$	1-tetradecylpyridinium
$[C_4C_{1\beta}py]^+$	1-butyl-3-methylpyridinium
$[C_4C_{1\gamma}py]^+$	1-butyl-4-methylpyridinium
$[C_6(dma)_\gamma py]^+$	1-hexyl-4-dimethylaminopyridinium
$[C_nC_1pyr]^+$	1-alkyl-1-methylpyrrolidinium
$[C_1C_1pyr]^+$	1,1-dimethylpyrrolidinium
$[C_1C_2pyr]^+$	1-ethyl-1-methylpyrrolidinium
$[C_2C_1pyr]^+$	1-ethyl-1-methylpyrrolidinium
$[C_3C_1pyr]^+$	1-propyl-1-methylpyrrolidinium
$[C_4C_1pyr]^+$	1-butyl-1-methylpyrrolidinium
$[C_5C_1pyr]^+$	1-pentyl-1-methylpyrrolidinium
$[C_6C_1pyr]^+$	1-hexyl-1-methylpyrrolidinium
$[C_1C_3pip]^+$	1-methyl-1-propylpiperidinium
$[C_2C_1pip]^+$	1-ethyl-1-methylpiperidinium
$[C_2C_6pip]^+$	1-ethyl-1-hexylpiperidinium
$[C_1C_1pip]^+$	1,1-dimethylpiperidinium
$[C_8quin]^+$	1-octylquinolinium
$[dabcoH]^+$	1,4-diazabicyclo[2.2.2]octan-1-ium(1+)
$[dabcoH_2]^{2+}$	1,4-diazabicyclo[2.2.2]octan-1-ium(2+)
$[(C_1)_2Phim]^+$	1,3-dimethyl-2-phenylimidazolium
$[(FcC_1)C_1im]^+$	1-(ferrocenylmethyl)-3-methylimidazolium
$[(H_2NC_2H_4)py]^+$	1-(1-aminoethyl)pyridinium
$[(H_2NC_3H_6)C_1im]^+$	1-(3-aminopropyl)-3-methylimidazolium
$[HN_{2\ 2\ 2}]^+$	Triethylammonium
$[H_2mor]^+$	Morpholinium
$[H_2pip]^+$	Piperidinium
$[Hpy]^+$	Pyridinium
$[H_2pyr]^+$	Pyrrolidinium
$[N_{0\ 1\ 1\ 1}]^+$	Trimethylammonium
$[N_{0\ 0\ 1\ 1}]^+$	Dimethylammonium
$[N_{0\ 0\ 0\ 1}]^+$	Methylammonium
$[N_{1\ 1\ 1\ 1}]^+$	Tetramethylammonium
$[N_{1\ 1\ 1\ 2OH}]^+$	Cholinium
$[N_{1\ 1\ 2\ 2OH}]^+$	Ethyl(2-hydroxyethyl)dimethylammonium

$[N_{1\ 1\ 1\ 4}]^+$	Trimethylbutylammonium
$[N_{1\ 4\ 4\ 4}]^+$	Methyltributylammonium
$[N_{1\ 8\ 8\ 8}]^+$	Methyltrioctylammonium
$[N_{0\ 2\ 2\ 2}]^+$	Trimethylammonium
$[N_{2\ 2\ 2\ 2}]^+$	Tetraethylammonium
$[N_{3\ 3\ 3\ 3}]^+$	Tetrapropylammonium
$[N_{3\ 3\ 3\ 11}]^+$	Tripropylundecylammonium
$[N_{3\ 3\ 6\ 8}]^+$	Dipropylhexyloctylammonium
$[N_{4\ 4\ 4\ 4}]^+$	Tetrabutylammonium
$[N_{5\ 5\ 5\ 5}]^+$	Tetrapentylammonium
$[N_{6\ 6\ 6\ 6}]^+$	Tetrahexylammonium
$[N_{6\ 6\ 6\ 14}]^+$	Trihexyl(tetradecyl)ammonium
$[N_{10\ 10\ 10\ 10}]^+$	Tetrakis(decyl)ammonium
$[N_{12\ 12\ 12\ 12}]^+$	Tetradodecylammonium
$[N_{n\ n\ n\ 0}]^+$	Trialkylammonium
$[PH_4]^+$	Phosphonium
$[P_{2\ 2\ 2(1O1)}]^+$	Triethyl(methoxymethyl)phosphonium
$[P_{4\ 4\ 4\ 3a}]^+$	(3-aminopropyl)tributylphosphonium
$[P_{4\ 4\ 4\ 4}]^+$	Tetrabutylphosphonium
$[P_{5\ 5\ 5\ 5}]^+$	Tetrapentylphosphonium
$[P_{6\ 6\ 6\ 14}]^+$	Trihexyl(tetradecyl)phosphonium
$[P_{8\ 8\ 8\ 14}]^+$	Tetradecyl(trioctyl)phosphonium
$[P_{10\ 10\ 10\ 10}]^+$	Tetrakis(decyl)phosphonium
$[P_{10\ 10\ 10}CH_2C(O)NH_2]^+$	Amidomethyl-tritetradecylphosphonium
$[P_{10\ 10\ 10}CH_2CO_2]^+$	Carboxymethyl-tritetradecylphosphonium
$[P_{18\ 18\ 18\ 18}]^+$	Tetraoctadecylphosphonium
$[pyH]^+$	Pyridinium
$[S_{2\ 2\ 2}]^+$	Triethylsulfonium
$[S_{2\ 2\ 16}]^+$	Diethylhexadecylsulfonium
$[(vinyl)C_1im]^+$	1-vinyl-3-methylimidazolium

Anions

$[Ace]^-$, $[OAc]^-$	Acetate, ethanoate
$[Ala]^-$	Alaninate
$[\beta Ala]^-$	β-alaninate
$[AlCl_4]^-$	Tetrachloroaluminate(III)
$[Al_2Cl_7]^-$	Heptachloroaluminate(III)
$[Al(hfip)_4]^-$	Tetrakis(hexafluoro*iso*propoxy)aluminate(III)
$[Arg]^-$	Arginate
$[Asn]^-$	Asparaginate
$[Asp]^-$	Asparatinate
$[B_{4\ 4\ 4\ 4}]^-$	Tetrabutylborate

$[BBB]^-$	Bis[1,2-benzenediolato(2-)-O,O']borate
$[BF_4]^-$	Tetrafluoroborate
$[Br]^-$	Bromide
$[BTA]^-$, $[(CF_3SO_2)_2N]^-$	Bis(trifluoromethylsulfonyl)imide
$[C_1CO_2]^-$	Ethanoate
$[C_1SO_4]^-$, $[O_3SOC_1]^-$	Methyl sulphate
$[C_8SO_4]^-$, $[O_3SOC_8]^-$	Octyl sulphate
$[C_nSO_4]^-$	Alkyl sulphate
$[(C_n)(C_m)SO_4]^-$	Asymmetrical dialkyl sulphate
$[(C_n)_2SO_4]^-$	Symmetrical dialkyl sulphate
$[CF_3SO_3]^-$, $[OTf]^-$	Trifluoromethanesulfonate, triflate
$[Cl]^-$	Chloride
$[ClCH_2COO]^-$	Chloroacetate
$[CTf_3]^-$	Tris{(trifluoromethyl)sulfonyl}methide
$[Cu_2Cl_3]^-$	Chlorocuprate(II)
$[CuBr_3]^-$	Bromocuprate(II)
$[Cys]^-$	Cysteinate
$[dbsa]^-$	Dodecylbenzenesulfonate
$[dca]^-$	Dicyanamide
$[FAP]^-$	Tris(perfluoroalkyl)trifluorophosphate
$[FSI]^-$	Bis(fluorosulfonyl)imide
$[Gln]^-$	Glutaminate
$[Glu]^-$	Glutamate
$[Gly]^-$	Glycinate anion
$[His]^-$	Histidinate
$[HSO_4]^-$	Hydrogen sulphate
$[Ile]^-$	Isoleucinate
$[lac]^-$	Lactate
$[Leu]^-$	Leucinate
$[Lys]^-$	Lysinate
$[Met]^-$	Methionate
$[Nle]^-$	Norleucinate
$[NH_4]^+$	Ammonium
$[NDf_2]^-$	Bis{bis(pentafluoroethyl)phosphinyl}amide
$[NMes_2]^-$	Bis(methanesulfonyl)amide
$[NO_3]^-$	Nitrate
$[NPf_2]^-$, $[BETI]^-$	Bis{(pentafluoroethyl)sulfonyl}amide
$[NTf_2]^-$, $[TFSI]^-$	Bis{(trifluoromethyl)sulfonyl}amide
$[O_3SOC_2]^-$	Ethyl sulphate
$[OMs]^-$	Methanesulfonate (mesylate)
$[PFBS]^-$	Perfluorobutylsulfonate
$[OTf]^-$, $[CF_3SO_3]^-$,	Trifluoromethanesulfonate, (triflate)
$[OTs]^-$, $[4\text{-}CH_3C_6H_4SO_3]^-$	4-toluenesulfonate, (tosylate)
$[PF_6]^-$	Hexafluorophosphate

[PFOS]$^-$, [C$_8$HF$_{17}$SO$_3$]$^-$	Perfluorooctanesulfonate
[Phe]$^-$	Phenylalaninate
[Pro]$^-$	Prolinate
[Sacc]$^-$	Saccharinate
[SCN]$^-$	Thiocyanate
[Ser]$^-$	Serinate
[Suc]$^-$	Succinate
[tfpb]$^-$	Tetrakis(3,5-bis(trifluoromethyl)phenyl)borate
[Thr]$^-$	Threoninate
[Tos]$^-$	Tosylate
[Trp]$^-$	Tryptophanate
[Tyr]$^-$	Tyrosinate
[Val]$^-$	Valinate
[XSO$_3$]$^-$, [C$_8$H$_{11}$O$_3$S]$^-$	Xylenesulfonate

Compounds

ACC	All-cellulose compounds
API	Active pharmaceutical ingredients
bpp	6,6'-[(3,3'-di-tert-butyl-5,5'-dimethoxy-1,1'-biphenyl-2,2'-diyl) bis(oxy)]bis(dibenzo[d,f][1,3,2]dioxaphosphepin), "biphephos"
bzp	2,2'-((3,3'-di-tert-butyl-5.5'-dimethoxy-[1,1'-biphenyl]-2,2'-diyl) bis(oxy)bis(4,4,5,5-tetraphenyl-1,3,2-dioxaphospholane), "benzopinacol"
CLA	Conjugated linoleic acid
DBNH	1,5-diazobicyclo[4,3,0]non-5-ene
DCPP	Dichlorophenylphosphine
DEPP	Diethoxyphenylphosphine
DHA	Docosahexanoic acid
DMF	Dimethylformamide
DMAc	N,N-dimethylacetamide
DMSO	Dimethylsulfoxide
dtbpmb	1,2-bis(di-tert-butylphosphinomethane)benzene
EPA	Eicosapentanoic acid
FA	Fatty acid
FAME	Fatty acid methyl ester
HMF	5-hydroxymethyl furfural
LCA	Lithocholic acid
LPG	Liquefied petroleum gas
MAA	Methyl acetoacetate
MeDBT	n-methyldibenzothiophene, n = 1,2,3,4,...
MIA	1-methylimidazole
MUFA	Monounsaturated fatty acid

NMMO	N-methylmorpholine-N-oxide
N_2O_4	Dinitrogen tetroxide
OB-CFA	Odd- and branched-chain fatty acid
OSC	Organic sulphur compound
PAC	Polycyclic aromatic compound
PAE	Phthalic acid esters, phthalates
PAH	Polycyclic aromatic hydrocarbon
PAN	Poly(acrylonitrile)
PASH	Polycyclic aromatic sulphur heterocycles
PCB	Polychlorinated biphenyl
PDMS	Phenyl polydimethylsiloxane
PEG	Polyethylene glycol $H(OCH_2CH_2)_nOH$, $n = 1,2,3,4,\ldots$
PEO	Poly(ethylene oxide)
PET	Polyethylene terephthalate
PLA	Polylactic acid
PSA	Poly(sulfonamide)
PTFE	Polytetrafluoroethylene
PUFA	Polyunsaturated fatty acid
PVA	Polyvinyl alcohol
PVC	Polyvinyl chloride
$scCO_2$	Supercritical carbon dioxide
SFA	Saturated fatty acid
sxp	4,5-bis(diphenylphosphino)-9,9-dimethylxanthene, "sulfoxantphos"
TBAF	Tetrabutylammonium fluoride
TCDD	2,3,7,8-tetrachlorobibenzo-p-dioxin
TCEP	1,2,3-tris(cyanoethoxy)propane
tppmim	Tri(m-sulfonyl)triphenylphosphine
	tri-1-butyl-2,3-dimethylimidazolium
tppts	Tri(m-sulfonyl)triphenylphosphine trisodium
VA	Vaccenic acid
VCM	Vinyl chloride monomer
VGO	Vacuum gas oil
VOC	Volatile organic compound

Part I
Introduction

Chapter 1
Important Developments in the History of Ionic Liquids from Academic Curiosity to Commercial Processes and Products

Mark B. Shiflett, Joe W. Magee and Dirk Tuma

Abstract Twenty years ago, research involving ionic liquids was a minor field of interest, and only a few chemists and even fewer engineers were interested in salts with melting points near room temperature. In April 2000, the first NATO advanced research workshop on ionic liquids was held in Heraklion, Crete. The conference was the first international meeting devoted to ionic liquids and attracted most of the active researchers at that time. Following that meeting, activity in the field began to flourish and the first books and international conferences devoted to ionic liquids began to appear. By the end of 2018, more than 80,000 scientific papers had been published, and 17,000 patents were applied for in the field of ionic liquids! This book provides an overview of the current and emerging industrial applications of ionic liquids covering the core processes and products, the practical implementation and technical challenges involved, and the potential future directions for research and development. The individual chapters were written by leading scientists in the field from industry and academia to address specific processes and products that are or will be soon commercialized. Examples include the use of a chloroaluminate ionic liquid as a next-generation alkylation catalyst to a new class of capillary gas chromatography (GC) columns with stationary phases based on ionic liquids. Over the past twenty years, there has been a growing realization that ionic liquids have moved from being mere academic curiosities to having genuine applications in fields as wide-ranging as advanced materials, biotechnology, catalysis, pharmaceuticals, renewable fuels, and sustainable energy. There are many optimistic indications that ionic liquids are on their way to becoming a commercial success story. This first book on "Commercial Applications of Ionic Liquids" provides over 50 applications that are either at the pilot scale or have been commercialized, which indicates that an exciting new chapter in the field of ionic liquids is about to begin!

M. B. Shiflett (✉)
Department of Chemical and Petroleum Engineering, University of Kansas, Lawrence, KS, USA
e-mail: mark.b.shiflett@ku.edu

J. W. Magee
National Institute of Standards and Technology, Boulder, CO, USA

D. Tuma
Federal Institute for Materials Research and Testing (BAM), Berlin, Germany

© Springer Nature Switzerland AG 2020
M. B. Shiflett (ed.), *Commercial Applications of Ionic Liquids*, Green Chemistry and Sustainable Technology, https://doi.org/10.1007/978-3-030-35245-5_1

Keywords Advanced materials · Commercialization · Ionic liquids · Ionic liquid history · Ionic liquid processes and products

1.1 Introduction

Who discovered the first ionic liquid? This honor is now almost unanimously attributed to the famous chemist Paul Walden (1863–1957), who was not only a renowned professor but also a scientific celebrity during his eventful lifetime. In an experimental study published in 1914 during his time as a professor at the Technical University of Riga in the Russian Empire (now Latvia), he reported data on density, capillarity, and electrical conductivity for 13 different alkyl- and aryl-substituted water-free and low melting ammonium salts [1]. The ethylammonium nitrate was found to have a melting point of about 13–14 °C, being a protic room-temperature ionic liquid (RTIL) in present terms. Walden determined the degree of association in the molten state using density and capillarity, a method commonly used in his time [2]. Today, almost every work that includes a chapter on the historical development of ionic liquids begins with this citation of a German paper in a Russian journal [1]. Nevertheless, Walden's paper is probably one of the least read and—notably during the beginning of modern-day research on ionic liquids—very often a wrongly cited work of scientific literature.

According to the literature, there are several substances that can be retrospectively classified as ionic liquids [3–5]. One candidate, first reported in 1877 and suggested by several authors, is the "red oil" observed as a separate phase during Friedel–Crafts reactions [4, 5]. This substance remained unspecified until 1976 when it was finally identified as a chloroaluminate with an alkylated aromatic cyclic cation [6]. Pernak et al. noted in their paper that Walden in his 1914 paper referred to an earlier work on the electrical conductivity of homologous alkylchinolinium iodides by Schall (like Walden a student of Wilhelm Ostwald at Riga) [7].

Other early ionic liquids found by Laus et al. were alkylpicolinium halides, first described by Ramsay in 1876 [8], and quaternary anilinium salts made at the turn of the twentieth century [3]. Even with their low melting points, these substances did not draw particular attention. The paper by Pernak et al. gave an explanation, namely that water was considered a universal solvent at that time, and salts with poor water solubility fell out of interest. The old papers often report that different batches of the same substance gave very divergent results. Purity and by-products could not be characterized as they are today due to the analytical means available at the time. These substances were also described as "intractable oils" and thus often ignored [9].

In the literature, there is considerable information about Paul Walden that is oddly if not wrongly interpreted. An autobiographical memoir was published in 1950 [10], a Russian paper followed in 2003 [11], and Everts had a biographical sketch in *Chemical & Engineering News* on the occasion of Latvia celebrating his 150th birthday [12]. The Latvian-born Walden started his academic career after graduation at

Leipzig (1891) as a professor of physical and analytical chemistry at the Riga Poly-technic Institute (1893/94). In 1899, already as an appointed professor, he received the Russian doctoral degree in chemistry working on stereochemistry at Saint Peters-burg. In 1910, Walden was given full membership in the Russian Imperial Academy of Sciences. Among his many landmark discoveries in various fields, the so-called Walden inversion in organic chemistry is generally considered his most outstanding achievement [13]. He was nominated six times (Stradiņš says seven times [14] but according to Morachevskii only two times [11]) for the Nobel Prize [15]. In 1919, Walden decided to leave for Germany and not continue his work in the capital of independent Latvia. From 1919 until his retirement in 1934, he was professor of chemistry at the University of Rostock. Walden remained active during his retire-ment, published works on the history of chemistry, and held—from 1934 on as a member of the National Socialist German Workers' Party—various influential posi-tions in academic organizations, such as the vice president of the German Chemical Society [15, 16]. After the war, he lived in the French-occupied zone of Germany and was given a guest professorship at the University of Tübingen where he gave lectures until the age of 90.

Contemporary German papers often refer to his name as "Paul von Walden," [17] and some Internet sources still use this name [18]. The tag "von" indicates nobility in the German style of writing names, and this fact convinced many that Paul Walden was a typical Baltic German (like his teacher and mentor Wilhelm Ostwald). Walden (who was educated in German schools and later entered a university where German was the language used, a peculiarity of the Baltic provinces) had opted for German citizenship after his emigration but kept strong ties to his homeland and saw himself primarily as a "chemist" [12]. In fact, Walden was granted hereditary peerage by the Russian Emperor in 1907 [10]. He did not use the "von" in his own works. Walden's genuine origin from Latvian peasantry was investigated, published, promoted, as well as popularized already in Soviet times by the academician Jānis Stradiņš [14]. The post-Soviet independent Latvia honored Walden by issuing a special commemorative stamp in 2013 [12], and, initiated by Stradiņš, a monument was erected near the University of Riga showing the Walden inversion [19].

At the University of Rostock, the Ludwig group conducted a study of ethylammo-nium nitrate to honor Walden's merits for the city and for the field of ionic liquids [20]. Walden had no conception of the ionic liquid as we have now. He was only interested in conductivity and molecular constitution. The term "ionic liquid" in its present sense was coined by Barrer in 1943, but he referred to classical high-temperature molten salts mostly by the term "ionic melts [21]."

From the remarkable amount of literature, the reader can choose from many detailed timelines that outline the development of ionic liquids. Four can be consid-ered representative and give the reader almost all corresponding key references. The paper by Wilkes in *Green Chemistry* was the first [22]. The most recently published historical overviews were contributed by Welton in *Biophysical Reviews* [23] and by Shiflett and Scurto as an introductory chapter of their book issued volume 1250 of the *Symposium Series of the American Chemical Society* [24]. The chapter on the history of ionic liquids by Freemantle in his tutorial book entitled *An Introduction to Ionic Liquids* and released in 2010 also provides key dates in ionic liquid research [4].

1.2 Ionic Liquid Generations

The classification of ionic liquids into different "generations" along a timeline is an established concept; however, there is no single criterion to apply except a general assessment on structure, properties, and potential applications.

During the 1960s and 1970s, chloroaluminate molten salts were believed to have a low melting temperature and to form a eutectic medium that could improve the overall performance in thermal batteries [22]. Further research and the employment of high-performance computers resulted in an emphasis on the dialkylimidazolium cations for their larger electrochemical window. The proper handling of these substances, however, was quite sophisticated because they required completely water-free conditions.

That is why a new generation (the second generation, but some researchers put the first generation here [5, 25]) became focused on ionic liquids composed of the air- and water-stable alkyl-substituted imidazolium cations with weakly coordinating anions, such as tetrafluoroborate $[BF_4]^-$ and hexafluorophosphate $[PF_6]^-$ [26]. Due to limited stability, these anions were later often replaced by the more stable tris(pentafluoroethyl) trifluorophosphate "FAP," triflate $[CF_3SO_3]^-$ or $[OTf]^-$, and particularly bis(trifluoromethanesulfonyl)amide $[(CF_3SO_2)_2N]^-$ or $[NTf_2]^-$, as well as by the halide-free anions methylsulfate $[C_1SO_4]^-$, acetate $[C_1COO]^-$ or $[Ace]^-$, thiocyanate $[SCN]^-$, and others. Typical cations were, among others, (substituted) ammonium, imidazolium, pyridinium, or pyrrolidinium.

Initially, ionic liquids were considered to be "green solvents" because of their very low volatility. Not surprisingly, the question soon arose whether these substances were really "green solvents." Studies on toxicity, persistence, and degradability were conducted [27, 28]. Ionic liquid research began to target ions with proven low toxicity while retaining the desired (and established) material properties. The methods of synthesis were also scrutinized to avoid toxic educts, to improve yields, and to minimize impurities. At the turn of the millennium, choline- and lactate-based ionic liquids were developed [29]. A new feature characterizing the third generation of ionic liquids was customized biological properties [25, 30]. Rogers and co-workers used the term "The Third Evolution of Ionic Liquids" in their paper on ionic liquids featuring active pharmaceutical ingredients [25]. The term "Ioliomics" has now been coined [31], and it is certain that the next step will focus on sustainability.

To sum up the evolution of ionic liquids in a single statement, unique tunable physical properties characterize generation one (or two by others), targeted chemical properties combined with chosen physical properties characterize generation two, and ultimately targeted biological properties while keeping the others characterize generation three. Upcoming ionic liquids of the fourth generation will have to be entirely sustainable.

The so-called deep eutectic solvents (DES) can be seen as "distant relatives" or a sideline of the ionic liquid family tree. DES share many typical features of ionic liquids, but the ultimate distinguishing criterion is that DES are always a mixture while an ionic liquid is a single substance (a salt formed from the combination of a cation and anion) [29, 32].

Most ionic liquids used in industrial processes and commercial applications still belong to the second generation. One reason is that many of them are universally and thoroughly characterized, like 1-hexyl-3-methylimidazolium bis(trifluoromethylsulfonyl)amide $[C_6C_1im][NTf_2]$, a reference material for an inter-laboratory thermophysical study with consolidated data available for design and development. The ionic liquids of the second generation do have drawbacks, but these drawbacks are controllable. A challenge comes from another direction, namely computational chemistry. As the number of possible ionic liquids is extraordinarily large, high-performance computing (using elaborate models that represent the molecule as realistically as possible) can assist in a preselection of substances that display some desired characteristics. In 2007, Maginn proposed a paradigm shift from "post-prediction" of properties to modeling systems that have not yet been synthesized [33].

1.3 Organization

This book is organized into four parts: Part I, Introduction (this chapter); Part II, Ionic Liquid Processes (Chaps. 2–5); Part III, Ionic Liquid Products (Chaps. 6–8); and Part IV, Future Ionic Liquid Applications (Chaps. 9–11).

Part II provides examples of processes using ionic liquids. Industry has a continuing long-range interest in advanced process technologies that have promise to save energy and materials and reduce waste. In Chap. 2, we highlight one such technology [34, 35] developed for next-generation alkylate gasoline manufacturing. Hye Kyung Timken, Huping Luo, and B. K. Chang[1] and Elizabeth Carter and Matthew Cole[2] provide a comparison of the new ionic liquid catalyst performance relative to the incumbent technologies. The first alkylation plant is scheduled for start-up in 2020 and will be one of the largest scale commercial applications of ionic liquids to date.[3][4]

---

[1] Chevron Energy Technology Company.

[2] Honeywell UOP.

[3] ISOALKY™, Chevron Energy Technology Company, and Honeywell UOP.

[4] Commercial suppliers, equipment, instruments, or materials are identified only in order to adequately specify certain procedures. In no case does such identification imply recommendation or endorsement by the National Institute of Standards and Technology, nor does it imply that the products identified are necessarily the best available for the purpose.

Professor Dr. Marco Haumann at Friedrich-Alexander-Universität (FAU) in Erlangen-Nürnberg, Germany, [36] wrote Chap. 3 on "Continuous Catalytic Processes with Supported Ionic Liquid Phase (SILP) Materials." Dr. Haumann and Professor Dr. Peter Wasserscheid, who is also at FAU and a director of the Helmholtz Institute Erlangen-Nürnberg (HI ERN) for Renewable Energy Production [37, 38], have pioneered the development of SILP materials as catalysts for numerous reaction chemistries.

Professor Robin Rogers and Dr. Julia Shamshina at The University of Alabama and their start-up company[5] [39] provided Chap. 4 entitled "Are Ionic Liquids Enabling Technology? Startup to Scale-up to Find Out." This insightful chapter provides the knowledge needed and the challenges associated with taking a new ionic liquid technology for a novel bio-based material made from chitin from an academic laboratory to commercialization.

Chapter 5, "Commercial Aspects of Biomass Deconstruction with Ionic Liquids," was written by Professor Jason Hallett and his group at Imperial College in London [40]. Dr. Hallett has pioneered the use of low-cost ionic liquids as a pretreatment step for biomass fractionation[6] that can dissolve the lignin and hemicellulose leaving a cellulose-rich pulp ready for saccharification.

Part III provides examples of products that have been developed using ionic liquids. Chapter 6 was written by Professor Daniel Armstrong and his group at the University of Texas in Arlington [41] and describes the development of a new class of capillary gas chromatography (GC) columns with stationary phases based on ionic liquids. His group has synthesized dicationic and polycationic ionic liquids that are stable to water and oxygen even at high temperatures, which is critical for their use as stationary phases in GC columns. The columns are now commercially available[7] [42].

Chapter 7 was provided by Dr. Natalia Plechkova, Dr. Martyn J. Earle, and colleagues at the Queen's University of Belfast in honor of the late Professor Kenneth Seddon (see Preface for dedication). Under Professor Seddon's leadership, the Queen's University Ionic Liquids Laboratory (QUILL) was founded in 1999 to explore, develop, and understand the role of ionic liquids and focuses on their synthesis, characterization, and applications [43]. The chapter describes a new ionic liquid–liquid chromatography (ILLC) instrument developed in collaboration with a commercial partner[8] [44] for performing novel separations. Compounds thought to be too insoluble or too immiscible with biphasic molecular solvent systems can now be separated using ILLC.

Chapter 8, entitled "Commercial Production of Ionic Liquids," was written by Dr. Thomas Schubert [45]. The company[9] is a commercial supplier of ionic liquids

---

[5]525 Solutions, Inc.

[6]IonoSolv, Imperial Innovations.

[7]Supelco/Sigma-Aldrich.

[8]AECS-QuikPrep™ Ltd/Quattro.

[9]IoLiTec GmbH.

for a variety of applications. This chapter provides important insights into several aspects of commercial ionic liquid production, such as synthesis, purity, and price.

Part IV provides examples of future products and processes that will use ionic liquids. Chapter 9, written by Dr. Luke Haverhals[10] [46] and Professors David Durkin and Paul Trulove (U.S. Naval Academy) [47], describes a revolutionary process for designing new high-performance composites. Their chapter entitled "Natural Fiber Welding" describes fabrication techniques for using natural materials to produce robust, functional, and biodegradable composites that can displace non-biodegradable plastics.

Chapter 10, entitled "Development of New Cellulosic Fibers and Composites using Ionic Liquids Technology," was written by Dr. Frank Hermanutz and Marc Philip Vocht at the German Institutes of Textile and Fiber Research [48] and Professor Dr. Michael Buchmeiser at the University of Stuttgart [49]. Several examples for processing cellulose to produce materials, such as super-microfibers, all-cellulose composites (ACCs), and carbon fibers, are discussed.

The final Chap. 11 written by Dr. Roland Kalb[11] [50] provides over 50 applications for ionic liquids that have been commercialized or are in pilot scale indicating that an exciting new era in the field is about to begin.

Next, we describe the development of the ionic compressor.

1.4   Ionic Compressor

The idea of an "ionic compressor" originates from the principle of communicating vessels. A liquid piston compression system for compressing low-pressure steam to recover waste heat was described in a 1984 patent document [51]. The compressing liquid was either water or glycol. A liquid with appropriate tribological properties can improve compression performance because the device can operate with fewer moving parts, produce less noise, work without sophisticated sealing systems, and the heat conductivity of the liquid can be used to perform a nearly isothermal compression. Wear is significantly reduced, and maintenance intervals can be lengthened compared to conventional techniques. In 2000, Pieperbeck [52] described a method to compress explosive gas mixtures using this principle. The working fluid was characterized by its low vapor pressure and low solubility of the gas in the fluid combined with non-reactivity at the operating conditions. As a typical feature, most ionic liquids have negligible vapor pressure.

The idea to use ionic liquids in compression technology was developed by Adler and co-workers [53]. Ionic liquids offer the possibility to combine anion and cation ("tailor-made solvents") that can specifically adapt to the medium. The handling of hydrogen as a fuel is challenging, because a fuel cell requires the absence of any contaminants, and storage and dispensing of hydrogen are often at high pressures up

---

[10]Natural Fiber Welding, Inc.

[11]Proionic GmbH.

to several hundred bars. Therefore, an ionic liquid suitable for this application must display good tribological behavior, minimum hydrogen solubility, suitable viscosity and density, as well as high thermal stability, conductivity, and heat capacity at the operational conditions. There must be no reactivity between the ionic liquid and hydrogen (or any other gas that may be present). This requires a certain defined quality of the ionic liquid. Ideally, the ionic liquid and hydrogen should be mutually immiscible.

Hydrogen solubility (in contrast to, for example, carbon dioxide) is extremely low in most ionic liquids. The temperature dependence of solubility is typically the opposite of that observed with carbon dioxide. Actually, hydrogen solubility increases with increasing temperature. This behavior was observed during the early stage of systematic solubility investigations on ionic liquids of the second generation [54]. Solubility data with a particularly low uncertainty over an extended p, T-range were reported for [C_4C_1im][PF_6] and the IUPAC-proposed reference [C_6C_1im][NTf_2] using a static-cell technique [55, 56]. Calculating the enthalpy $\Delta_{sol}H$ and entropy $\Delta_{sol}S$ of gas dissolution provides information about the molecular interaction between the gas and the ionic liquid and about the degree of ordering that determines the solubility. In the case of hydrogen, the Henry's Law constant decreases at higher temperatures which results in a positive dissolution enthalpy [57, 58]. From a thermodynamic consideration, the solubilities of different gases converge to the same value at the solvent critical temperature, an imaginary point for most ionic liquids due to their limited stability [59]. Figure 1.1 illustrates this particular behavior of the Henry's Law constant based on solubility data for [C_6C_1im][NTf_2]. The low-soluble hydrogen shows the inverse temperature dependence in the region relevant to technical applications.

In 2002, a company[12] started to pursue the idea of developing a commercial ionic compressor with an ionic liquid as the operational fluid [61]. A patent for this

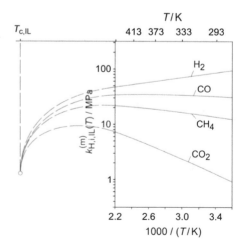

Fig. 1.1 Gas solubility as represented by Henry's Law constant (approaching zero pressure and on the molality scale) of various gases in the ionic liquid [C_6C_1im][NTf_2] with extrapolation to a hypothetical critical temperature. For the corresponding data, see the paper by Kumełan et al. and the references cited therein [60]

[12]Linde GmbH.

technique was granted in 2006 [62]. At the same time, a patent was obtained by another company[13] [63] on liquids for compressing a gaseous medium. The claims comprise liquids with a vapor pressure $<10^{-3}$ mbar, preferably ionic liquids. Another patent for a modified ionic compressor followed in 2008 [64]. The first commercial installation started in July 2005 in Austria.[14] The unit supported a fuel station for natural gas and operated at 25 MPa with a capacity of 500 m^3 natural gas per hour [53].

The ionic compressor is now part of some hydrogen fueling stations. The flagship model that qualifies for the fueling protocol SAE J2601-A70 [65] is a five-stage ionic compressor with a stage compression ratio of 1:2.8, an input pressure from 0.5 to 20 MPa, a maximum output pressure of 100 MPa, a stroke frequency of 5.8 Hz, a maximum delivery rate of about 33.6 kg hydrogen per h (as single line, double line is optional), and a specific energy consumption of 2.7 kWh per kg hydrogen. Noise emission is lower than 75 dB at 5 m. The target fueling pressure is 70 MPa at 288 K [66]. Compared with conventional piston pumps, energy consumption is reduced by about 20%, and the equipment can run about 500 days without maintenance—a factor of 10 longer. The costs for maintenance were also reduced by about 50% [53, 61]. In 2014, the compressor[15] entered production [61].

In a recently released brochure entitled *The Driving Force,* some reference projects on hydrogen fuels describe the entire supply chain from production to the final customer [67]. Ionic compressors supply hydrogen to a bus fleet in the towns of Aberdeen, Scotland, and Hamburg, Germany, at the power-to-gas installation at the Energiepark in Mainz, Germany, at the first hydrogen fueling station in the USA (opened in October 2014) in Sacramento, California, and at Japan's first commercial hydrogen fueling station in Iwatami.

In a presentation from November 18, 2014, which was later displayed on the web, Beckman outlined a corporate strategy in the hydrogen fueling business [68]. As a result of an infrastructure project with the government of California, ionic compressors are working at nine different hydrogen fueling stations. Beckman gives an estimate of 87 stations in California by the year 2020. An initiative for hydrogen mobility in Germany forecasts the construction of about 400 hydrogen fueling stations in the country by 2023.

Notably, the corporate documents released on an ionic compressor do not explicitly specify an ionic liquid. That information is proprietary, but it can be assumed that the best substance to be employed matches the above-mentioned criteria for an operating fluid. It is also likely that the company uses additives that improve the performance.

The Ph.D. thesis by Nasrin Arjomand Kermani completed at the Technical University of Denmark (DTU) and published in 2017 gives an indication of potential ionic liquid candidates [69]. In her work on the design of an ionic compressor, Arjomand

---

[13] Proionic GmbH.

[14] Wien Energie GmbH.

[15] Model IC90 v 1.3, Linde GmbH.

Kermani applied elimination criteria on five ionic liquids that had already passed several preliminary exclusion criteria for the anion and cation. The five candidates that made the final round were $[C_2C_1im][CF_3SO_3]$, $[C_2C_1im][NTf_2]$, $[P_{6\,6\,6\,14}][NTf_2]$, $[N_{1\,1\,1\,4}][NTf_2]$, and $[N_{1\,8\,8\,8}][NTf_2]$. In the final analysis, $[C_2C_1im][NTf_2]$ was the ultimate choice over $[C_2C_1im][CF_3SO_3]$ due to lower water miscibility, higher thermal stability, higher heat capacity, and a lower melting point. In addition, $[C_2C_1im][NTf_2]$ and $[C_2C_1im][CF_3SO_3]$ are available at industrial scale and both are classified by REACH [70]. Consolidated data for hydrogen solubility are only available for $[C_2C_1im][NTf_2]$ [71], which is an ionic liquid of the second generation. Currently, the database *ILThermo* 2.0 contains solubility data of ten binary systems (hydrogen + ionic liquid) [72] as will be highlighted in the next section.

1.5 IUPAC Projects with Impact on Commercial Applications

A historical perspective by Joe Magee, (NIST, Boulder, Colorado, U.S.A.).

1.5.1 IUPAC Project 2002-005-1-100 Thermodynamics of Ionic Liquids, Ionic Liquid Mixtures, and the Development of Standardized Systems

The utilization of ionic liquids in both chemical research and in industrial chemistry requires a systematic study of their thermodynamic and thermophysical properties that are required for chemical process design. For these reasons, Professor Kenneth (Ken) Marsh (formerly with the University of Canterbury, New Zealand) formed an international task group under the auspices of IUPAC with the goal of providing reliable data for a range of properties that could be used to check methods and calibrations for experimental instruments. Professor Marsh called the first task group meeting in Rostock, Germany, at the IUPAC International Conference on Chemical Thermodynamics [73]. Professor Marsh announced the intentions of this project at a workshop co-chaired with Dr. Joe Magee (NIST). Choosing a good reference material was paramount to this work. It would have to have a low melting point, a low viscosity, hydrophobic properties, be unrestricted by patents, and readily synthesized by users. After lengthy deliberations, the task group chose the reference material 1-hexyl-3-methylimidazolium bis(trifluoromethylsulfonyl)amide ($[C_6C_1im][NTf_2]$). A total of one liter of reference material was synthesized, purified, and characterized by Dr. Mark Muldoon of Professor Joan Brennecke's laboratory at the University of Notre Dame, Indiana. That sample was shipped to Dr. Magee at NIST, where it was further dried and characterized by ^1H and ^{19}F NMR spectroscopy and Karl Fischer titration. Dr. Jason Widegren (NIST) divided the sample into aliquots that were prepared for shipping to the participants in the measurement program. Each sample was handled inside a dry nitrogen glovebox where it was added to chemically cleaned, hermetically sealed Schlenk tubes. Most of the 1 L sample was divided and distributed to participants. NIST asked each participating laboratory to perform a Karl Fischer titration for water as soon as they received it and to report this information to Dr. Widegren. All samples were received intact with no water contamination due to an airtight seal that was maintained during shipping.

Professor Marsh selected task coordinators in consultation with the full task group and chair, and laboratories were selected to measure a specific property. Over several years, various measurements were conducted by the participating laboratories. Participants e-mailed their data sets to their task coordinator. Coordinators handled a report from a data owner as confidential information, and each was promptly forwarded to Dr. Robert (Rob) Chirico (NIST, Boulder, Thermodynamics Research Center). During the data evaluation phase, only Dr. Chirico had access to the experimental measurements. Authors of the specific data sets were allowed to publish their own studies at any time, and some chose to do so.

Dr. Chirico collected all data sets in an unpublished archive. Professor Marsh et al. summarized them in a report published by IUPAC [74]. This report described the experimental methods, results, and uncertainties for each data set. It surprised no one that, with a large international effort, there were challenges. Probably the central challenge was that each participating laboratory was overcommitted. That is to say, they were already busy processing their own measurements, which took priority. Since this work was not explicitly funded by the IUPAC Task Group, it was naturally given a lower priority in the experimental queue for some participants. Lower priority meant that NIST had to hold up the data evaluation phase for about three years until all samples were shipped. Another challenge was that the experimental results were posted to files with very different formats that varied from a text table with no ancillary description to a reprint of a published paper. Luckily, Dr. Chirico could contact authors directly for the missing details, mostly related to standard uncertainties and experimental methods. Finally, all the experimental data and metadata were collected. After all data sets had been submitted to Dr. Chirico, he then applied an array of assessment tools described in a recent paper [75] to critically evaluate the experimental data. Dr. Chirico et al. provided a final report and recommended property values in *Pure and Applied Chemistry* (2009) [76].

1.5.2 IUPAC Project 2003-020-2-100 Ionic Liquids Database

Critically evaluated experimental data are essential to carry out innovative chemical process design or to improve material and energy efficiencies of existing chemical processes. Developed under the auspices of IUPAC, *ILThermo* (https://ilthermo.boulder.nist.gov/) is a data archive of experimental thermophysical properties (including 120 thermodynamic, transport, and thermochemical properties) of ionic liquids and mixtures that contain them. Version 1.0 was announced at COIL-1 [77], and detailed capabilities were described [78].

Creation of a broadly scoped database required a sustained effort at a high level, so it is useful to reflect on how *ILThermo* was created. As mentioned earlier, in April 2000, the first international meeting devoted to ionic liquids, *Green Industrial Applications of Ionic Liquids*, was held in Heraklion, Crete; the 50 or so attendees issued recommended research needs. Among those needs was "a verified, web-based database of physical, thermodynamic, and related data (not process specific)." It was quite clear that a database was needed; the remaining issues, such as when and its scope, would be decided after a task group was formed. Such a project required minimizing risk factors. Risk factors include: (1) committing to an insufficient scope, (2) failing to understand customers' needs, and (3) devoting insufficient scientific talent. Within a burgeoning research field, it seemed that boundaries were being pushed back in all directions. In addition, announcements of new discoveries were commonplace. It was clear that the work needed to get underway. Figure 1.2 illustrates why time was of the essence. Because the research was growing at a power-law rate, postponing the work to build a database would clearly be a mistake. The project was starting from scratch, the backlog of work was already formidable, and it was growing. Dr. Magee recognized that industrial needs always include mixtures, so mixtures were added to the project scope from the beginning. Realizing there was a great fit to NIST's mission, Dr. Magee organized a NIST group capable of the work and secured new internal funding.

To build consensus that NIST should do the work to create a database, a new IUPAC project was proposed. Shortly afterward, while attending an ACS meeting (Boston, August 2002), Dr. Magee met with Professor Kenneth (Ken) Seddon (QUILL, UK) where they launched a collaboration to build a database. In the weeks that followed, Dr. Magee wrote an IUPAC proposal. Professor Seddon recruited an international task group, presented the proposal to IUPAC, and guided it through approval.

Fig. 1.2 Ionic liquid publications per year (1990–2005)

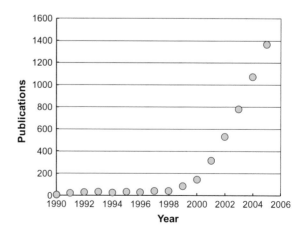

A few months later in January 2004, the task group met in Delft, The Netherlands, to discuss a work plan. Dr. Magee led that meeting with a presentation of NIST's vision and approach, which was approved by the task group. After the task group meeting, a NIST team was assembled and led by Dr. Magee and Dr. Michael Frenkel. Thermodynamics Research Center (TRC) group members (Qian Dong, and Drs. Robert Chirico, Andrei Kazakov, Chris Muzny, and Vladimir Diky) and Experimental Properties of Fluids group members Professor Ken Marsh (guest scientist) and Dr. Widegren were responsible for different aspects of the work. The team met about once per week and, between meetings, new data (and metadata) were captured from about ten primary journals and about twelve secondary journals, which were queried monthly for data in the scope of the project. The team also ensured that classic published papers had been well covered. NIST/TRC's in-house archive, known as SOURCE, had been conceived for molecular substances and their mixtures, not ionic substances. However, the NIST team modified SOURCE by expanding its scope to include ionic compounds by, for example, permitting searches to proceed by a specific ion and by adding impurity specifications for water and for halides. The team decided to retain the structure of SOURCE in the ionic liquids database, called *ILThermo*, shorthand for *ionic liquids thermophysical properties*. Dr. Kazakov wrote a simple computer code that would port data from SOURCE to *ILThermo*. The porting step was conducted on a frequent basis. In July 2006, NIST hosted THERMO International 2006, the world's most comprehensive conference on thermophysical properties, in Boulder, Colorado. It was natural that the time was right to publicly release *ILThermo*. On July 31, 2006, *ILThermo* Version 1.0 was released with holdings of 18,000 data points.

ILThermo is best described as a web-based, open-access database that provides data and metadata from published experimental studies of ionic liquids, including numerical values of thermophysical properties, chemical structures, measurement methods, sample purity, critically evaluated standard uncertainties of property values, and other significant measurement details. Since its public release in 2006, the quantity of data stored in *ILThermo* has doubled five times, and in May 2018, the database stored 561,000 data points. In addition to this prodigious growth, *ILThermo* has been released in Version 2.0 [72], which is keyed on the chemical structure of each chemical species curated. This feature brings inherent advantages to users of the database. At this writing, *ILThermo* continues to see significant usage (~15,000 annual users) by practitioners in this field. In addition to researchers and educators, usage data indicate that engineers in industry are using *ILThermo* to search and retrieve thermophysical properties and phase behavior data for use in engineering applications.

The successful outcomes of both IUPAC projects were, in no small part, due to the task group leaders, the late Professor Kenneth Marsh and the late Professor Kenneth Seddon, whose intellect, leadership, and collegiality are deeply missed by all who worked with them.

1.6 Industrial Activities at the Turn of the Century in Europe

How could the new (second generation) RTILs find their way into industrial processes? A key premise was the availability on a large scale, so that such processes could be successful. Another important factor that promoted the use of ionic liquids was a strong cooperation between academic consortia and industry.

In 1999, the QUILL Research Centre at the Queen's University, Belfast, Northern Ireland, was founded by the late Professor Seddon OBE and Professor Swindall. QUILL's statutes are similar to the US NSF Industry-University Cooperative Research Center model, where research is supported by an industrial advisory board [43, 79, 80]. As previously discussed, one year later in 2000, a NATO-sponsored advanced research workshop in Greece further promoted the interest in ionic liquids [81]. In 2004, a document that displayed a roadmap for ionic liquids in industrial innovation followed in which both scientific and business issues were addressed [82]. Dr. Plechkova and Professor Seddon gave a comprehensive overview of industrial applications of ionic liquids in their 2008 review with 148 listed references. Many items represent a cluster of book chapters, journal papers, technical reports, and patents [83]. Their work portrays an account of early industrial applications of ionic liquids in great detail. Notably, they pointed out—in contrast to the general notion at that time—that the developments of ionic liquids regarding academic research and industry were parallel from the very beginning.

In Germany, DECHEMA (originally "Deutsche Gesellschaft für chemisches Apparatewesen") did a similar job to that of QUILL in the UK, namely in promoting IL science. DECHEMA is a non-profit organization with 5800 members, both individual persons and institutional members, which defines itself as a hub of interdisciplinary knowledge exchange [84]. Under the auspices of DECHEMA and organized by Professor Wasserscheid from the RWTH Aachen University, the first international "Green Solvents Conference" was held in 2002 that was followed by seven additional meetings in a biennial interval. One-third of the approximately 220 participants of the first meeting came from industry [85]. Discussions were not restricted to only ionic liquids, but also included the so-called alternative solvents, meaning alternatives to traditional volatile organic solvents. In 2005, DECHEMA hosted "COIL 1," the First International Congress on Ionic Liquids, at Salzburg, Austria. A section entitled "Alternative Solvents" was established within the organization in 2003, and a policy paper entitled "Advanced Fluids—New Solvent Concepts for Process and Product Optimization" came out in 2006 [86]. The following Green Solvent Conferences kept the interdisciplinary orientation. Notably, the section "Advanced Fluids" was dissolved by 2018 (with 160 members at that time). The acting organizing committee did this because the "mission of establishing alternative solvents is accomplished." As ionic liquids or alternative solvents gained importance within the scientific community, the organizers of other conferences in relevant fields also began to establish sessions explicitly dedicated to them, often labeled with terms like "Green Chemistry." For example, the European Conference on Molten Salts that started in 1966 was joined by the ionic liquid community in 2006; the resulting 21st EUCHEM Conference on Molten Salts and Ionic Liquids organized by Marcelle Gaune-Escard and Kenneth R. Seddon was held at Hammamet, Tunisia. Following these events, the decision by DECHEMA can be seen as consequential and a manifestation of the fact that the field of ionic liquids has been established.

1.7   The Startup Companies

A historical perspective by Dirk Tuma (BAM, Berlin, Germany).

How did the production of ionic liquids begin on a commercial scale? In contrast to most laboratory experiments, industrial applications require substances that have well-defined qualities. The substances also have a completely different body of rules regarding licensure, handling, and risk management. Major European chemical companies[16] began to develop a portfolio of ionic liquids [83]. Additionally, ionic liquids created a market niche for highly specialized small enterprises.

The performance of newly founded companies specializing in ionic liquids was the subject of a scientific study on technology entrepreneurship conducted by Runge at the Karlsruhe Institute of Technology (KIT), Germany, and published in 2014 [87]. The case study comprised two German companies[17] as well as two British businesses.[18] According to Runge, their business model is a so-called technology push approach which is based on a new technology (i.e., ionic liquids). A disruptive innovation based on a platform technology with a potential for broad applicability in diverse markets but with little or no venture capital was launched. The "technology push" strategy involves introducing new materials to the marketplace as an optimized solution in order to gain a competitive market share with established technologies.

In 1999, a corporate spin-off[19] of the Rheinisch-Westfälische Technische Hochschule (RWTH) Aachen University was founded by Claus Hilgers and Peter Wasserscheid, who later recruited a minority partner.[20] In 2002, the company was awarded the Innovationspreis der deutschen Wirtschaft [88]. From 2004 onwards at their new location at Cologne, the company had an annual production capacity greater than 5 metric tons. By 2005, the spin-off changed its business model from a research-driven solvent maker to providing integrated systems and solutions for customers. Twenty-two ionic liquids were available on a multi-kilogram scale; among them a substituted quaternary ammonium sulfate,[21] labeled by the company as "Peg-5-cocomonium methosulfate," is the first substance produced on ton scale and registered by the European Inventory of Existing Commercial Chemical Substances

---

[16]BASF SE, Degussa AG, and Merck KGaA.

[17]Solvent Innovation GmbH and IoLiTec GmbH.

[18]Scionix Ltd. and Bioniqs Ltd.

[19]Solvent Innovation GmbH.

[20]Degussa AG (Evonik Industries AG since 2007).

[21]ECOENG™ 500, Solvent Innovation GmbH.

(EINECS). In addition, a family of ammonium-based ionic liquids[22] was designed as dispersing agents for homogenization of pigment formulations. In January 2008, the spin-off was acquired by a large chemical company[23], and the portfolios were integrated.

A company[24] was founded in 2003 by Thomas Schubert (who came from another ionic liquids company[25]) at Freiburg [45]. The company was profitable from the beginning due to a supporting number of clients and financially strong joint projects, as the company entered an already consolidated market. Notably, the company was a direct competitor in sales and was also engaged in contract R&D services and developed several trademarks to enlarge its portfolio.[26] In March 2005, the company launched a free newsletter. A major upgrade came in 2008 with the introduction of microreactor systems for scaling-up. This technique enabled variable configurations to produce different ionic liquids on a multi-kilogram-per-day scale. The engagement of a local investment firm[27] with 30% stake resulted in the company's transfer to Heilbronn [89], and by 2010 a branch office operated at Tuscaloosa, Alabama, USA. In 2010, the company opened a rental service for ionic liquids. The business figures attested to a continuous and healthy growth and its shares were sold back to the founders of the firm in 2017. Presently, the firm supplies the industry with more than 325 ionic liquids, 180 nanomaterials, and 50 key intermediates. The company is currently preparing an additional production site in the eastern part of Germany (Bitterfeld-Wolfen), where they will produce ionic liquids in a 1 metric ton reactor (autoclave) and several continuous-flow production units. A new technique has also been developed for drying ionic liquids at larger scale, and production is planned to begin in the third quarter of 2019.

A British company[28] started in 1999 as a joint venture between the University of Leicester (Professor Andrew Abbott) and a corporate venture company[29] [90]. In their business model, research is done at the university, while production, licensing, and marketing remains with the company. The ionic liquids are manufactured by another chemical manufacturer[30] in Huddersfield. The ionic liquids from this company are formulations of quaternary ammonium salts with either a metal salt or a hydrogen-bond donor giving deep eutectic solvents.[31] A significant number of formulations are based on choline chloride (i.e., vitamin B4).

---

[22] AMMOENG™, Solvent Innovation GmbH.

[23] Merck KGaA.

[24] IoLiTec GmbH.

[25] Solvent Innovation GmbH.

[26] Including IoLiSens® for sensor technology, IoLiTherm® for thermofluids, IoLiLyte® for electrolytes, and IoLiTive® for additives, IoLiTec GmbH.

[27] ZFHN Zukunftsfonds Heilbronn GmbH & Co. KG.

[28] Scionix Ltd.

[29] Genacys Ltd., a subsidiary of the Whyte Group Ltd.

[30] Grosvenor Chemicals Ltd.

[31] Scionix Ltd.

A second British company,[32] started in 2004, was a spin-off from the University of Cambridge and later moved with its founders to the University of York. This firm targeted primarily biochemical and biocatalytical processes, and the ionic liquids were designed to replace water. The company pursued a more science-driven approach and generated its revenue by consultancy and process design. In 2006, a local Yorkshire investment company invested in this firm, and a partnership with a larger chemical company[33] was established for the production and distribution of the products. Subsequently, the production of environmentally friendly solvents in accordance with the respective legislation began, and this firm promoted a label[34] that addressed sustainability and anticipated ionic liquids of the third generation. The qualification criteria were that all precursor materials used in manufacture came from renewable feedstocks, were biodegradable within 14 days (>98% by mixed community of natural soil organisms), were non-mutagenic as determined by the Ames test (mutagenicity index < 2, mutagenic activity ratio <2.5), and had low toxicity to both Daphnia magna (EC50 > 250 mg/L) and the green algae *Selanstrum capricornutum* (ErC50 > 100 mg/L) [91]. The company strongly resorted to in-silico design to develop these systems. Another success was the extraction of the anti-malaria drug artemisinin, a sesquiterpene lactone, from plant material with ionic liquids [92]. In 2007, however, the firm faced financial problems and ultimately went bankrupt in 2009.

From an economic point of view, Runge came to the conclusions that the company founded in 1999 as a corporate spin-off from RWTH Aachen had difficulty transforming scientific achievements into a business providing sufficient revenues and had to cope with a close competitor founded in 2003 with the same market segments. The 2003 firm had a better start due to specialized people, a consolidating market, a broader network, and an effective business strategy. In addition, the 2003 firm serves a niche market and volumes are currently too small for larger chemical manufacturers, so there is minimal mutual competition. The British firm founded in 1999 as a joint venture had less competitive strength compared to the 2003 company due to its constitution and served a different market sector. Ultimately, the other British firm, founded in 2004, was curiosity-driven and had a remarkable scientific record. The company operated continuously at a high-risk level, and an irreversible market entry did not succeed. These were the main factors that contributed to bankruptcy [87].

There are two additional companies of similar background that also have had success, the first in France and the second in Austria.

The French start-up,[35] founded in 2003 by François Malbosc and located at Toulouse, produces ionic liquids and develops applications in the fields of catalysis, surface preparation, and energy storage [93]. This firm is strongly engaged in energy-related research projects funded by the European Union, the most recently launched

---

[32]Bioniqs Ltd.
[33]Merck KGaA.
[34]ECONIQS™, Bioniqs Ltd.
[35]Solvionic SA.

being MAGENTA, that is, magnetic nanoparticle-based liquid energy materials for thermoelectric device application, within the framework Horizon 2020 [94].

The Austrian company[36] was started at Grambach in 2004 [50]. The founders Roland Kalb and Michael Kotschan straightforwardly pursued the cost-efficient production of ionic liquids on an industrial scale in an entirely sustainable way while maintaining high purity. The company introduced a product line[37] that uses a halide- and waste-free synthesis route for production. This technology was developed by Kalb in 2003 and is shown in Scheme 1. Additional details and the underlying patent documents are given in the references [95, 96].

The proprietary method follows a modular conception. The cation of the desired ionic liquid enters the reaction as a hydrogen carbonate or methyl carbonate, the precursor (see Scheme 1), and the anion joins as a Brønsted acid. The reaction proceeds quantitatively with the chemical equilibrium continuously shifted by the generation of CO_2. Alternatively, when the corresponding Brønsted acid is not available, an ammonium salt can be used.

The new process enabled the Austrian firm to produce multi-ton quantities of ionic liquids (today's annual capacity is 150 tons, by mid-2019, 500 tons). Large chemical suppliers offer selected precursors [97], and some may be the license holder of the methods [98]. Following the customers' requirements, the company's best-sellers cover ionic liquids composed of the following ions: $[C_2C_1im]^+$, $[C_4C_1im]^+$, and $[C_4C_1pyr]^+$ as the cations, and $[Ace]^-$, $[BF_4]^-$, $[CH_3SO_3]^-$, $[OTf]^-$, $[NTf_2]^-$, $[FSI]^-$, and $[C_1CO_3]^-$ as the anions. In January 2010, a group of international engineering companies[38] focusing on process plants with its headquarters in Austria took a 70% share of the company [99].

$$(Cation)^+CH_3CO_3^- + H^+(Anion)^- \rightarrow (Cation)^+(Anion)^- + CH_3OH +$$
$$CO_2\uparrow$$
$$(Cation)^+HCO_3^- + H^+(Anion)^- \rightarrow (Cation)^+(Anion)^- + H_2O + CO_2\uparrow$$
$$(Cation)^+CH_3CO_3^- + NH_4^+(Anion)^- \rightarrow (Cation)^+(Anion)^- + CH_3OH$$
$$+ CO_2\uparrow + NH_3\uparrow$$

Scheme 1 Synthesis of an ionic liquid [$(Cation)^+(Anion)^-$] by the CBILS® method

---

[36]Proionic GmbH.

[37]Carbonate Base Ionic Liquid Synthesis, CBILS®, Proionic GmbH.

[38]VTU Group GmbH.

1.8   The Major Chemical Companies

In their review of the industrial use of ionic liquids, Dr. Plechkova and Professor Kenneth Seddon reported the situation as of 2007 [83]. There were four major European chemical companies with ionic liquids in their portfolio.[39] In fact, one such chemical company[40] was involved in the foundation of QUILL [79], and a second[41] had ionic liquids on display at the first Green Solvents Conference. Runge also portrayed the engagement of major players in his market study [87].

The company that was a founding member of QUILL was the first large company to go into ionic liquids. Runge mentioned a project with ionic liquids for battery applications as early as the 1980s, but the program was discontinued. Ionic liquids were started at this firm in 2002 at a facility that was also used for the production of liquid crystals. This company not only offered a broad selection of ionic liquids (about 250 by 2004) but also provided a searchable web-based database containing physical properties. Notably, to overcome the problem of hydrolytic decomposition of perfluorinated anions, this company developed and patented the so-called FAP (i.e., tris(perfluoroalkyl)trifluorophosphate) anion, which also displayed a high electrochemical and thermal stability [100–102]. By 2008, this company had acquired another firm[42] and integrated the business. A chemical supplier[43] became a subsidiary in 2015. Today, ionic liquids can be found within the portfolio of their company brand[44] [103].

Another major company,[45] known as one of the biggest producers of bulk chemicals, performance products, and formulations in the world, turned to ionic liquids around 2002. A competitive portfolio of diverse chemicals as well as processes was established within a short time [83, 98]. The ionic liquids were classified[46] within the category of "intermediates," an arrangement that remains today [104]. The ionic liquids in this classification are intended to qualify for applications and processes, such as separation enhancers, reaction mediums and lubricants, polymer additives, and eventually as reaction solvents. This firm supplied 23 substances in 2005, most cations being alkyl-substituted imidazolium, such as $[C_4C_1im]^+$ or $[C_2C_1im]^+$, with $[Cl]^-$, $[HSO_4]^-$, $[C_1SO_4]^-$, $[Ace]^-$, or $[SCN]^-$ as anions [105]. Notably, choline acetate and choline salicylate were also on display. Smaller quantities were provided in cooperation with a chemical supplier.[47] In addition to a high patent record, this company cultivated strong ties with academic research, such as sponsoring the activities of DECHEMA. It paid off, as the biphasic acid scavenging utilizing ionic liquids

---

[39]BASF SE, Merck KGaA, Degussa AG, and Acros.

[40]Merck KGaA.

[41]BASF SE.

[42]Solvent Innovation GmbH.

[43]Sigma-Aldrich.

[44]Performance Materials, Merck KGaA.

[45]BASF SE.

[46]BASIONICS™, BASF SE.

[47]Sigma-Aldrich.

process,[48] in operation since 2002, is generally recognized as the first example of a commercially successful process using ionic liquids (although it was probably not the first, as another chemical company[49] claimed priority with an isomerization process using ionic liquids [80, 83]) [106, 107]. This reaction served as a prime example for an alternative that could outperform an established method causing the old process to be abandoned. In 2004, the new process won the Innovation for Growth Award of the magazine *European Chemical News* [83] as well as the Innovation Award of the company's Board of Executive Directors [105]. In fact, many scientific or popular tracts on the industrial uses of ionic liquids describe this biphasic acid scavenging process. In this regard, we refer to three standard monographs: (1) *Ionic Liquids in Synthesis* by Wasserscheid and Welton [108], (2) the volume on ionic liquids in the *Handbook of Green Chemistry* edited by Anastas [109], and (3) the book *Multiphase Homogeneous Catalysis* by Cornils et al. [110]. Scheme 2 illustrates the reaction principle. Alkoxyphenylphosphines are precursors in the production of the company's substances that are used as photoinitiators[50] to cure coatings and printing inks by exposure to UV light. A process had been established that removed the hydrochloric acid by a tertiary amine resulting in the generation of an ammonium salt that precipitated as a sticky slurry. The idea was to employ 1-methylimidazole (MIA), another existing bulk product, instead of the tertiary amine that luckily generated a separate liquid phase, a genuine ionic liquid, at operating conditions.

The lower ionic liquid phase could easily be removed and later retransformed into 1-methylimidazole. The reaction was significantly improved and enabled the company to replace the batch vessel with a jet reactor. The company also explored the possibilities of applying this principle to other similar reactions. Presently, the company supplies 18 ILs [111] and 8 have a REACH registration.[51]

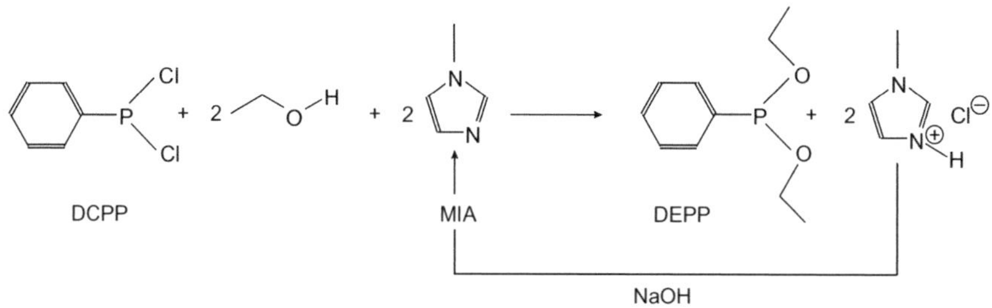

Scheme 2 Acid quench reaction of dichlorophenylphosphine (DCPP) to diethoxyphenylphosphine (DEPP) with a base that forms an ionic liquid, 1-methylimidazolium chloride

---

[48] BASIL™, BASF SE.

[49] Eastman Chemical Company.

[50] LUCIRINES®, BASF SE.

[51] Basionics™ BC01 $[C_2C_1im][Ace]$, HP01 $[C_2C_1im][NTf_2]$, LQ01 $[C_2C_1im][C_2SO_4]$, ST70 $[C_4C_1im][Cl]$, ST80 $[C_2C_1im][Cl]$, VS03 $[C_2C_1im][N(CN)_2]$, FS01 $[N_{1\ 2OH\ 2OH\ 2OH}][C_1SO_4]$, and the formulation $[C_2C_1im][C_1CO_3]$ in methanol, respectively.

Another chemical supplier[52] developed ionic liquids that were somehow different [83, 112]. As early as the 1990s, quaternary ammonium salts were investigated for their performance as, among others, emulsifiers, dispersing agents, and antistatic agents. Paint additives, so-called compatibilizers, were developed under a company brand name[53] (named after Theodor Goldschmidt, who founded a chemical company in 1847). Notably, the use of quantum-chemical methods played a significant role in experimental design. This firm also established a hydrosilylation process of a polydimethysiloxane with an ionic liquid acting as a catalyst. In the biphasic operation, the catalyst was dispersed in the ionic liquid phase, from which the pure product separated as a new liquid phase with the ionic liquid phase remaining active. In 2015, this firm reported a breakthrough in the application of SILP catalysts in a commercial large-scale hydroformylation plant [113]. The catalyst consisted of a rhodium complex with one ligand based on a polycyclic anthracenetriol structure. The corresponding ionic liquid was an imidazolium cation with an anion based on a binary amine. This SILP catalyst exhibited long-term stability of more than 2000 h [113].

1.9 Conclusions

The tremendous research efforts over the past two decades focused on the field of ionic liquids which has yielded a wealth of fundamental knowledge. Many applications in chemistry, engineering, and material science are being studied; thousands of papers and patents are being published; and new products and processes utilizing ionic liquids are being commercialized. This book has brought together the contributions by some of the foremost researchers to help scientists and engineers in the area obtaining knowledge and understanding of the field's broad interest and potential.

Ionic liquids have successfully entered various fields of industrial application. There are not only big stakeholders in the ionic liquid market, but there are also considerable segments for smaller and more specialized enterprises. Referring to the enterprises discussed here, the market research report "Technology Advancements in Ionic Liquids" issued by Frost and Sullivan and another market analysis by Grand View Research specified seven key stakeholders and key players[54] [114, 115].

Finally, in this age where digital information is available in the palm of one's hand, the editors and authors still experience the joy and serendipity of reading a physical book. We hope the readers will delve into all the chapters!

---

[52]Degussa AG (Evonik Industries AG since 2007).

[53]TEGO Dispers®, Degussa AG (Evonik Industries AG since 2007).

[54]BASF SE, IoLiTec GmbH, Merck KGaA, Proionic GmbH, Solvionic SA, Evonik Industries AG, and QUILL.

References

1. Walden P (1914) Ueber die Molekulargrösse und elektrische Leitfähigkeit einiger geschmolzener Salze. Bull Imp Acad Sci St.-Pétersbourg 8:405–422
2. Vaubel W (1903) Lehrbuch der theoretischen Chemie. Erster Band. Materie und Energie – Molekül und Lösung. Springer, Berlin, Heidelberg. https://doi.org/10.1007/978-3-642-50729-8
3. Laus G, Bentivoglio G, Schottenberger H, Kahlenberg V, Kopacka H, Röder T, Sixta H (2005) Ionic liquids: current developments, potential and drawbacks for industrial applications. Lenzinger Berichte 84:71–85. https://www.lenzing.com/fileadmin/content/PDF/03_Forschung_u_Entwicklung/Ausgabe_84_2005.pdf. Accessed on 2019-06-12
4. Freemantle M (2010) An introduction to ionic liquids. RSC Publishing, Cambridge
5. Pernak J, Rzemieniecki T, Materna K (2016) Ionic liquids "in a nutshell" (history, properties and development). Chemik 70:471–480
6. Nambu N, Hiraoka N, Shigemura K, Hamanaka S, Ogawa M (1976) A study of the 1,3,5-trialkylbenzenes with aluminum chloride-hydrogen chloride systems. Bull Chem Soc Jpn 49:3637–3640. https://doi.org/10.1246/bcsj.49.3637
7. Schall C (1908) Über organische und geschmolzene Salze (Eine Leitfähigkeitsstudie). Z Elektrochem 14:397–405. https://doi.org/10.1002/bbpc.19080143002
8. Ramsay W (1876) On picoline and its derivatives. Phil Mag 2(5th series):269–281. https://doi.org/10.1080/14786447608639105
9. MacFarlane D, Kar M, Pringle JM (2017) Fundamentals of ionic liquids: from chemistry to applications. Wiley-VCH, Weinheim. https://doi.org/10.1002/9783527340033
10. Walden P (1950) Aus den Erinnerungen eines alten chemischen Zeitgenossen. Naturwiss 37(4):73–81. https://doi.org/10.1007/BF00631950; Walden P (1951) Notes from the life of a chemist. J Chem Educ 28:160–163. https://doi.org/10.1021/ed028p160
11. Morachevskii AG (2003) Academician Pavel Ivanovich Walden (on 140th anniversary of his birthday). Russ J Appl Chem 76:1186–1190. https://doi.org/10.1023/A:1026399420965
12. Everts S (2013) Latvia celebrates Paul Walden. Chem Eng News 91(29):28–29. https://doi.org/10.1021/cen-09129-scitech
13. Walden P (1896) Ueber die gegenseitige Umwandlung optischer Antipoden. Ber Dt Chem Ges 29(1):133–138. https://doi.org/10.1002/cber.18960290127
14. Stradiņš J (2011) Pauls Valdens – latviešu nācijas pazaudētais un Ķīmijas gadā jaunatrastais dēls, Latvijas Zinātņu akadēmija (Latvian Academy of Sciences), Zinātnes Vēstnesis - 2011.g. 26.septembris, 11(422)
15. University of Rostock. http://purl.uni-rostock.de/cpr/00002667. "Paul Walden" in Catalogus Professorum Rostochiensium. Accessed on 2019-01-16
16. Maier H (2015) Chemiker im "Dritten Reich". Wiley-VCH, Weinheim. https://doi.org/10.1002/9783527694631
17. Lockemann G (1953) Paul von Walden, dem Nestor der Chemie, zum 90. Geburtstage am 26. Juli 1953, Naturwiss 40(14):373–374. https://doi.org/10.1007/BF00589294
18. Institute of Organic Chemistry, University of Tübingen. http://www.oc2.chemie.uni-tuebingen.de/history/paul_walden/paul_walden.htm. Accessed on 2019-01-16
19. The Riga Tourism Development Bureau Foundation. https://www.liveriga.com/en/3370-denkmal-fur-paul-walden. Accessed on 2019-01-16
20. Emel'yanenko VN, Boeck G, Verevkin SP, Ludwig R (2014) Volatile times for the very first ionic liquid: understanding the vapor pressures and enthalpies of vaporization of ethylammonium nitrate. Chem Eur J 20:11640–11645. https://doi.org/10.1002/chem.201403508
21. Barrer RM (1943) The viscosity of pure liquids. II. Polymerized ionic melts, Trans Faraday Soc 39:59–67. https://doi.org/10.1039/TF9433900059
22. Wilkes JS (2002) A short history of ionic liquids—from molten salts to neoteric solvents. Green Chem 4:73–80. https://doi.org/10.1039/b110838g

23. Welton T (2018) Ionic liquids: a brief history. Biophys Rev 10:691–706. https://doi.org/10.1007/s12551-018-0419-2
24. Shiflett MB, Scurto AM (2017) Ionic liquids: current state and future directions. ACS Symp Ser 1250:1–13. https://doi.org/10.1021/bk-2017-1250.ch001
25. Hough WL, Smiglak M, Rodríguez H, Swatloski RP, Spear SK, Daly DT, Pernak J, Grisel JE, Carliss RD, Soutullo MD, Davis JH Jr, Rogers RD (2007) The third evolution of ionic liquids: active pharmaceutical ingredients. New J Chem 31:1429–1436. https://doi.org/10.1039/b706677p
26. Wilkes JS, Zaworotko MJ (1992) Air and water stable 1-ethyl-3-methylimidazolium based ionic liquids. J Chem Soc Chem Commun 965–967. https://doi.org/10.1039/C39920000965
27. Ranke J, Stolte S, Störmann R, Arning J, Jastorff B (2007) Design of sustainable chemical products—the example of ionic liquids. Chem Rev 107:2183–2206. https://doi.org/10.1021/cr050942s
28. Jordan A, Gathergood N (2015) Biodegradation of ionic liquids—a critical review. Chem Soc Rev 44:8200–8237. https://doi.org/10.1039/C5CS00444F
29. Pena-Pereira F, Namieśnik J (2014) Ionic liquids and deep eutectic mixtures: sustainable solvents for extraction processes. ChemSusChem 7:1784–1800. https://doi.org/10.1002/cssc.201301192
30. Ferraz R, Branco LC, Prudêncio C, Noronha JP, Petrovski Ž (2011) Ionic liquids as active pharmaceutical ingredients. ChemMedChem 6:975–985. https://doi.org/10.1002/cmdc.201100082
31. Egorova KS, Gordeev EG, Ananikov VP (2017) Biological activity of ionic liquids and their application in pharmaceutics and medicine. Chem Rev 117:7132–7189. https://doi.org/10.1021/acs.chemrev.6b00562
32. Smith EL, Abbott AP, Ryder KS (2014) Deep eutectic solvents (DESs) and their applications. Chem Rev 114:11060–11082. https://doi.org/10.1021/cr300162p
33. Maginn EJ (2007) Atomistic simulation of the thermodynamic and transport properties of ionic liquids. Acc Chem Res 40:1200–1207. https://doi.org/10.1021/ar700163c
34. Chevron Energy Technology Company, San Ramon, CA. https://www.chevron.com/technology. Accessed on 2019-02-10
35. Honeywell UOP, Des Plaines, IL. https://www.uop.com/. Accessed on 2019-02-10
36. Privatdozent Dr. Marco Haumann, Friedrich-Alexander-Universität (FAU), Erlangen-Nürnberg. https://www.crt.tf.fau.eu/person/pd-dr-marco-haumann/. Accessed on 2019-02-10
37. Professor Dr. Peter Wasserscheid, Friedrich-Alexander-Universität (FAU), Erlangen-Nürnberg. https://www.crt.tf.fau.eu/person/prof-dr-peter-wasserscheid/. Accessed on 2019-02-10
38. Helmholtz Institute Erlangen-Nürnberg (HI ERN) for Renewable Energy Production. https://www.hi-ern.de/EN/HOME/node.html. Accessed on 2019-02-10
39. Solutions, Inc., Tuscaloosa AL. http://www.525solutions.com/. Accessed on 2019-02-10
40. Professor Jason P. Hallett, Imperial College London. https://www.imperial.ac.uk/people/j.hallett. Accessed on 2019-02-10
41. Professor Daniel W. Armstrong, University of Texas at Arlington. https://www.uta.edu/chemistry/faculty/dan_armstrong.php. Accessed on 2019-02-10
42. Sigma-Aldrich Ionic Liquid Gas Chromatography Columns. https://www.sigmaaldrich.com/analytical-chromatography/analytical-products.html. Accessed on 2019-02-10
43. Queen's University Ionic Liquids Laboratory (QUILL), Belfast. https://www.qub.ac.uk/schools/SchoolofChemistryandChemicalEngineering/Research/QUILL/. Accessed on 2019-02-10
44. AECS-QuikPrep Ltd/Quattro, London. http://www.quattroprep.com/. Accessed on 2019-02-10
45. IoLiTec Ionic Liquids Technologies GmbH, Heilbronn. https://www.iolitec.de/. Accessed on 2019-02-10
46. Natural Fiber Welding, Inc., Peoria IL. http://www.naturalfiberwelding.com/. Accessed on 2019-02-10

47. Professor David Durkin and Professor Paul Trulove, U.S. Naval Academy, Annapolis MD. https://www.usna.edu/ChemDept/faculty/index.php. Accessed on 2019-02-10
48. German Institutes of Textile and Fiber Research (DTIF), Denkendorf. https://www.ditf.de/en/index/ditf.html. Accessed on 2019-02-10
49. Professor Dr. Michael R. Buchmeiser, University of Stuttgart. https://www.ipoc.uni-stuttgart.de/institute/team/Buchmeiser-00003/. Accessed on 2019-02-10
50. Proionic GmbH, Grambach. https://www.proionic.com/. Accessed on 2019-02-10
51. Cowan B (1984) Liquid piston compression systems for compressing steam. United States Patent 4,566,860
52. Pieperbeck B (2000) Verfahren und Verdichter zum Komprimieren von Gasen. Offenlegungsschrift DE 198 48 234 A1
53. Kömpf M (2006) Mobility under high pressure. In: Linde AG (ed) Linde technology, Linde reports on science and technology, pp 24–29. ISSN 1612-2232
54. Dyson PJ, Laurenczy G, Ohlin CA, Vallance J, Welton T (2003) Determination of hydrogen concentration in ionic liquids and the effect (or lack of) on rates of hydrogenation. Chem Commun 2418–2419. https://doi.org/10.1039/B308309H
55. Kumełan J, Pérez-Salado Kamps Á, Tuma D, Maurer G (2006) Solubility of H_2 in the ionic liquid [bmim][PF_6]. J Chem Eng Data 51:11–14. https://doi.org/10.1021/je050362s
56. Kumełan J, Pérez-Salado Kamps A, Tuma D, Maurer G (2006) Solubility of H_2 in the ionic liquid [hmim][Tf_2N]. J Chem Eng Data 51:1364–1367. https://doi.org/10.1021/je060087p
57. Finotello A, Bara JE, Narayan S, Camper D, Noble RD (2008) Ideal gas solubilities and solubility selectivities in a binary mixture of room-temperature ionic liquids. J Phys Chem B 112:2335–2339. https://doi.org/10.1021/jp075572l
58. Anderson JL, Anthony JL, Brennecke JF, Maginn EJ (2008) Gas solubilities in ionic liquids. In: Wasserscheid P, Welton T (eds) Ionic liquids in synthesis, 2nd edn. Wiley-VCH, Weinheim, pp 103–129. https://doi.org/10.1002/9783527621194.ch3
59. Beutier D, Renon H (1978) Gas solubilities near the solvent critical point. AIChE J 24:1122–1125. https://doi.org/10.1002/aic.690240628
60. Kumełan J, Pérez-Salado Kamps Á, Tuma D, Maurer G (2009) Solubility of the single gases carbon monoxide and oxygen in the ionic liquid [hmim][Tf_2N]. J Chem Eng Data 54:966–971. https://doi.org/10.1021/je8007556
61. Mayer M (2014) From prototype to serial production. Manufacturing hydrogen fueling stations. Presented at Eco-Mobility 2014, Austrian Association for Advanced Propulsion Systems A3PS, Vienna, Austria, 20–21 Oct 2014
62. Adler R, Siebert G (2006) Method and device for compressing a gaseous medium. WO 2006/034748 A1
63. Kotschan M, Kalb R (2006) Liquid for compressing a gaseous medium and use of the same. WO 2006/120145 A1
64. Adler R, Mayer H (2008) Pistonless compressor. WO 2008/031527 A1
65. SAE International, Warrendale PA. www.sae.org. Accessed on 2019-02-10
66. Linde AG (2014) Hydrogen technologies. The Ionic Compressor 90 MPa–IC 90. Datasheet 358503.
67. Linde AG (2015) The driving force. Managing hydrogen projects with Linde. Brochure tcm14-233488.
68. Beckman M (2014) Linde hydrogen fueling overview. Presentation at Washington DC, 18 Nov 2014
69. Arjomand Kermani N, Rokni M, Elmegaard B (2017) Design and prototyping of an ionic liquid piston compressor as a new generation of compressors for hydrogen refueling stations. Technical University of Denmark (DTU), DCAMM Special Report No. S229
70. European Chemicals Agency (ECHA), Helsinki. https://echa.europa.edu/; [C_2C_1im][NTf_2] is filed under the EC List no. 700-235-5, [C_2C_1im][OTf] under 680-002-1
71. Raeissi S, Schilderman AM, Peters CJ (2013) High pressure phase behavior of mixtures of hydrogen and the ionic liquid family [cnmim][Tf_2N]. J Supercrit Fluids 73:126–129. https://doi.org/10.1016/j.supflu.2012.09.003

72. Kazakov A, Magee JW, Chirico RD, Paulechka E, Diky V, Muzny CD, Kroenlein K, Frenkel M (2019) NIST standard reference database 147: NIST ionic liquid database (ILThermo), Version 2.0. National Institute of Standards and Technology, Gaithersburg MD, 20899. https://ilthermo.boulder.nist.gov. Accessed on 2019-01-16

73. Magee JW (2003) Papers presented at the workshop on ionic liquids, ICCT, Rostock, Germany, July 28 to August 2, 2002. J Chem Eng Data 48:445. https://doi.org/10.1021/je0304771

74. Marsh KN, Brennecke JF, Chirico RD, Frenkel M, Heintz A, Magee JW, Peters CJ, Rebelo LPN, Seddon KR (2009) Thermodynamic and thermophysical properties of the reference ionic liquid: 1-Hexyl-3-methylimidazolium bis[(trifluoromethyl)sulfonyl]amide (including mixtures) Part 1. Experimental methods and results. Pure Appl Chem 81:781–790. https://doi.org/10.1351/PAC-REP-08-09-21

75. Chirico RD, Frenkel M, Magee JW, Diky V, Muzny CD, Kazakov AF, Kroenlein K, Abdulagatov I, Hardin GR, Acree WE Jr, Brennecke JF, Brown PL, Cummings PT, De Loos TW, Friend DG, Goodwin ARH, Hansen LD, Haynes WM, Koga N, Mandelis A, Marsh KN, Mathias PM, McCabe C, O'Connell JP, Padua A, Rives V, Schick C, Trusler JPM, Vyazovkin S, Weir RD, Wu J (2013) Improvement of quality in publication of experimental thermophysical property data: challenges, assessment tools, global implementation, and online support. J Chem Eng Data 58:2699–2716. https://doi.org/10.1021/je400569s

76. Chirico RD, Diky V, Magee JW, Frenkel M, Marsh KN (2009) Thermodynamic and thermophysical properties of the reference ionic liquid: 1-Hexyl-3-methylimidazolium bis[(trifluoromethyl)sulfonyl]amide (including mixtures) Part 2. Critical evaluation and recommended property values. Pure Appl Chem 81:791–828. https://doi.org/10.1351/PAC-REP-08-09-22

77. Magee JW, Widegren JA, Frenkel M, Dong Q, Muzny C, Chirico RD, Diky VV (2005) Comprehensive data retrieval system for ionic liquids. In: 1st International Congress on Ionic Liquids, Salzburg, Austria, 19–22 June 2005

78. Dong Q, Muzny CD, Kazakov A, Diky V, Magee JW, Widegren JA, Chirico RD, Marsh KN, Frenkel M (2007) ILThermo: a free-access web database for thermodynamic properties of ionic liquids. J Chem Eng Data 52:1151–1159. https://doi.org/10.1021/je700171f

79. Seddon KR (1999) QUILL rewrites the future of industrial solvents. Green Chem 1:G58–G59. https://doi.org/10.1039/GC990G58

80. Plechkova NV, Seddon KR (2007) Ionic liquids: "designer" solvents for green chemistry. In: Tundo P, Perosa A, Zecchini F (eds) Methods and reagents for green chemistry: an introduction. Wiley, Hoboken NJ, pp 105–130. https://doi.org/10.1002/9780470124086.ch5

81. Rogers RD, Seddon KR, Volkov S (2002) Green industrial applications of ionic liquids. NATO Sci Ser 92. https://doi.org/10.1007/978-94-010-0127-4

82. Atkins MP, Davey P, Fitzwater G, Rouher O, Seddon KR, Swindall J (2004) Ionic liquids: a map for industrial innovation, Report Q001, January 2004. QUILL, Belfast, UK

83. Plechkova NV, Seddon KR (2008) Applications of ionic liquids in chemical industry. Chem Soc Rev 37:123–150. https://doi.org/10.1039/b006677j

84. DECHEMA Gesellschaft für Chemische Technik und Biotechnologie e. V. (Society for Chemical Engineering and Biotechnology), Frankfurt am Main. https://dechema.de/en/. Accessed on 2019-01-16

85. Leitner W, Seddon KR, Wasserscheid P (2003) Foreword: green solvents for catalysis. Green Chem 5:G28. https://doi.org/10.1039/B302757K

86. Behr A, Hoff A, Leitner W, Wasserscheid P (2006) Advanced Fluids – neue Lösungsmittelkonzepte für die Prozess- und Produktoptimierung, Positionspapier des DECHEMA-Arbeitskreises „Alternative Lösungsmittelsysteme für technische Anwendungen". DECHEMA e. V., Frankfurt am Main

87. Runge W (2014) Technology entrepreneurship. A treatise on entrepreneurs and entrepreneurship for and in technology ventures, vols 1 and 2. Karlsruher Institut für Technologie, KIT Scientific Publishing, Karlsruhe. http://dx.doi.org/10.5445/KSP/1000036459 (vol 1); http://dx.doi.org/10.5445/KSP/1000036460 (vol 2)

88. Initiative Der Deutsche Innovationspreis. http://www.der-deutsche-innovationspreis.de. Accessed on 2019-01-16
89. ZFHN Zukunftsfonds Heilbronn GmbH & Co. KG. https://www.zf-hn.de. Accessed on 2019-01-16
90. Scionix Ltd., London. http://www.scionix.co.uk. Accessed on 2019-01-16
91. Hembury GA, Sullivan N, Tate L, Fairless G, Newton R (2009) Standards for Green Solvents: An Example Using Ionic Liquids. In: 13th Annual Green Chemistry & Engineering Conference, College Park MD, 23–25 June 2009
92. Extraction of artemisinin using ionic liquids (2008) Bioniqs project report 003-003/2. Bioniqs Ltd., Heslington, York
93. Solvionic SA, Toulouse. http://en.solvionic.com/presentation. Accessed on 2019-01-16
94. Solvionic SA, Toulouse. http://en.solvionic.com/research-projects; https://www.magenta-h2020.eu. Accessed on 2019-01-16
95. Kalb RS, Stepurko EN, Emel'yanenko VN, Verevkin SP (2016) Carbonate based ionic liquid synthesis (CBILS®): thermodynamic analysis. Phys Chem Chem Phys 18:31904–31913. https://doi.org/10.1039/C6CP06594E
96. Kalb RS, Damm M, Verevkin SP (2017) Carbonate based ionic liquid synthesis (CBILS®): development of the continuous flow method for preparation of ultra-pure ionic liquids. React Chem Eng 2:432–436. https://doi.org/10.1039/C7RE00028F
97. Sigma-Aldrich Chemie GmbH, Munich. https://www.sigmaaldrich.com/chemistry/chemical-synthesis/technology-spotlights/cbils.html. Accessed on 2019-01-16
98. Massonne K (2010) Ionic liquids at BASF SE, Leuven Ionic Liquids Summer School, 23–27 Aug 2010. Leuven/Louvain, Belgium
99. VTU Group GmbH, Raaba-Grambach. https://www.vtu.com/en. Accessed on 2019-01-16
100. Freemantle M (2004) Ionic liquids in organic synthesis. Room-temperature ionic liquids provide unique environment for organic reactions. Chem Eng News 82(45):44–49. https://doi.org/10.1021/cen-v082n045.p044
101. Ignat'ev NV, Welz-Biermann U, Kucheryna A, Bissky G, Willner H (2005) New ionic liquids with tris(perfluoroalkyl)trifluorophosphate (FAP) anions. J Fluorine Chem 126:1150–1159. https://doi.org/10.1016/j.jfluchem.2005.04.017
102. Ignat'ev NV, Pitner W-R, Welz-Biermann U (2007) Synthesis and application of new ionic liquids with tris(perfluoroalkyl)trifluorophosphate anions. ACS Symp Ser 950:281–287. https://doi.org/10.1021/bk-2007-0950.ch022
103. Merck KGaA, Darmstadt. https://www.merckgroup.com/en/performance-materials.html; https://www.ionic-liquids.com. Accessed on 2019-01-16
104. BASF SE, Ludwigshafen. http://www.intermediates.basf.com/chemicals/ionic-liquids/index. Accessed on 2019-01-16
105. Basionics™ and Basil™. Ionic liquids—Solutions for Your Success (2005) CZ 0510-06/05, BASF SE, Ludwigshafen
106. Freemantle M (2003) BASF's smart ionic liquid. Process scavenges acid on a large scale without producing solids. Chem Eng News 81(13):9. https://doi.org/10.1021/cen-v081n013.p009
107. Maase M (2004) Erstes technisches Verfahren mit ionischen Flüssigkeiten. Chem unserer Zeit 38:434–435. https://doi.org/10.1002/ciuz.200490093
108. Maase M (2008) Industrial applications of ionic liquids. In: Wasserscheid P, Welton T (eds) Ionic liquids in synthesis, 2nd edn. Wiley-VCH, Weinheim, pp 663–687. https://doi.org/10.1002/9783527621194.ch9
109. Saling P, Maase M, Vagt U (2010) Eco-efficiency analysis of an industrially implemented ionic liquid-based process—the BASF BASIL process. In: Anastas PT (ed) Handbook of green chemistry, green solvents, vol 6; Wasserscheid P, Stark A (eds) Ionic liquids, Wiley-VCH, Weinheim, pp 299–314. https://doi.org/10.1002/9783527628698.hgc070
110. Maase M (2005) BASIL™ process. In: Cornils B, Herrmann WA, Horváth IT, Leitner W, Mecking S, Olivier-Bourbigou H, Vogt D (eds) Multiphase homogeneous catalysis, vol 2. Wiley-VCH, Weinheim, pp 560–566. https://doi.org/10.1002/9783527619597.ch5c

111. BASF intermediates product catalogue: We create chemistry. Building blocks and reagents for our customers' needs (2012) CZ 1203-02/12-20.000, BASF SE, Ludwigshafen. http:// www.intermediates.basf.com/chemicals/brochures/intermediates. Accessed on 2019-01-16

112. Hoff A, Jost C, Prodi-Schwab A, Schmidt FG, Weyershausen B (2004) Ionic liquids. New designer compounds for more. Elements (Degussa Science Newsletter) 09:10–15

113. Franke R, Hahn H (2015) A catalyst that goes to its limits. Elements 51(2):18–23. https:// corporate.evonik.com/en/products. Accessed on 2019-01-16

114. Technology advancements in ionic liquids (TechVision) (2016). Market research report, CM00755-GL-TR_16372, Frost & Sullivan, Mountain View CA.

115. Ionic liquids market size and forecast by application (Solvents and catalysts, extractions and separations, bio-refineries, energy storage), by Region (North America, Europe, Asia Pacific, Latin America and Middle East & Africa) and Trend analysis from 2018 to 2025 (2016) Market research report, Report ID GVR-1-68038-283-9, Grand View Research Inc., San Francisco CA.

Part II
Ionic Liquid Processes

Chapter 2
ISOALKY™ Technology: Next-Generation Alkylate Gasoline Manufacturing Process Technology Using Ionic Liquid Catalyst

Hye Kyung Timken, Huping Luo, Bong-Kyu Chang, Elizabeth Carter and Matthew Cole

Abstract Chevron and Honeywell UOP have developed a next-generation alkylate gasoline manufacturing process, ISOALKY Technology, which has performance benefits (yield, product octane number, safety, and environmental) over the conventional acid-based processes, while the economical costs for capital plant construction and operation are comparable to the conventional processes. The ISOALKY Catalyst is a non-volatile ionic liquid that operates efficiently via on-line regeneration. ISOALKY Technology is for greenfield plants as well as for retrofit and expansion of existing alkylation plants. Chevron has entered into an alliance agreement with Honeywell UOP, proven experts in alkylation technology, to license this technology to the industry. UOP is the exclusive licensor of the ISOALKY Technology. Chevron made the final investment decision to retrofit and convert its approximately 5,000 barrel per day (BPD), that is, 190 kton/year, alkylate production plant in its refinery from a hydrofluoric acid technology to the ISOALKY Technology. Construction for the conversion project began in 2018, with the ISOALKY Plant due for startup in 2020. Chevron will own the first ISOALKY Plant in the world.

Keywords Alkylation · Commercialization · Ionic liquid catalyst · ISOALKY™ Technology · Refinery

2.1 Introduction

Worldwide, oil refineries produce over two million barrels of alkylate gasoline per day as a part of their gasoline production. Alkylate gasoline is a desirable blending component for cleaner-burning motor gasoline, with its high octane number, low sulfur and nitrogen levels, and near-zero olefin and aromatic contents. Table 2.1

H. K. Timken (✉) · H. Luo · B.-K. Chang
Chevron Energy Technology Company, Richmond, CA, USA
e-mail: htimken@chevron.com

E. Carter · M. Cole
Honeywell UOP, Des Plaines, IL, USA

© Springer Nature Switzerland AG 2020
M. B. Shiflett (ed.), *Commercial Applications of Ionic Liquids*, Green Chemistry and Sustainable Technology, https://doi.org/10.1007/978-3-030-35245-5_2

Table 2.1 World's leading gasoline specifications versus alkylate properties

Specification	RBOB Tier III	CARBOB phase 3	Euro VI	China V	Alkylate
$(R + M)/2$, [RON][a]	87/89/91	87/89/91	[91/95]	[95]	[95]+
Sulfur, ppm max.	10	20	10	10	< 5
Benzene, vol% max.	0.62	0.8	1.0	1.0	0
Aromatics, vol% max.	–	25	35	40	0
Olefins, vol% max.	–	6	18	24	0
Oxygen, wt% max.	2.7	2.2	2.7	2.7	0

[a][$R + M$]/2 is the average octane number where R is research octane number (RON) and M is motor octane number (MON)

shows the world's leading gasoline specifications versus the typical properties of the alkylate gasoline.

Motor alkylate gasoline, primarily a blend of C_7–C_8 isoparaffins, has become an increasingly important blending component in the production of environmentally mandated clean fuels, both in the USA and abroad. Ever-tightening gasoline and LPG specifications worldwide, increasingly demanding environmental regulations, and an abundant supply of low-cost isobutane in the USA and elsewhere have promoted the expansion of alkylate gasoline production in the refining industry. This trend is expected to continue in the next decade.

Refineries currently use either concentrated sulfuric acid (H_2SO_4) or hydrofluoric acid (HF) catalyst-based technologies for alkylate gasoline manufacturing. There are over 300 alkylation plants worldwide. Roughly one half of the plants use HF alkylation technologies, and the other half use H_2SO_4 alkylation technologies. In the USA alone, there are about 100 alkylation plants and about 50 of them use HF alkylation technologies.

For the last several decades, there have been numerous efforts to develop alternate alkylation technologies that could compete with the current conventional technologies, but no significant commercial success has been achieved. Steady improvements of the existing technologies coupled with the difficult chemistry of paraffin alkylation have hindered the emergence of a new, alternate technology.

Recently, commercial implementations of a couple of solid alkylation plants [1, 2] and one ionic liquid alkylation plant [3] in China were reported. Long-term performance data of these plants are not available yet.

Chevron and Honeywell UOP have committed significant time and resources to develop a new alkylation technology that can compete with the existing, conventional acid-based alkylation technologies, with goals to develop a new alkylation process technology that has significant performance and operational advantages, acceptable new technology risks, and favorable economics. We finally reached these goals and began commercialization efforts a few years ago. Here, we report key features of the ISOALKY Technology and introduce overall development and commercialization efforts.

2.2 Discovery and Development History

Chevron discovered in 2004 in their strategic research program that ionic liquid catalysts are very effective in alkylating LPG olefins and that the ionic liquid catalyst-based paraffin alkylation process has several performance advantages over the conventional alkylation technologies. These discoveries led to more than a decade of extensive efforts to scale up the ISOALKY Technology to a commercially viable alternative to the traditional alkylation processes that require hydrofluoric acid or sulfuric acid.

Chevron initially scaled up the ISOALKY Technology from bench-scale R&D units to a 0.1 BPD laboratory pilot plant. Our laboratory pilot plant has been in continuous operation since 2005. This unit was used initially for reactor concept testing, alkylation process chemistry study, and optimization of the alkylation process. Extensive modifications were then made to the pilot plant to incorporate other necessary "sub-processes", such as effluent separation, on-line regeneration of the ionic liquid catalyst, and alkylate product treating. Currently, this unit is used to support the commercialization of the ionic liquid catalyst as well as licensing support.

In 2006, we started further scale-up of the ISOALKY Technology and designed a demonstration (demo) plant with a 100-fold scale-up factor to 10 BPD (Fig. 2.1). Construction of the demo plant was finished in late 2009. The demo plant was operated for over 5 years, from 2010 until mid-2015. Various process data, design data, fundamental knowledge on ionic liquid catalysis, and know-how for optimization of the integrated sub-processes were collected to develop the ISOALKY Technology. The information collected was used to address all key technology risks and to support a financial decision to implement the technology at a full commercial scale.

2.3 ISOALKY Technology Simplified Description

Feeds to the paraffin alkylation process are isobutane (iC_4) and LPG olefins $\left(C_n^=\right)$. The olefin feed sources are typically supplied from the fluid catalytic cracking (FCC) C_3–C_5 olefin stream. The predominant chemistry for the alkylate process can be described as:

$$iC_4 + C_4^= \rightarrow C_8 \text{ alkylate}$$

$$iC_4 + C_3^= \rightarrow C_7 \text{ alkylate}$$

Fig. 2.1 ISOALKY Demonstration Plant in a refinery that consists of demo plant skids, cooling unit, and analytical shed (HF plant columns are shown in the background)

$$iC_4 + C_5^= \rightarrow C_9 \text{ alkylate}$$

The ISOALKY Technology process scheme, shown in Fig. 2.2, has many features of the conventional alkylation processes, as well as many unique features.

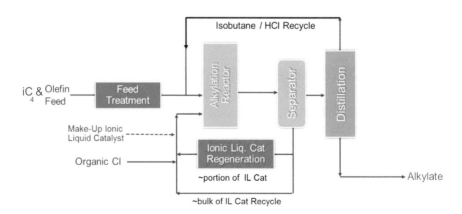

Fig. 2.2 Overall process scheme of ISOALKY Technology

The overall ISOALKY Technology can be roughly divided into five sections:

- Feed treating—conventional technology
- Alkylation reactor and effluent separation—unique
- Ionic liquid catalyst regeneration—unique
- Product distillation and recycling—conventional technology
- Product finishing—unique.

For the front feed-treating section and the back-end of product distillation section, the ISOALKY Technology uses conventional technologies that are commonly used in current refineries. Alkylation reactor, reactor effluent product separation, ionic liquid catalyst regeneration, and alkylate product treating are unique innovations specific for the ISOALKY Technology. Performance of these new components were extensively developed and tested into optimized sub-processes in order to minimize the scale-up technical risks.

2.4 ISOALKY Ionic Liquid Catalyst

The ISOALKY Technology uses a state-of-the-art chloroaluminate-based ionic liquid catalyst that is more environmentally sustainable than conventional mineral acid catalysts. A trace amount of anhydrous HCl co-catalyst is added in situ by addition of an organic chloride. The combined ISOALKY Catalyst system can be described as an anhydrous $AlCl_3$-based Lewis acid catalyst promoted with anhydrous HCl Brønsted acid. Together, the catalyst system exhibits the characteristics of a superacid (see Sect. 2.6).

The ionic liquid alkylation catalyst can be characterized by the general formula Q^+A^-, wherein Q^+ is an ammonium or phosphonium cation and A^- is a negatively charged ion, such as $AlCl_4^-$, $Al_2Cl_7^-$, $GaCl_4^-$, $Ga_2Cl_7^-$, or $Ga_3Cl_{10}^-$. Various ionic liquids, selected from the group consisting of hydrocarbyl-substituted pyridinium chloroaluminate, hydrocarbyl-substituted imidazolium chloroaluminate, quaternary amine chloroaluminate, trialkyl amine hydrogen chloride chloroaluminate, and alkyl pyridine hydrochloride chloroaluminate, were evaluated [4]. Our study found that phosphonium-containing ionic liquid catalysts [5] and gallium chloride-based ionic liquids are also effective for the paraffin alkylation process.

Figure 2.3 illustrates common ammonium cation structures that showed excellent performance for our paraffin alkylation process where the anion (X^-) is $Al_2Cl_7^-$.

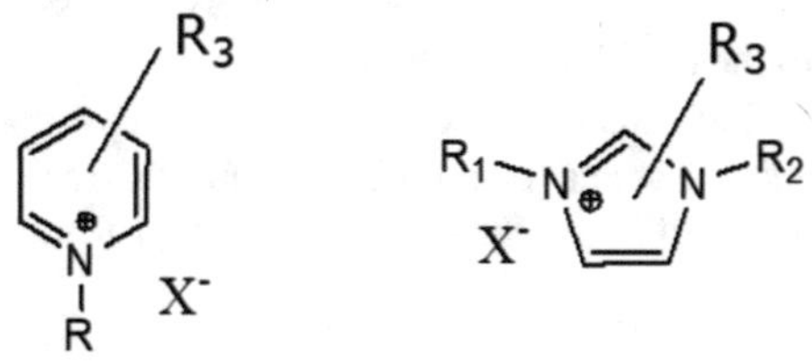

Fig. 2.3 Structures of ionic liquid catalysts that performed well for alkylation process

The paraffin alkylation process requires a very high acidity catalyst. An ionic liquid catalyst with mole ratio of AlCl$_3$ to amine chloride close to 2:1 was selected for enhanced performance in producing higher octane number alkylate gasoline. The ionic liquid catalyst shows the highest Lewis acidity when the mole fraction of AlCl$_3$ in the ionic liquid, x(AlCl$_3$) is 0.67 (2:1 mole ratio of AlCl$_3$ to amine chloride). The catalyst for the commercial process was selected for its overall performance, i.e., activity and selectivity for alkylation process as well as for long-term stability.

In addition, a trace amount of anhydrous hydrogen chloride (HCl) co-catalyst is needed for the high activity of the ISOALKY Catalyst system, and a constant supply of HCl is required to maintain the reactivity of the catalyst system for extended, continuous operation of the alkylation process [6]. For the commercial process, the HCl is generated in situ by the addition of an organic chloride promoter.

The fundamental aspects of chloroaluminate ionic liquid and the nature of its acidity were studied by Johnson et al. [7, 8] and Smith et al. [9]. Other characteristics of halometallate ionic liquids have been well reviewed by Estager et al. [10]. Catalysis for paraffin alkylation using an ionic liquid catalyst and the effect of HCl addition for the process were studied by Jess et al. [11–13].

Independently, the China University of Petroleum developed a composite ionic liquid catalyst containing CuCl, trialkyl ammonium hydrochloride, and anhydrous aluminum chloride for the paraffin alkylation process and commercialized the technology in China [3, 14, 15].

Other researchers have studied many different ionic liquid catalysts for the paraffin alkylation process. A recent review article by Wong et al. has a comprehensive list [16].

2.5   Characteristics of Ionic Liquid Catalyst and Its Use in a Refinery

The ionic liquid catalyst is a *liquid* salt (consisting of positive and negative ions only) at ambient temperature. It has the typical properties of a salt, such as no measurable vapor pressure, stability for long-term storage, and low solubility in hydrocarbon. These properties allow for easier handling of the catalyst in a refinery. Ionic liquid catalyst spills can be contained and managed with no concern of the catalyst vaporizing.

In addition, hydrocarbon product separation from the ionic liquid is facile due to low solubility of ionic liquid in the hydrocarbon. Efficient separation and recovery of the ionic liquid catalyst minimize the loss of the ionic liquid catalyst via carry-over to the distillation section.

The ISOALKY Catalyst is very hydrophilic and easily hydrolyzed in the presence of water. To protect the catalyst from hydrolysis, high-performance dryers are installed to dry the feeds to less than 1 ppm moisture level.

Refinery-standard personal protective equipment (PPE) gives adequate protection for the operation of the ISOALKY Technology. This is a significant advantage of the ISOALKY Technology compared with HF technology where extensive PPE is required due to high volatility and toxicity of HF.

2.6 ISOALKY Technology Performance Advantages

The trace HCl added to the ionic liquid catalyst leads to coordination of the HCl with $[Al_2Cl_7]^-$ anions and formation of "free" superacidic protons [8, 9] per the following equation where B is a base or a proton acceptor, which could be an olefinic species for our alkylation process.

$$HCl + B + [Al_2Cl_7]^- \rightarrow BH^+ + 2\,[AlCl_4]^-$$

To compare Brønsted acidities in non-aqueous media, Hammett and Deyrup developed the concept of Hammett acidity H_o (pK_{BH+} at infinite dilution) where a more negative number indicates stronger acidity using a log10 exponential scale [17].

Smith et al. [9] and Ma and Johnson [7] each measured the Hammett acidity of the protons in ethylmethylimidazolium chloroaluminate ionic liquids with 0.55 mol fraction of $AlCl_3$ ($x(AlCl_3)$) in the ionic liquid. Johnson reported a H_o of -14 [7], while Smith reported -18 [9]. Concentrated sulfuric acid has a H_o of -10.6 [17] and concentrated HF a H_o of -10.7 [7]. These results suggest that the acidity of HCl dissolved in ionic liquid is at least 3 orders of magnitude greater than the acidity of conventional alkylation catalysts.

Due to a significantly higher activity of the ISOALKY Catalyst than the conventional acid catalysts, a much smaller catalyst volume and a shorter residence time are used in the ISOALKY Alkylation sub-process. Our alkylation reactor runs with droplets of the ionic liquid catalyst (3–6 vol%) dispersed in a continuous phase of hydrocarbon. This contrasts with the conventional mineral acid-based alkylation process where the continuous phase is the acid. The acid occupies over 50 vol% in the sulfuric acid alkylation reactor and 60–75 vol% in the HF acid alkylation reactor. Based on the catalyst volume and residence time, the ionic liquid catalyst is about 60 times more active than the H_2SO_4 catalyst.

In addition to the higher activity, the ISOALKY Ionic Liquid Catalyst has several other performance advantages.

The ionic liquid catalyst has a very wide operational temperature window of 30–120 °F, unlike the other conventional alkylation processes.

The ISOALKY Technology shows about a two (2) research octane number (RON) advantage in alkylate gasoline product over the H_2SO_4 processes [18] at comparable reactor temperature with a mixed C_4 olefin feed as shown in Fig. 2.4. A higher RON of 98 can be achieved with the ISOALKY Technology at 40 °F alkylation reactor temperature with a typical mixed C_4 olefin feed from a refinery fluid catalytic cracking (FCC) unit. For the ISOALKY Technology, a selective hydroisomerization

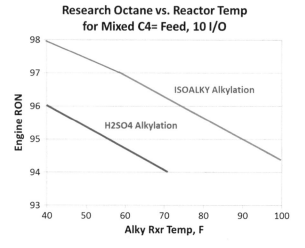

Fig. 2.4 Operating temperature window and impact of alkylation reactor temperature on research octane number (RON) of ISOALKY (our data) versus H_2SO_4 alkylation [18] with mixed $C_4^=$ feed and 10:1 ratio of $iC_4/C_4^=$

of C_4 olefin feed, which converts 1-butene to 2-butene, is necessary to achieve the high octane number.

Other benefits of the ISOALKY Technology include its feed flexibility and robustness of the process. ISOALKY Technology is equally effective for a wide range of different olefin feeds. Alkylation of 100% pure propene feed, 100% pure isobutene feed, 50/50 mol% mix of $C_3^=/C_4^=$ feed from a refinery, and 20/80 mol% mix of $C_4^=/C_5^=$ feed from a refinery were tested, and all showed comparable satisfactory performance. Thus, no complicated feed segregation or dilution of the olefin feed is necessary to optimize the alkylation sub-process.

Olefin may react with acid to form saturated hydrocarbons via H-transfer. HF has high H-transfer tendency, where the HF catalyst facilitates transfer of hydrogens from isobutane and/or HF to the olefins, causing conversion of some propene and n-butene olefins in the feed to the corresponding propane and n-butane paraffins [19]. Albright estimated 15–20% of the propene is converted to propane and 6–10% of n-butene to n-butane. The ionic liquid catalyst has low H-transfer tendency. Production of propane and n-butane from the conventional C_3–C_4 olefin feed during the alkylation step are nearly zero. Formation of isopentane from a mixed $C_5^=$ feed during the alkylation step is also low, and the C_9 selectivity is high, suggesting that the ISOALKY Technology is well suited to alkylate $C_5^=$ with isobutane to produce C_9 alkylate gasoline.

All alkylation processes generate "conjunct polymer," also known as "red oil" or simply "polymer," that builds up in the catalyst. The conjunct polymer is made via undesirable side reactions of olefins and acid catalyst. The conjunct polymer is made of unsaturated heavy hydrocarbon with heteroatom functionality depending on the nature of alkylation catalyst (i.e., $-SO_4$ for H_2SO_4 processes, $-F$ for HF processes, and $-Cl$ for ISOALKY Technology). The amount of conjunct polymer in a liquid catalyst is carefully controlled in an alkylation plant as it affects the catalyst performance and the alkylate product quality.

In conventional alkylation plants, rapid buildup of conjunct polymer in an acid catalyst may occur in case of a severe unit upset. In this case, the acid catalyst with high-conjunct polymer may form an inseparable emulsion with the hydrocarbon, and a massive amount of acidic emulsion can be carried to the distillation section causing plant downtime and equipment corrosion. This situation is called acid runaway [20]. Conventional alkylation processes are carefully controlled to prevent this situation from occurring.

The ISOALKY Technology has no risk of acid runaway due to the low volume of the ionic liquid catalyst in the process and the limited solubility of hydrocarbon in the ionic liquid catalyst. Impacts of severe process upsets and continuous buildup of conjunct polymer in the ionic liquid were alleviated by reducing the olefin feed and continuing on-line regeneration of the ionic liquid catalyst.

2.7 ISOALKY Technology Sub-processes and Key Features

2.7.1 Alkylation Reactor

The alkylation reaction occurs at the interface of the ionic liquid catalyst droplets and the hydrocarbon phase, i.e., a biphasic reaction system.

Typical process conditions are as follows:

Reaction medium	Hydrocarbon continuous phase with droplets of ionic liquid catalyst
Temperature	30–120 °F operable window
Pressure	40–250 psig operational window
Isobutane/Olefin mole ratio	8–10 external I/O
Ionic liquid catalyst volume	3–6 vol% of ionic liquid catalyst
Olefins conversion	>99.9%
Conjunct polymer formation	0.3–0.5 wt% of olefins.

To optimize the alkylation reactor design, years of work in computer modeling, cold-flow unit testing, and demonstration plant testing were conducted. Our reactor design includes injection nozzles for efficient dispersion of the ionic liquid catalyst and an external heat exchanger. For reliable operation, there are no moving parts (such as impellers) in the reactor. Our reactor can operate in a wide range of olefin feed rates and 50% turn down is feasible.

Fig. 2.5 Alkylate reactor
effluent containing ionic
liquid catalyst (inlet) and
hydrocarbon stream after
coalescer (outlet)

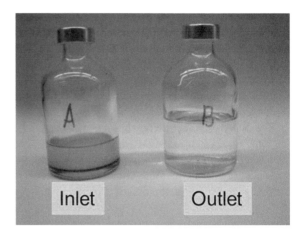

2.7.2 Alkylation Reactor Effluent Separation

Complete separation of alkylation reactor effluent into the pure hydrocarbon stream
and pure ionic liquid catalyst stream is critical for efficient operation of the overall process. The hydrocarbon stream is sent to the distillation section for further
separation into hydrocarbon product streams (propane, *n*-butane, and alkylate). The
ionic liquid catalyst stream is sent back to the alkylation reactor and to the regeneration unit. Complete separation will minimize the loss of ionic liquid catalyst via
carry-over to the distillation section. In addition, poor separation of catalyst may
contaminate and/or degrade the hydrocarbon products.

After extensive research, a proprietary liquid–liquid coalescing technology was
developed that allows full separation of the ionic liquid catalyst from hydrocarbons
[21]. We achieved a water-clear effluent stream to the distillation unit as shown in
Fig. 2.5.

The ionic liquid is separated from hydrocarbons using a coalescer element pad
material having a stronger affinity for the ionic liquid than for the hydrocarbons. The
coalescer element is made of a high surface area material to provide a large contact
area to which ionic liquid droplets dispersed in the hydrocarbons may adhere. After
the capturing and coalescence steps, the ionic liquid droplets fall via gravity from
the material to separate the ionic liquid from the hydrocarbons and provide a clean
hydrocarbon effluent.

2.7.3 Regeneration of ISOALKY Catalyst

Formation of conjunct polymer (Sect. 2.6) deactivates the catalyst; therefore, removal
of conjunct polymer is required to maintain the catalyst activity. A portion of used
ionic liquid catalyst is sent to the regeneration unit to balance the conjunct-polymer

formation rate by the alkylation step. The unique on-line regeneration sub-process converts conjunct polymer into saturated hydrocarbon "regen naphtha" in gasoline boiling range and liquefied petroleum gas (LPG). The regeneration unit effluent is sent to the alkylation reactor effluent separator where the regen naphtha and LPG are transferred to the hydrocarbon stream and then sent to the distillation section. Reuse of the ISOALKY Catalyst with on-line regeneration enables very low ionic liquid catalyst consumption.

2.7.4 Product Finishing and Chloride Control

Our process design was optimized to minimize chloride formation in the product streams. Recycling of HCl co-catalyst is incorporated to minimize the makeup of organic chloride promoter.

The ISOALKY Technology does not wash its product streams (propane, n-butane, and alkylate gasoline), and the process does not produce any waste caustic solutions from the product treating. Solid adsorbents are used to remove the residual chlorides from the finished alkylate, n-butane, and propane products. Treating with these commercially available adsorbents removes residual chlorides and meets all product specifications.

2.8 Compared to Conventional Alkylation Technologies

Table 2.2 compares key features of conventional alkylation technologies with ISOALKY Technology. The sulfuric acid alkylation technologies operate at 60 °F or lower alkylation reactor temperature with cryogenic cooling to control the conjunct-polymer formation rate. The HF technology typically operates at about 95 °F with cooling water. The ISOALKY Technology can operate at a very wide temperature window, and the process can be designed to fit the needs of customers for product quality and/or their constraint in cooling capability.

The ISOALKY Technology can produce higher octane number alkylates than other alkylation technologies by operating at about 60 °F reactor temperature or lower. Alternatively, the unit can be designed to operate at 95 °F with cooling water to lower the capital cost.

Alkylate yields by the three alkylation processes are similar. The ISOALKY Technology is estimated to have about 2 wt% alkylate yield advantage with the lower conjunct polymer formation, on-line regeneration, and conversion of conjunct polymer back to regen naphtha compared to the sulfuric acid process.

The efficient regeneration process allows for a much lower makeup rate for the catalyst. The makeup rate is about 400 times less than the H_2SO_4 alkylation process and about 2 times less than the HF alkylation process.

Table 2.2 Comparison of ISOALKY Technology to conventional alkylation technologies

	H_2SO_4	HF	ISOALKY
Alkylation temperature, °F	30–60	95	30–120
Alkylation pressure, psi	60	200	40–250
Catalyst volume in reactor (%)	50	50–80	3–6
Isobutane to olefin molar ratio (I/O) to the alkylation reactor	8–10	10	8–10
Feed moisture requirement	Not critical	<10 ppm	<1 ppm
Alkylate quality w/mixed $C_4^=$ at 10 I/O and typical operating temp. (RON)	95–96	95	94–99
Alkylate yield per bbl of $C_4^=$ (bbl/bbl)	1.8	1.8	1.8[a]
Conjunct-polymer formation rate, wt% olefin (%)	1–1.5	~0.5	0.3–0.5
Handling of conjunct polymer	Incineration	Incineration	Converted to naphtha & LPG
Catalyst makeup rate, lb IL/bbl alkylate	~400 × base, off-plot regen	~2 × base, on-line regen	base, on-line regen
Safety and environmental impact	Large acid inventory, acid transport for off-plot regen facility, SO_x emission during regeneration	Smaller acid inventory, volatile HF requires engineering controls and special PPE	Smallest catalyst inventory, non-volatile catalyst, integrated regeneration, reduction of caustic solution waste

[a]ISOALKY Technology has about a 2 wt% yield advantage compared to H_2SO_4 technology

Significantly higher activity of the ISOALKY Catalyst coupled with efficient on-line regeneration resulted in a significantly lower catalyst inventory in a refinery. The inventory volume of the ionic liquid catalyst is an order of magnitude less than that required for the sulfuric acid process.

The unique properties of the ISOALKY Catalyst and the sub-processes allow efficient design of an ISOALKY Alkylation plant. Advantages of the ISOALKY Technology for process safety include reduction in acid catalyst inventory, elimination of emissions during acid catalyst regeneration, elimination of waste caustic solutions from the product washing, and elimination of the engineering controls needed to handle volatile HF acid.

2.9 UOP's Efforts in Alkylation Process Development Using Ionic Liquid Catalyst

Honeywell UOP started exploratory work on ionic liquid catalytic processes in 2005. By 2009, UOP launched a research and development program on an alkylation process using ionic liquid catalysts. The efforts led to the construction and operation of an alkylation pilot plant and a regeneration pilot plant. The studies led to the design of a demonstration plant in 2014 for scale-up of the UOP alkylation technology. Since then, the project direction was changed as Honeywell UOP and Chevron initiated discussion on an alliance formation to produce a better alkylation technology through synergy between the companies.

2.10 Formation of Chevron-UOP Alliance for ISOALKY Licensing to Industry

The alliance was formed in March 2016 to license the ISOALKY Technology broadly to the refining industry, and the two companies have been working closely ever since to further improve the technology. The two companies have complementary strengths in that both are key technology developers for the industry. Chevron brought in operational experiences of the ISOALKY Technology and commercial plant design, and UOP brought in new technology launch experiences and process engineering expertise.

UOP is the exclusive licensor of the ISOALKY Technology and for sales of relevant catalyst and materials. Process designs of a revamp plant as well as grassroots plant for licensing have been developed jointly.

2.11 ISOALKY Technology Commercialization

In September 2016, Chevron Corporation made a final investment decision to convert the existing approximately 5,000 BPD HF alkylation unit at its 53,000 BPD refinery in Salt Lake City, Utah, into ISOALKY Technology, the first-ever retrofit conversion of a HF alkylation plant into an alternative technology. Construction on the conversion project began in 2018 with the ISOALKY Plant due for a startup in 2020.

The refining industry has shown strong interests in ISOALKY Technology and has kept watchful eyes on the progress of the first commercial Chevron project. Due to the scale of refinery operations, this ISOALKY Alkylation Catalyst is believed to be the largest volume industrial application of ionic liquids to date.

2.12 Conclusions

After many years of R&D and scale-up efforts, the ISOALKY Alkylation Technology has become a commercially viable alternative that offers a compelling economic solution compared to conventional liquid acid technologies. The ISOALKY Catalyst exhibits superior performance with a wide range of olefin feeds compared to conventional acid catalysts. The ionic liquid catalyst has negligible vapor pressure and can be regenerated on-site, resulting in a lower environmental impact compared to other technologies.

With the advancement of fracking, abundant supply of low-cost LPG becomes available in the industry and this gives new opportunities for refiners. It is a revolutionary new technology that offers refiners the ability to upgrade low-value butanes and refinery olefins to high-value alkylate and to improve the quality of their gasoline pool. The ISOALKY Technology is expected to make a significant impact on global production of clean fuels in the years to come.

Acknowledgements We thank many of our colleagues who contributed to the development of ISOALKY Technology ranging from researchers for strategic research and sub-process development, operators and constructors of various units, chemists for fundamental data collection, and process engineers for design and operation of various scale-up plants. We thank the Chevron Salt Lake City refinery for hosting the demo plant, providing operational resources, and leading the commercial plant implementation. Finally, we thank the Chevron and Honeywell UOP management for strong support and commitment throughout the development and commercialization process.

References

1. Medina J, Chuanhua Z, van Broekhoven EH (2016) Successful start-up of the first solid catalyst alkylation unit. AFPM annual meeting presentation report, AM-16-22
2. KBR New release of first licensing contract for K-SAAT solid alkylation technology. In: Hydrocarbon Processing, 17 Feb 2016. https://www.hydrocarbonprocessing.com/news/2016/02/kbr-licenses-new-alkylation-technology-in-china
3. Liu Z, Zhang R, Meng X, Liu H, Xu C, Zhang X, Chung W (2018) Composite ionic liquid alkylation technology gives high product yield and selectivity. Hydrocarbon Processing, March 2018, p 35
4. Elomari S, Trumbull S, Timken HKC, Cleverdon R (2008) Alkylation process using chloroaluminate ionic liquid catalysts. US Patent 7,432,409
5. Martins S, Nafis D, Bhattacharyya A (2015) Alkylation process using phosphonium-based ionic liquids. US Patent 9,156,028
6. Timken HKC, Elomari S, Trumbull S, Cleverdon R (2008) Integrated alkylation process using ionic liquid catalysts. US Patent 7,432,408
7. Ma M, Johnson KE (1995) Carbocation formation by selected hydrocarbons in trimethylsulfonium bromide–AlCl$_3$/AlBr$_3$–HBr ambient temperature molten salts. J Am Chem Soc 117:1508. https://pubs.acs.org/doi/abs/10.1021/ja00110a007
8. Campbell JLE, Johnson KE (1995) The chemistry of protons in ambient-temperature ionic liquids: solubility and electrochemical profiles of HCl in HCl:ImCl:AlCl$_3$ ionic liquids as a function of pressure. J Am Chem Soc 117:7791. https://pubs.acs.org/doi/abs/10.1021/ja00134a026

9. Smith GP, Dworkin AS, Pagni RM, Zingg SP (1989) Brønsted superacidity of HCl in a liquid chloroaluminate. AlCl₃–1-Ethyl-3-methyl-1*H*-imidazolium chloride. J Am Chem Soc 111:525. https://pubs.acs.org/doi/abs/10.1021/ja00184a020

10. Estager J, Holbrey JD, Swadźba-Kwaśny M (2014) Halometallate ionic liquid – revisited. Chem Soc Rev 43:847. https://pubs.rsc.org/en/content/articlelanding/2014/CS/C3CS60310E

11. Aschauer S, Schilder L, Korth W, Fritschi S, Jess A (2011) Liquid-phase isobutane/butene-alkylation using promoted Lewis-acidic IL-catalysts. Catal Lett 141:1405. https://link.springer.com/article/10.1007/s10562-011-0675-2

12. Pöhlmann F, Schilder L, Korth W, Jess A (2013) Liquid phase isobutane/2-butene alkylation promoted by hydrogen chloride using Lewis-acidic ionic liquids. ChemPlusChem 78:570. https://onlinelibrary.wiley.com/doi/abs/10.1002/cplu.201300035

13. Schilder L, Maas S, Jess A (2013) Effective and intrinsic kinetics of liquid-phase isobutane/2-butene alkylation catalyzed by chloroaluminate ionic liquids. Ind Eng Chem Res 52:1877. https://pubs.acs.org/doi/abs/10.1021/ie3028087

14. Liu Z, Zhang R, Xu C, Xia R (2006) Ionic liquid alkylation process produces high-quality gasoline. Oil Gas J 104(40):52. https://www.tib.eu/en/search/id/olc%3A174772467X/PROCESSING-Ionic-liquid-alkylation-process-produces/

15. Cui J, de With J, Klusener PAA, Su X, Meng X, Zhang R, Liu Z, Xu C, Liu H (2014) Identification of acidic species in chloroaluminate ionic liquid catalysts. J Catal 320:26. https://doi.org/10.1016/j.jcat.2014.09.004

16. Wong H, Meng X, Zhao G, Zhang S (2017) Isobutane/butene alkylation catalyzed by ionic liquids: a more sustainable process for clean oil production. Green Chem 19:1462. https://doi.org/10.1039/C6GC02791A

17. Hammett LP, Deyrup AJ (1932) A series of simple basic indicators. I. The acidity functions of mixtures of sulfuric and perchloric acids with water. J Am Chem Soc 54:2721. https://pubs.acs.org/doi/abs/10.1021/ja01346a015

18. Kranz K (2008) Intro to alkylation chemistry—mechanisms, operating variables and olefin interactions. DuPont Stratco report, Sept 2008. http://www.dupont.com/content/dam/dupont/products-and-services/consulting-services-and-process-technologies/consulting-services-and-process-technologies-landing/documents/AlkylationChemistry_RU.pdf

19. Albright LF (2009) Present and future alkylation processes in refineries. Ind Eng Chem Res 48:1409. https://pubs.acs.org/doi/abs/10.1021/ie801495p

20. Liolios G (2001) Acid Runaways in a sulfuric acid alkylation unit. DuPont Stratco report, Nov 2001. http://www2.dupont.com/Clean_Technologies/es_MX/assets/downloads/AcidRunaway2001.pdf

21. Luo H, Ahmed M, Parimi K, Chang BK (2011) Liquid-liquid separation process via coalescers. US Patent 8,067,656

Chapter 3
Continuous Catalytic Processes with Supported Ionic Liquid Phase (SILP) Materials

Marco Haumann

Abstract The concept of supported ionic liquid phase (SILP) materials offers many attractive features for continuous catalytic processes. Due to the extremely low volatility of the ionic liquid, homogenous catalysts can be immobilized and applied in gas-phase reactions without the need for elaborating separations or recycling strategies. The concept has been developed since 2003, and this chapter highlights the current status of the technology. No large-scale commercial process has been established, but several examples exist at the demonstration level for hydroformylation, water-gas shift, and even asymmetric hydrogenation.

Keywords Catalysis · Supported ionic liquid phase · Continuous processes · Industrialization · Pilot plants

3.1 Introduction

Catalysts can help achieve sustainable chemical production by preventing undesired by-product formation and lowering the energy consumption due to milder reaction conditions. Well-defined homogeneous transition metal complexes and biocatalysts allow high selectivity and mild reaction conditions [1]. However, the often tedious and energy-consuming separation of these catalysts from the reaction mixture hampers the implementation of these benign systems in industry. Numerous techniques have been developed to immobilize or heterogenize these liquid catalysts [2, 3]. An interesting field of immobilization is the use of supported ionic liquid phase (SILP) materials [4]. In these systems, the ionic liquid is dispersed on a solid support, and the dissolved catalyst complex can act truly homogeneous on a microscopic level [5–8]. On the macroscopic level, the material is a powder or pellet and can easily be separated from the reaction mixture. Since the ionic liquid has a negligible vapor pressure under reaction conditions, it allows continuous gas-phase operation. The

M. Haumann (✉)
Friedrich-Alexander-Universität Erlangen-Nürnberg (FAU), Lehrstuhl für Chemische Reaktionstechnik (CRT), Egerlandstr. 3, 91058 Erlangen, Germany
e-mail: marco.haumann@fau.de

© Springer Nature Switzerland AG 2020   49
M. B. Shiflett (ed.), *Commercial Applications of Ionic Liquids*, Green Chemistry and Sustainable Technology, https://doi.org/10.1007/978-3-030-35245-5_3

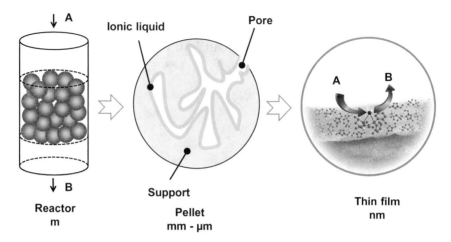

Fig. 3.1 Schematic representation of the supported ionic liquid concept for continuous catalytic processing

SILP concept was independently introduced by Mehnert et al. and Riisager et al. in 2002–2003 [9, 10] (Fig. 3.1).

While the SILP technology was introduced over 15 years ago, no large-scale application in catalysis has been reported yet. This is mainly due to the complexity of catalysis and the challenge to compete against established and depreciated processes. There are, however, several examples of SILP catalysis on a demonstration level in continuously operated liquid phase and gas phase mini-plants. These examples include hydroformylation, water-gas shift reaction, asymmetric hydrogenation, and hydrochlorination. This chapter will highlight the major outcomes of these studies, especially with respect to intensification of the future processes.

3.2 SILP-Catalyzed Gas-Phase Hydroformylation of Short-Chain Olefins

The first example of slurry phase SILP hydroformylation (see Scheme 3.1) was reported in 2003 by Mehnert et al. describing the conversion of 1-hexene as a batch reaction [9b]. A surface-modified silica gel was used in this work as the solid support. Its surface has been previously modified by covalently anchored imidazolium fragments. The motif of the classical water-soluble ligand tppts (see Fig. 3.2) had been modified with ionic tags, and this tpptim ligand (see Fig. 3.2) showed high activity with a turnover frequency (TOF) exceeding 3900 h^{-1}. Due to the miscibility of the ionic liquids used with the aldehyde product, severe and fast deterioration of the thin film was observed. Gas-phase SILP-catalyzed hydroformylation of propylene with a

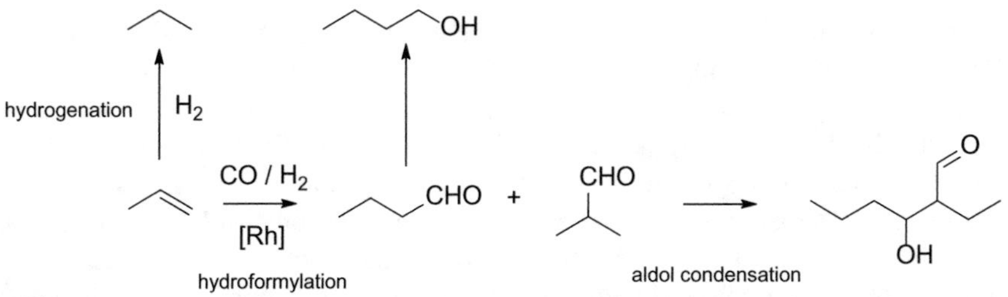
hydrogenation
H_2
OH
CO / H_2
[Rh]
CHO
CHO
O
OH
hydroformylation
aldol condensation

Scheme 3.1 General reaction network for Rh-catalyzed propylene hydroformylation including undesired side reactions

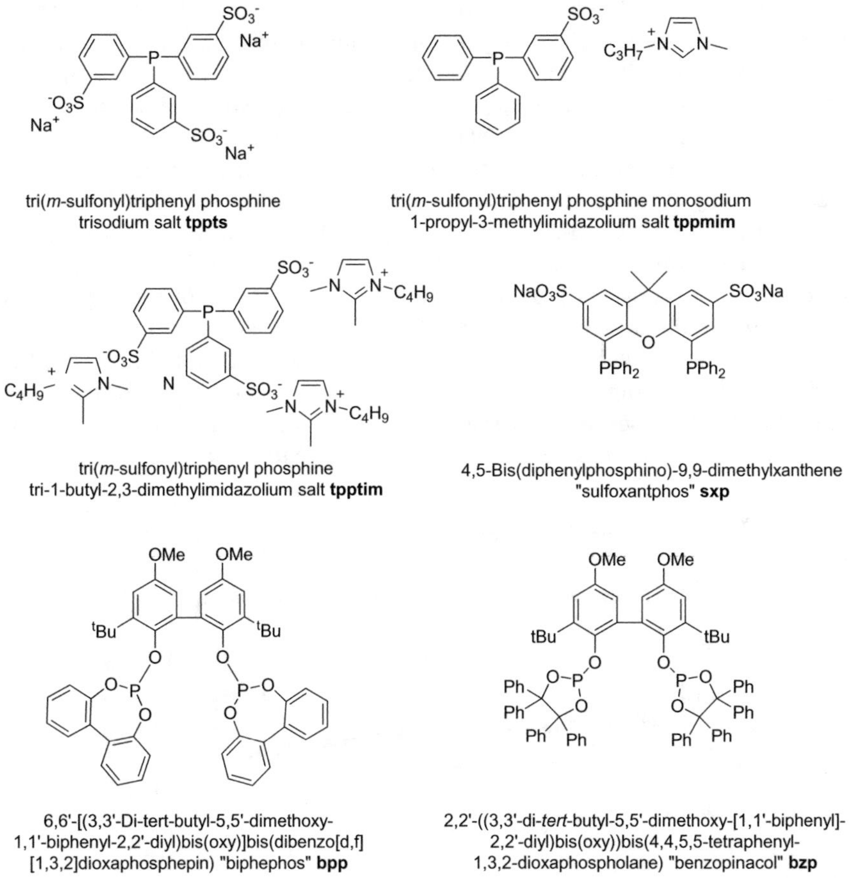
tri(*m*-sulfonyl)triphenyl phosphine
trisodium salt **tppts**
tri(*m*-sulfonyl)triphenyl phosphine monosodium
1-propyl-3-methylimidazolium salt **tppmim**
tri(*m*-sulfonyl)triphenyl phosphine
tri-1-butyl-2,3-dimethylimidazolium salt **tpptim**
4,5-Bis(diphenylphosphino)-9,9-dimethylxanthene
"sulfoxantphos" **sxp**
6,6'-[(3,3'-Di-tert-butyl-5,5'-dimethoxy-
1,1'-biphenyl-2,2'-diyl)bis(oxy)]bis(dibenzo[d,f]
[1,3,2]dioxaphosphepin) "biphephos" **bpp**
2,2'-((3,3'-di-*tert*-butyl-5,5'-dimethoxy-[1,1'-biphenyl]-
2,2'-diyl)bis(oxy))bis(4,4,5,5-tetraphenyl-
1,3,2-dioxaphospholane) "benzopinacol" **bzp**

Fig. 3.2 Selected ligands applied in SILP hydroformylation studies. Reprinted with permission [8]

homogeneous transition metal catalyst was able to overcome this drawback of ionic liquid and catalyst leaching [10].

A sulfoxantphos (sxp, see Fig. 3.2) modified Rh complex was dissolved in $[C_4C_1im][PF_6]$ and $[C_4C_1im][n\text{-}C_8H_{17}OSO_3]$. Commercially available porous silica gel was used as support material, providing a large internal surface area around $300\ m^2\ g^{-1}$. The activity was low with TOF values of only $18\ h^{-1}$ but could be maintained for at least 4 h. In a follow-up study, the reaction was conducted in a mini-plant setup that allowed higher conversion levels, yielding initial TOF values around $70\ h^{-1}$ [11]. Hydroformylation of propylene was found to be first order in substrate partial pressure, slightly positive order in hydrogen, and negative order in carbon monoxide. These data are in accordance with the established Wilkinson mechanism for modified rhodium-catalyzed hydroformylation [12]. The Rh-sxp-SILP-catalyzed gas-phase reaction was operated for approximately 210 h (time on stream) during which a slight decline in conversion was observed as shown in Fig. 3.3. The n/iso selectivity remained unchanged, which indicated that the Rh-sxp complex itself remained intact. It was concluded that the decline in activity was mainly due to blocking of transport pores, and short pressure swing scenarios (vacuum of approximately 70 mbar at 100 °C for 10 min) proved beneficial to regain the activity.

The activity was more than twice as high when the less reactive 1-butylene was used instead of propylene [13]. Interestingly, this difference in catalyst activity closely reflects the difference in molar solubility between the two olefins in the applied ionic liquid, with 1-butylene being 2.4 times more soluble in $[C_4C_1im][n\text{-}C_8H_{17}OSO_3]$ than propylene. Calculating space-time yields (STY) for these two non-optimized SILP scenarios leads to values of $110\ kg\ m^{-3}\ h^{-1}$ for propylene and $250\ kg\ m^{-3}\ h^{-1}$ for 1-butylene. The industrial biphasic Ruhrchemie/Rhône-Poulenc

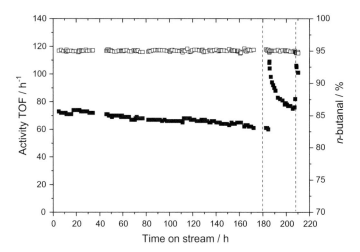

Fig. 3.3 Gas-phase hydroformylation of propylene using Rh-sxp SILP catalysts at 100 °C and 10 bar syngas pressure ($p_{propylene} = 1.8$ bar), $n_{rhodium} = 353\ \mu mol$, residence time = 0.4 s. Dashed lines represent 10 min vacuum (70 mbar at 100 °C) treatment. Data taken from Ref. [11]

(RCH/RP) process, utilizing a water-soluble Rh-tppts catalyst (see Fig. 3.2), operates at a STY of 200 kg m^{-3} h^{-1} for propylene at 120 °C and 50 bar [14]. It should be noted that the gas-phase process pressure was only 10 bar, while the RCH/RP process requires 50 bar. These values indicate the potential of a new SILP gas-phase process, especially considering the transformation from propylene to 1-butylene feedstock due to better product properties. Mass transport limitations were estimated and found to be negligible on overall reaction kinetics up to pellet diameters of 2.5 mm [15].

An industrially relevant reaction would be the transformation of internal olefins in an isomerization–hydroformylation sequence [16–18]. EVONIK reported a biphephos-modified (bpp, see Fig. 3.2) Rh-SILP system that can transform a mixed C4 feedstock in continuous gas-phase hydroformylation (see Fig. 3.4), yielding almost exclusively the desired linear pentanal [19].

The feedstock used in this study represents the light (non-converted) fraction of a commercial C4 dimerization plant and consists of only 1.5 wt% 1-butene, 28.5 wt% cis/trans-2-butenes, and 70 wt% inert n-butane (butane crude). The added value of such an olefin-depleted feed is of high technical relevance since it is typically not subjected to further chemical transformation due to its low concentration of reactive olefins.

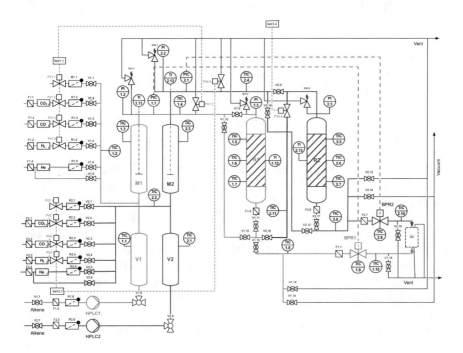

Fig. 3.4 Detailed flow scheme of the continuous gas-phase mini-plant used for SILP-catalyzed hydroformylation of C4 and C4 feed. The plant contains two parallel reactors (orange and blue) that can be operated independently in a parallel fashion or as a reactor cascade

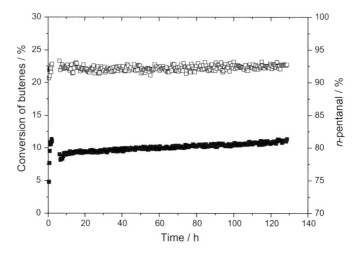

Fig. 3.5 Conversion (filled square) and selectivity (empty square) in the Rh-SILP-catalyzed hydroformylation of mixed 1-butene and 2-butene, highly diluted in butane (butane crude). $T = 140$ °C, $p_{total} = 20$ bar, 6 g Rh-bpp-SILP, $w_{Rh} = 0.4$ wt%, ligand biphephos, $\alpha_{IL} = 10$ vol%, $\tau = 80$ s, $CO:H_2 = 1$, molar flow (CO) = molar flow (H_2) = 2.2 mmol min^{-1}, molar flow (butane crude) = 1 mmol min^{-1} [19]. Reprinted with permission [8]

At elevated temperatures and pressures, the SILP catalyst converted up to 81 wt% of the reactive butylenes while achieving an n-pentanal selectivity higher than 93%. The Rh-bpp-SILP catalyst was stable for more than 100 h time on stream without significant loss of activity and selectivity (see Fig. 3.5).

Under optimized conditions, a STY close to 100 kg$_{n\text{-pentanal}}$ m^{-3} h^{-1} was reached, a value that is remarkably high in view of the very low concentration of terminal olefins in the applied feed.

Another active and highly selective SILP system for the conversion of a second technical C4 feedstock, coined Raffinate 1, was reported by EVONIK and academic partners in 2011 [20]. This C4 mixture originates from the steam cracker after the removal of 1,3-butadiene and consists of iso-butene (43.1 wt%), 1-butene (25.6 wt%), cis/trans-2-butene (16.1 wt%), inert butanes (14.9 wt%), and 1,3-butadiene (0.3 wt%). For high n-pentanal yield, the hydroformylation of iso-butene must be suppressed, and both 2-butenes must be isomerized prior to hydroformylation. A novel benzopinacol-based diphosphite ligand developed by LIKAT and EVONIK was used to modify the active Rh species [21]. The same mini-plant as shown in Fig. 3.4 was used. All internal butenes were converted only after isomerization totaling 99.5% n-pentanal selectivity. Undesired by-product 3-methylbutan-1-al formed by aldol condensation was below the detection limit of the online gas chromatograph. The TOF reached 3600 h^{-1} at 120 °C and 25 bar, which corresponded to a STY of 850 kg$_{n\text{-pentanal}}$ m^{-3} h^{-1}. Such high values would be acceptable for industrial applications in fixed-bed reactors (Fig. 3.6).

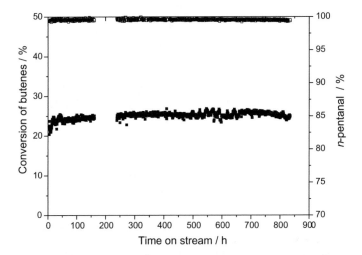

Fig. 3.6 Hydroformylation of an industrial C4 mixture (Raffinate 1, <16 ppm H_2O) in the presence of Rh-bzp-SILP catalyst showing conversion (filled square) and n-pentanal selectivity (empty square). $T = 100$ °C, $p_{total} = 10$ bar, $p_{raffinate\ 1} = 2$ bar, $p_{H_2} = p_{CO} = 4$ bar. Total volume flow = 29.2 mL min^{-1}, residence time = 15 s, $m_{SILP} = 3$ g, $w_{Rh} = 0.2$ wt%, L/Rh = 10, Stabilizer/L = 4, [C_2C_1im][NTf$_2$], ionic liquid loading = 10 vol%. Reprinted with permission [8, 20]

In 2015, Walter et al. reported the combined sequence of butane dehydrogenation and hydroformylation in a mini-plant cascade [22]. An industrial Cr-oxide catalyst from EVONIK was used for butane dehydrogenation at temperatures of 450 °C and ambient pressure. The as-formed mixture of 1-butylene and internal butylenes was condensed in a storage tank filled with glass beads at −20 °C. Hydrogen and lighter alkanes were released via a vent valve. When the liquid reached a certain level, an HPLC pump transferred the mixture into an evaporator mixing unit. The gaseous syngas-C4 feed passed a Rh-bpp-SILP fixed bed at 100 °C and 10 bars, thereby transforming all butylenes (ca. 20 wt% 1-butene, ca. 30 wt% cis-2-butene, and ca. 43 wt% trans-2-butene, 7 wt% butane) into n-pentanal. The STY of this combined dehydrogenation–hydroformylation process ranged between 10–170 kg$_{n\text{-pentanal}}$ m$_{SILP}^{-3}$ h^{-1} depending on the reaction conditions. Reducing the syngas pressures and the syngas to C4 ratio in the reactor resulted in better STY in general.

The latest development for process optimization in gas-phase hydroformylation catalysis was recently reported by EVONIK and academic partners [23]. The concept aims at combining the SILP reaction concept with membrane separation. The design principle is the creation of a catalytically active membrane as schematically depicted in Fig. 3.7.

A porous monolithic support structure will be coated by a thin film of ionic liquid, containing the active catalyst plus ligand. On the outside of the cylindrical monolith, a coating layer will be placed that has the potential to separate the formed aldehydes from the hydroformylation feed components. In such "two-in-one-reactors,"

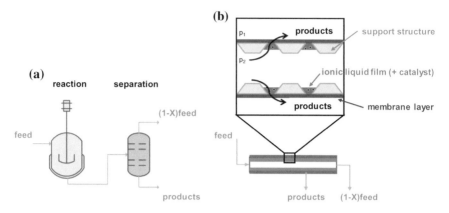

Fig. 3.7 Schematic representation of **a** classical sequential homogeneously catalyzed hydroformylation followed by separation steps and **b** the combined reaction–separation using a membrane reactor with supported ionic liquid films according to the EU H2020 Reactor Optimization by Membrane Enhanced Operation (ROMEO) project approach. Reprinted with permission [8]

the lifetime of the catalysts could be enhanced via removal of the products because unwanted aldol condensation will be minimized.

For the hydroformylation of olefins larger than C4, Hintermair et al. suggested combining SILP with supercritical carbon dioxide (scCO$_2$) as the mobile phase in order to remove products that could condense inside the pore network [24] (Fig. 3.8).

The SILP catalyst was prepared from [C$_3$C$_1$im][Ph$_2$P(3-C$_6$H$_4$SO$_3$)] and [Rh(CO)$_2$(acac)] (acac = acetylacetonate). It was dissolved in [C$_8$C$_1$im][NTf$_2$] and immobilized on microporous silica gel. The continuous-flow hydroformylation of 1-octene was carried out at 100 °C and 100 bar, and the Rh-SILP catalyst exhibited high catalytic rates (TOF of up to 800 h^{-1}). The SILP/scCO$_2$ system showed no sign of deactivation over 40 h time on stream, even though the formed nonanals have higher boiling points than the aldol condensation products obtained as side products in C3 or C4 hydroformylation. Rhodium and ionic liquid leaching were very low (<0.5 ppm) as measured by ICP-MS and NMR analysis. The combination of scCO$_2$ and SILP allows for excellent substrate diffusion, while the scCO$_2$ at the same time extracts heavy products from the supported ionic liquid film. This is a significant advantage compared to the classical SILP approach in continuous gas contact and broadens the application window of the SILP technology to higher boiling substrates.

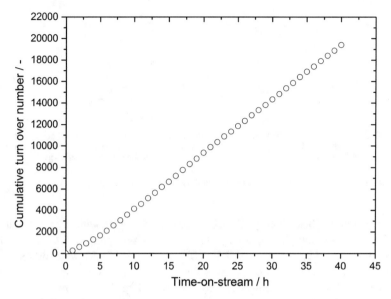

Fig. 3.8 Cumulative turnover number for the Rh-SILP-catalyzed hydroformylation of 1-octene in scCO$_2$. 100 °C, 100 bar, 14 wt% IL; Rh 0.146 mmol; P:Rh = 10, silica = 7.5 mL; substrate flow 0.42 mL min^{-1}, CO/H$_2$:substrate = 10, total flow = 854 mL min^{-1}. Data from [24]. Reprinted with permission [8]

3.3 Gas-Phase Selective Asymmetric Hydrogenation

Chiral hydrogenation catalysts that have been developed in the last decades for asymmetric transformations can also be applied in the form of SILP systems. The continuous gas-phase hydrogenation of methylacetoacetate (MAA) with Ru-SILP catalysts was reported in 2011 [25] (Scheme 3.2).

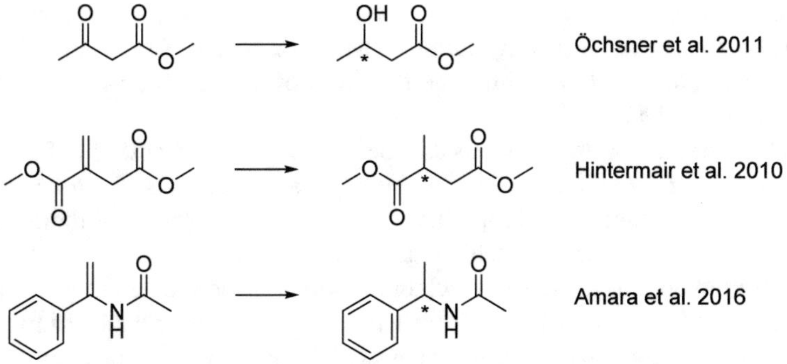

Öchsner et al. 2011

Hintermair et al. 2010

Amara et al. 2016

Scheme 3.2 Asymmetric gas-phase hydrogenation reactions catalyzed by SILP materials [25, 27, 30]

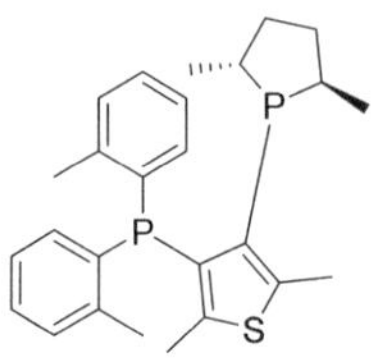

3-(2,5-(2R,5R)-dimethylphospholanyl-1)-
4-di-o-tolylphosphino-2,5-dimethyl-
thiophene

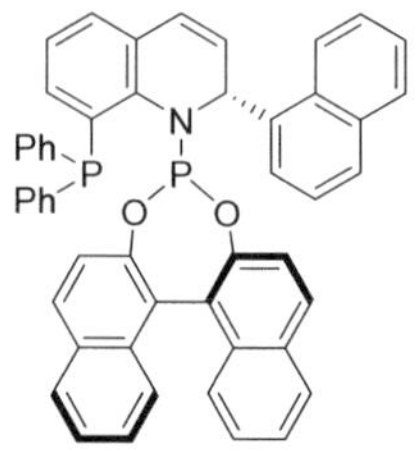

(2R)-1-((11bS)-dinaphtho[2,1-d:1',2'-f][1,3,2]dioxaphosphepin-
4-yl)-8-(diphenylphosphaneyl)-2-(naphthalen-1-yl)-1,2-
dihydroquinoline "quinaphos" **qp**

Fig. 3.9 Selected ligands used in SILP-catalyzed hydrogenation reactions

The precursor [bis(2-methylallyl)(1,5-cyclooctadiene)ruthenium and the ligand 3-(2,5-(2R,5R)-dimethylphospholanyl-1)-4-di-o-tolylphosphino-2,5-dimethylthio-phene (see Fig. 3.9, property of EVONIK, now SOLVIAS) were dissolved in [C_2C_1im][NTf_2], and this solution was dispersed onto porous silica. In order to keep the high-boiling compounds in the gas phase, a large helium flow was adjusted through a gradient-free recycle reactor. No activity was observed when the Ru complex was dispersed onto the support without ionic liquid. The activity of the SILP increased linearly with higher ionic liquid loading to reach its maximum at ionic liquid loading of 100% of the support's volume. Rinker's group did a detailed study for SILP systems in 1973 and explained such behavior by a slow chemical reaction that is not limited by diffusion effects [26]. Therefore, the more liquid volume containing catalyst complexes are present, the higher becomes the conversion.

Supercritical CO_2 as the mobile phase has been used for the asymmetric hydrogenation of dimethylitaconate by Hintermair et al. [27]. The chiral bidentate phosphine–phosphoramidite ligand quinaphos-modified (qp, see Fig. 3.9) Rh-qp-SILP catalyst was immobilized in imidazolium-based ionic liquids, [C_2C_1im][NTf_2], [C_4C_1im][BF_4], and [C_6C_1im][OTf], dispersed on silica gel. A reactor setup (see Fig. 3.10) was established to allow continuous flow reaction with scCO_2 flow [28]. The SILP catalysts achieved full conversion shortly after the start of the reaction and did not show a decrease in activity during 65 h time on stream. The enantioselectivity was extremely high (>99%) within the first 10 h of the run, followed by a steady decline (see Fig. 3.11).

The loss in enantioselectivity was ascribed to partial decomposition of the active species, resulting in unselective rhodium hydrogenation catalysts. The total turnover number was remarkable with 115,000 moles of substrate per mole of rhodium. The STY was calculated to be reasonably high with 0.3 kg m^{-3} h^{-1}.

In a follow-up study, the same authors reported a more detailed investigation of scCO_2/Rh-SILP asymmetric hydrogenation [29]. The water content in the feed was found to be a crucial factor for catalyst stability. Drying of incoming feed is well established in industry (e.g., by using redundant guard beds) and would allow implementation of the combined scCO_2-SILP catalysis for industrial asymmetric hydrogenations.

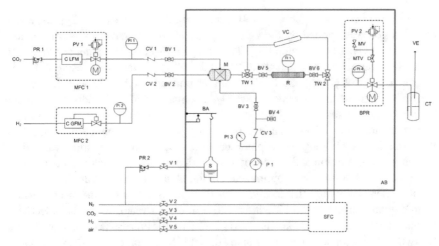

Fig. 3.10 Flow scheme of the continuous reaction system for supercritical fluid processing (PR = pressure-reducing valve, MFC = mass flow controller, LFM = liquid-flow meter, GFM = gas-flow meter, PI = pressure indicator, PV = proportional valve, CV = check valve, BV = ball valve, AB = air bath, V = screw-down valve, M = mixing chamber, S = substrate reservoir, BA = balance, P = piston pump, TW = three-way valve, R = reactor, VC = high-pressure view cell, TI = temperature indicator, BPR = back pressure regulator, MTV = magnetic trigger valve, MV = metering valve, CT = cooling trap, VE = vent, SFC = supercritical fluid chromatograph). Reprinted with permission [28]

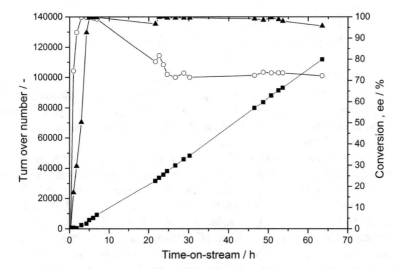

Fig. 3.11 Asymmetric hydrogenation using chiral Rh-qp-SILP ([EMIM][NTf$_2$]) on silica (α_{IL} = 0.35) catalysts. 40 °C, 120 bar, F(substrate) = 0.01 mL min^{-1}, F(CO$_2$) = 85 mL min^{-1}, F(H$_2$) = 10 mL min^{-1}, m_{SILP} = 1.56 g. Conversion (filled triangle), enantioselectivity (empty circle), and cumulative TONs (filled square). Data from [27]

The Rh-qp complex (see Fig. 3.9), immobilized in $[C_2C_1im][NTf_2]$ dispersed on silica gel, was applied by the Leitner group for the asymmetric hydrogenation of *N*-(1-phenylvinyl)acetamide [30]. The compound serves as model substrate for active pharmaceutical ingredient (API) synthesis of the target molecule *N*-(1-(5-fluoropyrimidin-2-yl)vinyl)acetamide from the portfolio of AstraZeneca. Due to the poor miscibility of the enamides in $scCO_2$, a helper solvent was added to facilitate the dissolution in the feed. Of the investigated solvents dichloromethane, toluene, and methanol, only toluene showed promising performance both with respect to the formation of a homogeneous phase under $scCO_2$ reaction conditions as well as preventing precipitation of the product. In a continuous $scCO_2$ reactor setup, the flow and process conditions were optimized. Again, water content within the feed was found to deactivate the SILP catalyst over time, leading to a decline of conversion from a quantitative level to 68% after 120 h time on stream, while the selectivity remained constant above 99%. The average STY of this process was calculated to be 21 kg m^{-3} h^{-1} with no Rh leaching being detected by means of ICP-OES (<1 ppm). Similar to the previous study, careful drying of the feed resulted in improved catalyst lifetime for more than 80 h time on stream and a cumulative turnover number (TON) exceeding 10,000. One kilogram of the starting material *N*-(1-phenylvinyl)acetamide could be converted by the $scCO_2$-SILP process within 18 h time on stream. These values would allow the $scCO_2$-SILP continuous flow process to be operated in an industrially viable scenario.

3.4 Gas-Phase Water-Gas Shift Reaction

The water-gas shift (WGS) reaction is the exothermic conversion of carbon monoxide and water to hydrogen and carbon dioxide. The reaction is of enormous industrial importance in the conversion of carbon monoxide to additional hydrogen, for example, in the Haber–Bosch process for ammonia production or in methane reforming for hydrogen production (Scheme 3.3).

The exothermic nature of this equilibrium allows the highest hydrogen purity (lowest level of remaining CO) at the lowest temperatures. Based on homogeneous Ru-chloro-carbonyl complexes dissolved in chloride-containing ionic liquids, Werner et al. developed active and stable Ru-SILP WGS catalysts [31, 32]. A mixed Ru-chloro-carbonyl $[Ru(CO)_3Cl_2]_2$ complex was identified as the resting state of the active catalyst under reaction conditions [33]. The basicity of both the support material and the ionic liquid was varied independently, since the literature data suggested a beneficial effect of the OH$^-$ concentration. The activity was increased by an order of magnitude when alumina was used instead of silica as the support [34]. The system stability could be further improved by changing the ionic liquid from

Scheme 3.3 Water-gas shift reaction

$$CO + H_2O \rightleftharpoons CO_2 + H_2$$

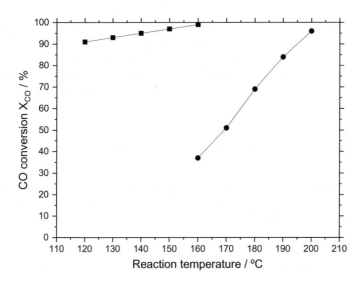

Fig. 3.12 Conversion in WGS reaction comparing a Ru-SILP (filled square) and an industrial Cu/ZnO (filled circle) catalyst. Conditions: GHSV = 12,000 h^{-1}, 200 mg catalyst (powder); p = 3 bar; syngas:H$_2$ 75 wt%, CO 8 wt%, CO$_2$ 13 wt%, N$_2$ 4 wt%; steam to gas 1:3, support alumina; ionic liquid [C$_4$C$_1$C$_1$im][Cl] (10 wt%); catalyst precursor Ru(CO)$_3$Cl$_2$ (2 wt% Ru). Data from [34]

[C$_4$C$_1$C$_1$im][OTf] to [C$_4$C$_1$C$_1$im][Cl]. For the latter IL, it is assumed that the IL's anion serves as a chloride reservoir for catalyst stabilization [35]. In the temperature range below 180 °C, the optimized SILP catalyst outperformed the industrial Cu/ZnO benchmark system (supplied by CLARIANT) by far, as shown in Fig. 3.12.

Based on the industrial relevance of the WGS, a scale-up study for SILP catalyst production was developed in 2012 [36]. Here, the established incipient wetness impregnation was replaced for larger scale synthesis by a spray-coating process. Reproducible coating of particles with an ionic liquid catalyst solution dissolved in a low-boiling solvent was demonstrated, allowing kilogram to ton-scale SILP preparation. The Ru-SILP catalyst was used to convert a premixed reformate synthesis gas mixture (7 wt% CO, 13 wt% CO$_2$, 5 wt% N$_2$, and 75 wt% H$_2$) to also evaluate the catalyst at elevated pressure. A TOF of around 40 h^{-1} was obtained at a gas hourly space velocity (GHSV) of 2000 h^{-1} corresponding to a STY of around 100 kg$_{CO_2}$ m^{-3}$_{SILP}$h^{-1}. In contrast to classical heterogeneous WGS catalysts, the Ru-SILP system allowed operation under fluctuating conditions, for example, when the feed flow was replaced for extended periods of time with nitrogen while continuing heating [37]. This robustness against dynamic operation is a strong benefit of the SILP WGS catalyst in comparison with its heterogeneous counterparts.

The Ru-SILP catalyst was tested by CLARIANT in an industrial WGS plant. The SILP catalyst (1.5 kg) was placed inside bypass tubing as a fixed-bed reactor [38]. The catalyst was able to completely convert the technical feed containing approximately 20 vol% CO. However, at these high conversion levels and high mass flows, the exothermicity of the WGS reaction caused a temperature increase inside the SILP bed above 200 °C. This probably led to the thermal decomposition of the Ru complex as well as the ionic liquid, resulting in irreversible catalyst deactivation. An industrial application would have to solve this heat management issue by an appropriate reactor design. As one possible option, LINDE, BIOENERGY2020, and academic partners have reported the idea to use membrane reactors for SILP WGS catalysis [23].

3.5   Miscellaneous Gas-Phase Applications

In addition to the continuous hydroformylation reactions, the related rhodium-catalyzed carbonylation of methanol has been reported as a SILP-catalyzed gas-phase reaction. The technical importance of this reaction is obvious for the so-called Monsanto process (see Scheme 3.4) [39].

The latter is the dominating technical process for the production of acetic acid (and methyl acetate) and is carried out on a large industrial scale as a homogeneous liquid-phase reaction.

Based on [Rh(CO)$_2$I$_2$]$^-$ anions as the catalytically active species, Riisager and coworkers developed a Monsanto-type SILP catalyst system, in which the active rhodium catalyst complex is part of the ionic liquid itself [40]. The SILP system was prepared by a one-step impregnation of the silica support using a methanolic solution

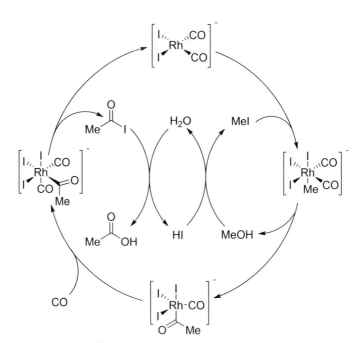

Scheme 3.4 Mechanism of Rh-catalyzed methanol carbonylation

of the ionic liquid $[C_4C_1im][I]$ and the dimeric precursor species $[Rh(CO)_2I]_2$ as depicted in Scheme 3.5. Under reaction conditions resembling typical industrial conditions ($T = 180$ °C, $p = 20$ bar), the water-free gas-phase carbonylation of methanol was carried out at full conversion of methanol reaching a TOF for the combined acetyl products (acetic acid plus methyl acetate) of $76.5\,h^{-1}$ and production rate of 21.0 mol $L^{-1}\,h^{-1}$. This productivity was comparable to that obtained with an analogous bubble-column reaction system containing about 100-fold volume of ionic liquid catalyst solution, clearly indicating the high efficiency of the designated SILP system. One drawback, however, was found to be the relatively low selectivity toward acetic acid (21% instead of 96% as in the large-volume reaction system).

The Riisager group recently reported the application of phosphine-modified palladium catalysts for the continuous gas-phase methoxycarbonylation of ethylene to methylpropanoate with promising results (see Scheme 3.6) [41]. The SILP catalyst was prepared by impregnation of a silica support with a solution containing a Brønsted acid IL, Pd(OAc)₂, and the ligand (1,2-bis(di-*tert*-butylphosphinomethane)benzene dtbpmb (see Scheme 3.6) that is well known to yield methylpropionate with 100% selectivity in the studied reaction.

Different reaction parameters were investigated, such as IL loading, Pd loading, and dtbpmb/Pd ratio. It was found that high IL loadings had a negligible effect on the catalytic activity probably due to mass transfer limitations in a thicker IL layer. On the other hand, the reaction seems to be nearly first order with respect to palladium at the tested reaction conditions; the higher the palladium content, the higher the conversion. Interestingly, a strong correlation was found between the dtbpmb to Pd ratio and the stability of the catalytic system (up to 50 h on stream) by increasing the dtbpmb to Pd ratio from 5 to 30. Further increase led to a very low conversion due to the excess of ligand, which prevents the substrates from approaching the metal center.

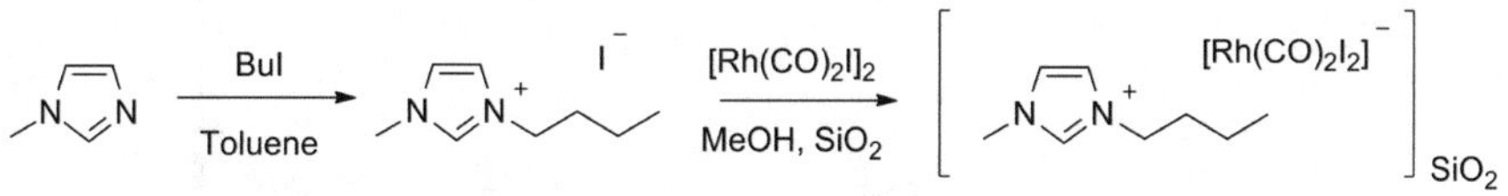

Scheme 3.5   Preparation of SILP $[C_4C_1im][Rh(CO)_2I_2]–[C_4C_1im][I]–SiO_2$ catalyst

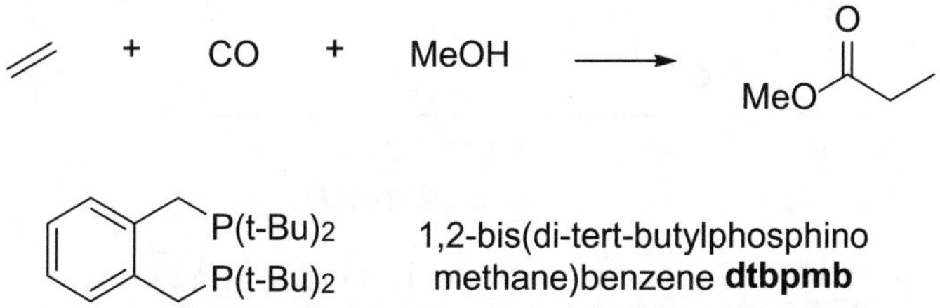

Scheme 3.6   Pd-catalyzed methoxycarbonylation of ethylene and dtbpmb ligand structure

In 2017, the reduction of toxic mercury compounds from vinyl chloride monomer (VCM) production was reported. High-valent Au(III) complexes have been immobilized in 1-propyl-3-methylimidazolium chloride $[C_3C_1im][Cl]$ supported on activated carbon [42]. In order to suppress reduction of the active Au(III) species to metallic Au^0, addition of $CuCl_2$ to the ionic liquid was found to be beneficial. Both metals were highly dispersed on the carbon surface, since no Au or Cu particles were detected by means of XRD and TEM in combination with EDS.

The as-prepared Au-SILP catalysts were tested in the gas-phase hydrochlorination of acetylene (see Scheme 3.7) in a tubular fixed-bed reactor at 180 °C and ambient pressure. While the pure Au/activated carbon (AC) and Cu/AC catalysts showed almost no activity, 40% acetylene conversion was observed for the Au-SILP system. The $CuCl_2$-modified Au-Cu-SILP showed 75% conversion and stable performance for at least 12 h. In an impressive long-term experiment, this catalyst was stable for more than 500 h as shown in Fig. 3.13.

The VCM selectivity was high with 99.8% over the total run time with a TON of more than 439,000. These results clearly mark an improvement toward more benign VCM production.

Scheme 3.7 Hydrochlorination of acetylene to yield vinyl acetate monomer

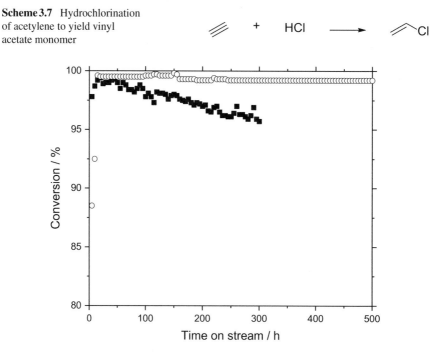

Fig. 3.13 Hydrochlorination of acetylene using two Au-SILP catalysts, Au(III)-SILP (filled square) and Au(III)-Cu(II)-SILP (empty circle). Reaction conditions: $T = 180$ °C, GHSV $= 50$ h^{-1}, feed ratio $V_{HCl}:V_{acetylene} = 1.2:1$

3.6 Conclusion

The examples in this chapter have summarized the state of the art in organometallic catalyst immobilization via thin-film technologies with SILP materials. The number of successful applications of SILP materials is constantly increasing. However, it should be noted that one needs to be careful when planning to apply a SILP material for a given reaction or process. While, on the one hand, a high solubility of substrate is desired, the high product solubility can lead to catalyst deactivation. Such accumulation must be taken into account, and pore flooding needs to be prevented upfront by adjusting flow rates and conversions appropriately. Despite these precautions, the surface coating of solid materials with ionic liquid thin films constitutes a versatile and broadly applicable technology.

References

1. Cornils B, Herrmann WA, Beller M, Paciello R (eds) (2017) Applied homogeneous catalysis with organometallic compounds: a comprehensive handbook in four volumes. Wiley-VCH, Weinheim
2. Cole-Hamilton D, Tooze R (eds) (2006) Catalyst, separation, recovery and recycling. Springer, Dordrecht
3. Cornils B, Herrmann WA, Horváth IT, Leitner W, Mecking S, Olivier-Bourbigou H, Vogt D (eds) (2005) Multiphase homogeneous catalysis. Wiley-VCH, Weinheim
4. Riisager A, Fehrmann R, Haumann M (eds) (2014) Supported ionic liquids—fundamentals and applications. Wiley-VCH, Weinheim
5. Gu Y, Li G (2009) Ionic liquids-based catalysis with solids: state of the art. Adv Synth Catal 351:817
6. Van Doorslaer C, Wahlen J, Mertens P, Binnemans K, De Vos D (2010) Immobilization of molecular catalysts in supported ionic liquid phases. Dalton Trans 39:8377
7. Selvam T, Machoke A, Schwieger W (2012) Supported ionic liquids on non-porous and porous inorganic materials—A topical review. Appl Catal A Gen 445–446:92
8. Marinkovic JM, Riisager A, Franke R, Wasserscheid P, Haumann M (2019) Fifteen years of supported ionic liquid phase-catalyzed hydroformylation: material and process developments. Ind Eng Chem Res 58:2409.
9. (a) Mehnert CP, Mozeleski EJ, Cook RA (2002) Supported ionic liquid catalysis investigated for hydrogenation reactions. Chem Commun 3010; (b) Mehnert CP, Cook RA, Dispenziere NC, Afeworki M (2002) Supported ionic liquid catalysis – a new concept for homogeneous hydroformylation catalysis. J Am Chem Soc 124:12932
10. (a) Riisager A, Wasserscheid P, van Hal R, Fehrmann R (2003) Continuous fixed-bed gas-phase hydroformylation using supported ionic liquid-phase (SILP) Rh catalysts. J Catal 219:452; (b) Riisager A, Eriksen KM, Wasserscheid P, Fehrmann R (2003) Propene and 1-octene hydroformylation with silica-supported, ionic liquid-phase (SILP) Rh-phosphine catalysts in continuous fixed-bed mode. Catal Lett 90:149
11. Riisager A, Fehrmann R, Haumann M, Gorle BSK, Wasserscheid P (2005) Stability and kinetic studies of supported ionic liquid phase catalysts for hydroformylation of propene. Ind Eng Chem Res 44:9853
12. (a) Young JF, Osborn JA, Jardine FA, Wilkinson G (1965) Hydride intermediates in homogeneous hydrogenation reactions of olefins and acetylenes using rhodium catalysts. Chem Commun (London) 131; (b) Evans D, Osborn JA, Wilkinson GJ (1968) Hydroformylation of alkenes

by use of rhodium complex catalysts. J Chem Soc A 3133; (c) Evans D, Yagupsky G, Wilkinson GJ (1968) The reaction of hydridocarbonyltris(triphenylphosphine)rhodium with carbon monoxide, and of the reaction products, hydridodicarbonylbis(triphenylphosphine)rhodium and dimeric species, with hydrogen. J Chem Soc A 2660

13. Haumann M, Dentler K, Joni J, Riisager A, Wasserscheid P (2007) Continuous gas-phase hydroformylation of 1-butene using supported ionic liquid phase (SILP) catalysts. Adv Synth Catal 349:425

14. Hibbel J, Wiebus E, Cornils B (2013) 75 Jahre Hydroformylierung – Oxoreaktoren und Oxoanlagen der Ruhrchemie AG und der Oxea GmbH von 1938 bis 2013. Chem Ing Tech 85:1853

15. Haumann M, Jakuttis M, Werner S, Wasserscheid P (2009) Supported ionic liquid phase (SILP) catalyzed hydroformylation of 1-butene in a gradient-free loop reactor. J Catal 263:321

16. van Leeuwen PWNM, Kamer PC, Reek JNH, Dierkes P (2000) Ligand bite angle effects in metal-catalyzed C−C bond formation. Chem Rev 100:2741

17. Klein H, Jackstell R, Wiese K-D, Borgmann C, Beller M (2001) Hoch selektive Katalysatoren für die Hydroformylierung interner Olefine zu linearen Aldehyden. Angew Chem 113:3505

18. Behr A, Obst D, Schulte C, Schosser T (2003) Highly selective tandem isomerization–hydroformylation reaction of *trans*-4-octene to *n*-nonanal with rhodium-BIPHEPHOS catalysis. J Mol Catal A 206:179

19. Haumann M, Jakuttis M, Franke R, Schönweiz A, Wasserscheid P (2011) Continuous gas-phase hydroformylation of a highly diluted technical C4 feed using supported ionic liquid phase catalysts. ChemCatChem 3:1822

20. (a) Jakuttis M, Schönweiz A, Werner S, Franke R, Wiese K-D, Haumann M, Wasserscheid P (2011) Rhodium–phosphite SILP catalysis for the highly selective hydroformylation of mixed C$_4$ feedstocks. Angew Chem Int Ed 50:4492; (b) Franke R, Brausch N, Fridag D, Christiansen A, Becker M, Wasserscheid P, Haumann M, Jakuttis M, Werner S, Schönweiz A (2012) Einsatz von supported ionic liquid phase (SILP) Katalysatorsystemen in der Hydroformylierung von olefinhaltigen Gemischen zu Aldehydgemischen mit hohem Anteil von in 2-Stellung unverzweigten Aldehyden. WO 2012041846 A1

21. Selent D, Franke R, Kubis C, Spannenberg A, Baumann W, Kreidler B, Börner A (2011) A new diphosphite promoting highly regioselective rhodium-catalyzed hydroformylation. Organometallics 30:4509

22. (a) Walter S, Haumann M, Wasserscheid P, Hahn H, Franke R (2015) *n*-butane carbonylation to *n*-pentanal using a cascade reaction of dehydrogenation and SILP-catalyzed hydroformylation. AIChE J 61:893; (b) Hahn H, Dyballa KM, Franke R, Fridag D, Walter S, Haumann M, Wasserscheid P (2016) SILP-Katalysator zur Hydroformylierung von Olefinen mit Synthesegas. DE 102015201560 A1

23. http://www.romeo-h2020.eu/. Accessed 10.01.2019

24. Hintermair U, Zhao G, Santini CC, Muldoon MJ, Cole-Hamilton DJ (2007) Supported ionic liquid phase catalysis with supercritical flow. Chem Commun 1462

25. (a) Öchsner E, Schneider MJ, Meyer C, Haumann M, Wasserscheid P (2011) Challenging the scope of continuous, gas-phase reactions with supported ionic liquid phase (SILP) catalysts—Asymmetric hydrogenation of methyl acetoacetate. Appl Cat A Gen 399:35; (b) Haumann M, Öchsner E, Wasserscheid P (2010) Hybridmaterialien zur heterogen katalysierten asymmetrischen Hydrierung im Gaskontakt und Verfahren zu ihrer Anwendung. DE 102009011815 A1

26. Abed R, Rinker RG (1973) Diffusion-limited reaction in supported liquid-phase catalysis. J Catal 31:119

27. Hintermair U, Höfener T, Pullmann T, Franciò G, Leitner W (2010) Continuous enantioselective hydrogenation with a molecular catalyst in supported ionic liquid phase under supercritical CO$_2$ flow. ChemCatChem 2:150

28. Hintermair U, Roosen C, Kaever M, Kronenberg H, Thelen R, Aey S, Leitner W, Greiner L (2011) A versatile lab to pilot scale continuous reaction system for supercritical fluid processing. Org Process Res Dev 15:1275

29. Hintermair U, Franciò G, Leitner W (2013) A fully integrated continuous-flow system for asymmetric catalysis: enantioselective hydrogenation with supported ionic liquid phase catalysts using supercritical CO_2 as the mobile phase. Chem Eur J 19:4538
30. Amara Z, Poliakoff M, Duque R, Geier D, Franciò G, Gordon CM, Meadows RE, Woodward R, Leitner W (2016) Enabling the scale-up of a key asymmetric hydrogenation step in the synthesis of an API using continuous flow solid-supported catalysis. Org Process Res Dev 20:1321
31. (a) Werner S, Szesni N, Fischer RW, Haumann M, Wasserscheid P (2009) Homogeneous ruthenium-based water–gas shift catalysts *via* supported ionic liquid phase (SILP) technology at low temperature and ambient pressure. Phys Chem Chem Phys 11:10817; (b) Szesni N, Kaiser M, Fischer RW, Haumann M, Werner S, Wasserscheid P (2011) Katalysatorzusammensetzung für die Umsetzung von Kohlenmonoxid in Gasströmen. WO 2011023368 A1
32. Werner S, Szesni N, Bittermann A, Schneider MJ, Härter P, Haumann M, Wasserscheid P (2010) Screening of Supported Ionic Liquid Phase (SILP) catalysts for the very low temperature water–gas-shift reaction. Appl Cat A Gen 377:70
33. Bauer T, Stepic R, Wolf P, Kollhoff F, Karawacka W, Wick CR, Haumann M, Wasserscheid P, Smith DM, Smith A-S, Libuda J (2018) Dynamic equilibria in supported ionic liquid phase (SILP) catalysis: *in situ* IR spectroscopy identifies $[Ru(CO)_xCl_y]_n$ species in water gas shift catalysis. Cat Sci Technol 8:344
34. Werner S, Szesni N, Kaiser M, Fischer RW, Haumann M, Wasserscheid P (2010) Ultra-low-temperature water–gas shift catalysis using supported ionic liquid phase (SILP) materials. ChemCatChem 2:1399
35. Stepić R, Wick CR, Strobel V, Berger D, Vučemilović-Alagić N, Haumann M, Wasserscheid P, Smith A-S, Smith DM (2019) Mechanism of the water–gas shift reaction catalyzed by efficient ruthenium-based catalysts: a computational and experimental study. Angew Chem Int Ed 58:741
36. Werner S, Szesni N, Kaiser M, Haumann M, Wasserscheid P (2012) A scalable preparation method for SILP and SCILL ionic liquid thin-film materials. Chem Eng Technol 35:1962
37. Werner S, Haumann M (2014) Ultralow temperature water–gas shift reaction enabled by supported ionic liquid phase catalysts. In: Riisager A, Fehrmann R, Haumann M (eds) Supported ionic liquids—fundamentals and applications, chap 16. Wiley-VCH, Weinheim
38. Szesni N Clariant Produkte Germany, private communication
39. Yoneda N, Kusano S, Yasui M, Pujado P, Wilcher S (2001) Recent advances in processes and catalysts for the production of acetic acid. Appl Catal A 221:253
40. (a) Riisager A, Jørgensen B, Wasserscheid P, Fehrmann R (2006) First application of supported ionic liquid phase (SILP) catalysis for continuous methanol carbonylation. Chem Commun 994; (b) Riisager A, Fehrmann R (2006) A process for continuous carbonylation by supported ionic liquid-phase catalysis. WO2006122563 A1
41. (a) Khokarale SG, García-Suárez EJ, Fehrmann R, Riisager A (2017) Highly selective continuous gas-phase methoxycarbonylation of ethylene with supported ionic liquid phase (SILP) catalysts. ChemCatChem 9:1824; (b) Riisager A, Fehrmann R, Garcia Suarez E, Xiong J (2017) Palladium catalyst system comprising zwitterion and/or acid-functionalyzed ionic liquid. US2017341067 A1
42. Zhao J, Yu Y, Xu X, Di S, Wang B, Xu H, Ni J, Guo LL, Pan Z, Li X (2017) Stabilizing Au(III) in supported-ionic-liquid-phase (SILP) catalyst using $CuCl_2$ via a redox mechanism. Appl Cat B Environ 206:175

Chapter 4
Are Ionic Liquids Enabling Technology?
Startup to Scale-Up to Find Out

Julia L. Shamshina and Robin D. Rogers

Abstract Commercialization of new sustainable technology from academia to industry is based on the technology-enabling innovation, the manufacturability, the implementation cost, and the technology's competitive advantage, such as functionality improvement(s) over the routine process or existing products. Future-minded thinking outside the accepted margins and innovative execution are involved in creating new markets. The majority of this chapter is dedicated to our experiences in pursuing the transition of ionic liquids (ILs)-based technology from academia to industry for the extraction of chitin ($(C_8H_{13}O_5N)_n$), the second most abundant biopolymer on the planet, directly from shrimp shells. While the dissolution and extraction of chitin was demonstrated as early as 2010, the necessity of using an IL presented hurdles for scaling the technology to a commercial level. The resultant chitin polymer could be extracted while maintaining its high-molecular weight and providing materials with high strength and unique control of the final form. In 2012, a Laboratory Demonstration Pilot Unit (LDPU) was built and tested, followed by further scale-up to a mini-pilot plant in 2014–2015 with funding from the U.S. Department of Energy. Currently, this mini-pilot plant provides the groundwork for the construction of a larger plant for a scaled-up chitin extraction by Mari Signum, Mid-Atlantic. This will allow the generation of sufficient supplies of chitin and create new markets for this polymer. The high quality of the polymer and the ability to produce high-value products from it will give Mari Signum, Mid-Atlantic a competitive advantage not only to enter multiple focused profitable markets but also to create new markets. Once the polymer becomes available on a large-scale not only will the price decrease, but it will become available for the invention of additional products. When large-scale supply is available, it will provide confidence to investors due to

RDR is president of 525 Solutions, Inc. and has partial ownership of 525 Solutions, Inc. RDR has a former ownership in Mari Signum, Mid-Atlantic, LLC. RDR and JLS are named inventors on related patents and applications. JLS is a former employee of 525 Solutions, Inc. and a former CTO of Mari Signum, Mid-Atlantic, LLC.

J. L. Shamshina (✉)
Mari Signum, Mid Atlantic, LLC, 8310 Shell Road, 23237 Richmond, VA, USA
e-mail: shams002@gmail.com

R. D. Rogers
525 Solutions, Inc., P.O. Box 2206, 35403 Tuscaloosa, AL, USA

© Springer Nature Switzerland AG 2020 69
M. B. Shiflett (ed.), *Commercial Applications of Ionic Liquids*, Green Chemistry and Sustainable Technology, https://doi.org/10.1007/978-3-030-35245-5_4

known and manageable marketing and supply costs. The tremendous potential of chitin will soon be exploited for a number of industrial applications utilizing the full potential of this IL-based platform.

Keywords Biomass · Biopolymers · Chitin · Commercialization · Ionic liquid · Process development · Product development · Renewable polymers

4.1 Introduction

4.1.1 More Plastic than Fish

With an increase in plastic production volume from about 2 million metric tons per year in 1950 to over 400 million metric tons per year in 2015, the global plastics market is expected to reach $654 billion by 2020 [1, 2]. At the same time, recent data on the lifecycle of plastics worldwide from production to utilization to recycle suggest that as much as 76% of all plastics produced to date have ended up as waste [3]. Only 9% of this waste has been re-processed, 12% has been incinerated, and 79% has accumulated worldwide. At the current plastic production rate, the amount of plastic waste accrued in the environment will practically double by 2050 compared with 2015 [2]. The Ellen MacArthur Foundation in its 2016 report [4] claimed that if plastic continues to be manufactured at current rates with irresponsible disposal, there will be more plastic than fish in our oceans by 2050 [5].

The enormous negative environmental impact of the plastics industry [6] has resulted in a major reconsideration of the role of renewables in sustainable product development. The classic definition of sustainable product development in value-added products' manufacture is the use of renewable resources, that is, resources that can be used repeatedly and are being naturally replaced. The potential of polymers sourced from nature, or *biopolymers* isolated from biomass as a by-product of agricultural, forestry, and marine ecosystems [7], are both vital components from a sustainability and economic value standpoint.

4.1.2 Taking Full Advantage of What Nature Creates

The definition of biopolymers has less to do with chemistry and more to do with semantics. Any chemist dealing with sustainable development in the last few years has run into the recurrence of the term "renewable", closely connected to "biorefinery". The well-established biorefinery model is focused on bio-based chemicals and products in which biomass is first converted into commodity building blocks and high-value chemicals for high-volume global markets. The production of lower-cost chemical building blocks is understandable, and ultimately, society must move towards bio-based substituents to reduce our dependence on petroleum. A biorefinery, however, is based on the idea that biomass needs to be "de-functionalized" at the cost of chemicals and energy to be made into the basic building blocks found in petroleum.

Oftentimes, these basic bio-based building blocks are utilized in polymer synthesis, and the resultant chemicals are also named "biopolymers". For instance, polylactic acid (PLA) is known as a biopolymer even though it is a polyester derived from renewably-sourced lactic acid. In this regard, the concept of using bio-based chemicals as precursors for polymer manufacturing is confusing, since often the same non-degradable plastic is produced, regardless of whether this plastic is oil-based or plant-based. The life cycle of a plastic product is more important than its origin. For example, recently, Coca-Cola advertised its new beverage container or "plant-bottle" [8] made of polyethylene terephthalate (PET), produced by condensation of bio-based monoethylene glycol with terephthalic acid. Laboratory experiments studying PET degradation predicted a life expectancy between 27 [9] and 93 [10] years, regardless of the starting material used for manufacture.

When we talk about biopolymers in the following chapter, we mean polymers isolated from naturally occurring biomass "as nature made them" [11], such as polysaccharides (e.g., cellulose, chitin, hemicellulose), proteins (e.g., spider silk), plants polyesters (e.g., lignin), and so forth, and not the ones produced from bio-based chemicals. These biopolymers are viewed not as a replacement for petroleum, but as a source of valuable chemicals and materials that cannot be obtained from petroleum. Instead of chemically modifying polymers obtained from nature and making synthetic analogs, we need to figure out how to take full advantage of what nature does so well.

4.1.3   Research and Development to Commercialization Constraints: Need of Economy of Scale

Research and development in the field of biopolymers is primarily small in scale and academic in nature. While renewable, biodegradable replacements for plastics are being actively developed, they have come nowhere close to completely replacing plastics. Plastics enjoy technological maturity and an entrenched economy of scale that have kept new technologies from competing with them.

Economic barriers include poor predictions for short-term profits, undefined demand in the marketplace because of prevailing inexpensive alternatives (synthetic polymers), and, most importantly, a lack of supply in needed volume. Plastics rose to dominance through the availability of cheap oil as a feedstock and open markets for its products, and for any replacement to be successful, those two factors must be addressed. Because biopolymers will undeniably be more expensive than synthetic analogs, much work will be required to lower the costs. Even if some biopolymers are shown to have advantageous properties when compared to conventional polymers, in only a few high-end applications, such as biomaterials (e.g., tissue engineering, plastic surgery, and drug delivery devices), will the relatively high costs of biopolymer precursors likely not interfere with market growth. In low-end and medium-end applications, it is hard to predict whether economy-of-scale manufacturing will be able to bring down the current high production costs.

Technical barriers are the same as for all new technologies, such as product design and long development periods, even more so taking into consideration a need for 'bench-to-scale' prototyping. Besides, any new process also requires development (or at least adaptation) of production equipment.

Finally, the enormous growth in the plastics economy during the twenty-first century reflects a large investment in the oil-based industry. While renewable resources are perfectly suited to provide the same rich variety of polymers and composites as that currently available from oil, there has not been the same extent of investments for renewables that have gone into plastics manufacturing. On the other hand, transition from non-biodegradable plastic materials to biodegradable biopolymeric products from renewable sources will be highly advantageous to society for development of new materials, new products, new unforeseen markets, and improvement in the environment.

4.2   Brief Foreword to Chitin and Current Isolation Technology

4.2.1   Chitin Polymer

Chitin, a linear carbohydrate made of N-acetyl-D-glucosamine units as shown in Fig. 4.1, is the second most abundant biopolymer on earth (after cellulose) [12] and a primary component of crustacean shells (e.g., crabs, lobsters, and shrimps), where it exists in a protein-mineral matrix. Chitin is a known wound-healing accelerator [13], has anti-inflammatory properties [14], is protein-regulating [15] and has cell-proliferating [16] properties, and demonstrates outstanding biocompatibility [17]. It is also biodegradable (12-weeks post-surgery degradation in the human body [18]), non-allergenic, and non-toxic. The polymer demonstrates high mechanical strength

Fig. 4.1 Structure of chitin

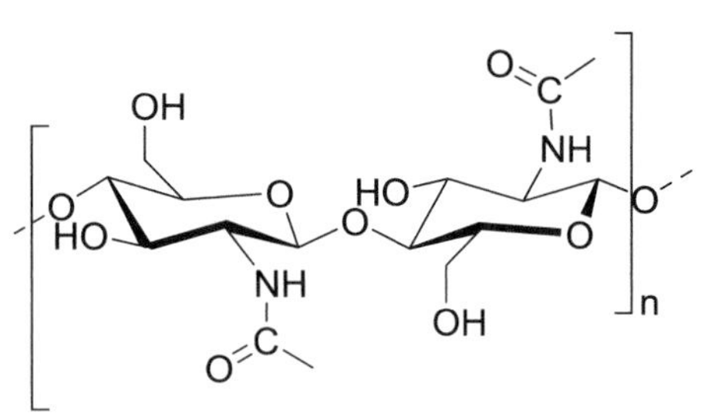

and provides the advantage of easy surface modifications [19]. Multiple reviews on chitin and its properties are available elsewhere [20–25].

4.2.2   Chitin Market

The global chitin market was worth $803 million in 2016, increased to $893 million in 2017, and is predicted to triple to $2,941 million by 2027 according to Global Industry Analysis [26]. The global chitin market is projected to have a compound annual growth rate (CAGR) of 12.7% throughout the period of 2017–2027. New biomedical products are the main drivers of this market. Overall, the healthcare segment, estimated at $309 million in 2017, is projected to grow at a CAGR of 14.2% throughout 2017–2027 and is anticipated to reach $1 billion by the end of 2027. For instance, chitin-based materials have been proposed for artificial organs [27], space-filling implants [28], drug delivery systems [29, 30], tissue engineering materials [31], wound dressings [32], treatment of burns [33], artificial skin, and plastic skin surgery [34].

Even though there is a high demand for chitin and its derivatives, currently there is no North American producer of chitin. In 2008, there were five companies that dealt with chitin derivatives (namely, chitosan and glucosamine) including Biothera Inc., CarboMer, Inc., HaloSource, Inc., V-Labs, Inc., and United Chitotechnologies [35]. Even though some of these companies claimed chitin production, they used a 100-year-old chemical and energy-intensive process that degraded chitin by reducing its molecular weight. This process also results in a large degree of deacetylation of chitin, producing chitosan, a polymer with different properties. Unfortunately, the lack of rigor in terminology has led to confusion in the markets about the actual properties of chitin as a material versus chitosan.

In 2018, there were only two functioning facilities identified: Tidal Vision, Inc. [36] and CarboMer, Inc. [37]. Tidal Vision is a company that manufactures chitin-based and chitosan-based products and serves various industries including textile, cosmetic, water treatment, agriculture, food, pharmaceutical, and so forth, and also sells raw material to research institutions and laboratories [36]. The company claims to use a "patent-pending, closed-loop processing system" to extract chitin from crab shells, although the company does not provide the technology of chitin isolation. Car-boMer, Inc. [37] is the second company that appears in numerous market reports as a producer of chitin; however, we were unable to find either chitin or its derivative, chitosan, in the company's product list. The company sells polyamino acids, polyglycolides and polylactides, collagens, poly(3-hydroxybutyrate), PLA, poly(estradiol phosphate), poly(ethylene adipate), poly(glycolic acid), and polyinosinic-polycytidylic acids as biopolymers [37].

Chitin is used in animal feed as a dietary supplement. It is reported to promote animal growth, to improve adsorption of nutrients, and to inhibit the effect of harmful microorganisms [38]. Its properties also make the polymer an ideal material for the following uses in agriculture: (1) as a fertilizer, (2) as a fungicide and pesticide in crop protection, (3) as an agent to improve seed quality (as well as crop yield and quality), and (4) as a plant growth stimulator; it also acts as amplifier of the beneficial chitinolytic microbes [39]. Chitin is used in environmental applications as an effective biosorbent, due to presence of both hydroxyl- and acetamide-moieties (easily modifiable into amine functionality), which demonstrate high adsorption potential for the removal of various metal ions from water sources [40]. Chitin, in the form of whiskers or nanofibers, can be utilized during manufacturing processes as an additive to reinforce existing materials (packaging, fibers, etc.) [41].

Key market players, however, are approaching the market with advanced high-quality *medical* products of higher efficacy. Unitika, Ltd. (Japan) [42] marketed a chitin-containing non-woven dressing (Beschitin W) for the treatment of burns and demonstrated its superior performance in speed of healing, wound adherence, exudate absorption, and scar minimization. In 1970, multiple new companies appeared on the market. Eisai Co., Ltd. produces wound dressings from chitin, Chitipack S® and Chitipack P® [43], which are used in the treatment of traumatic wounds preventing the formation of scar tissue [44]. Syvek-Patch® produced by Marine Polymer Technologies, Inc. is made of microfiber chitin [45], as is Excel Arrest® dressing from Hemostasis, LLC [46]. Numerous opportunities for chitin products can be found in selected reviews [20–25].

4.2.3 Current Chitin Isolation Methods

Chitin isolation targets a biomass source generated by U.S.-based fisheries as a costly waste that can be turned into valuable products. Yet, currently, chitin is isolated from crustacean biomass via a pulping process. Pulping typically includes three steps: (1) demineralization to remove calcium carbonate present in a shell matrix (using acids (e.g., HCl)), (2) deproteinization to remove proteins (conducted using hydroxides (e.g., NaOH)), and (3) bleaching/discoloration (using organic solvents [47] or oxidation agents [48]). Because pulping is conducted at relatively high temperatures (70–100 °C) and usually for a prolonged time [49, 50], the process adequately removes both proteins and shell inorganics but generates a large amount of waste.

Manufacturing 1 kg of chitin using the pulping method requires 10 kg of biomass, 300 L of freshwater, 9 kg of HCl, 8 kg of NaOH, and 1.2 kWh of electricity. The liquid waste generated is equal to the input freshwater volume plus the process water; the overall amount of waste per 1 kg of chitin exceeds 500 L [26]. In addition, the emission of CO_2 is estimated to be 0.9 kg/kg of chitin [26] Such high cost involved in the production of chitin and the huge quantity of generated waste resulted in the pulping process raising public and governmental concerns. As a result, there is no chitin producing plant that uses acid/base treatment in the United States [51]. In addition, crustacean shells contain a host of potentially valuable components in addition to chitin, including other biopolymers, such as proteins, small molecules, such as astaxanthin that have medical value, and minerals, such as calcite that may be useful as construction materials; all of them get destroyed during the pulping process.

Lastly, harsh conditions used in isolation were found to decrease the quality of the isolated chitin, to promote deacetylation and depolymerization, and to result in a lack of reproducible high-quality polymer product [52]. However, many applications require specific polymer properties, and the strength of materials has been proven to be governed by the molecular weight and the degree of acetylation. We have shown (and will detail it below in this chapter) that high-molecular weight chitin is critical for the preparation of materials with different shapes (fibers, films, packages, hydrogels, beads, and electrospun mats). This host of new materials and the preparation of chitin composites and blends [53–68] are made possible by an ionic liquid [69] solution-based process.

4.3   Startup to Scale-up

4.3.1   The Beginning: Business Opportunity

In 2010, Rogers demonstrated the dissolution and extraction of the biopolymer chitin directly from shrimp shells [70]. Before that, no one had reported the direct dissolution of crustacean biomass or the extraction of chitin polymer from it using ionic liquids. The IL 1-ethyl-3-methylimidazolium acetate ($[C_2C_1im][OAc]$) was shown to be an excellent solvent for chitin [71]. Using this IL, microwave irradiation facilitated the dissolution and demineralization of crustacean biomass and resulted in the extraction of all available chitin in minutes. The polymer maintained its high-molecular weight resulting in a material with high strength and unprecedented high quality. At that point, this extraction method was demonstrated on a 100 mL scale using a domestic microwave.

The quantities of isolated polymers are critical to many materials applications and must be produced at a larger scale. For instance, co-dissolved with alginic acid, high-molecular weight chitin–IL solutions were shown to be suitable for the preparation of spun chitin–calcium alginate fibers [63, 70, 72], for intracutaneous biocompatibility testing, and wound-healing studies. Using a domestic microwave and a lab-scale fiber pulling setup, it took 3 weeks to prepare only 6 g of bandages [63]. Using a somewhat larger setup, with a small manufacturing, custom-made fiber extruder, required a minimum loading of 1 kg of the solution to produce 1,000 m of monofilament fiber. Clearly, a prototype was necessary to make scaled-up quantities for the preparation of sustainable, high-value chitin materials.

4.3.2 Baby Steps: Laboratory Demonstration Pilot Unit (Alabama Innovation Grant)

Seeking funding to scale up chitin production, we turned our attention to the Alabama Department of Commerce's "Accelerate Alabama" Program, particularly to the Alabama Innovation Fund (AIF) that was established "to maximize the use of the State's economic development resources by leveraging annual research and development expenditures by public institutions and generate high technology resources which can be used to support economic development activities" [73].

The AIF Fund mainly supported research collaborations among universities and industries working on research that could be economically beneficial to the state. If chitin extraction at scale was successful, there would be new and better business options for the seafood industries by utilizing shrimp shells, an industrial waste product. A collaboration was established with the Gulf Coast Agricultural and Seafood Cooperative [74] in Bayou La Batre, AL. They had built a seafood waste drying/pulverizing facility with the support of the Alabama Farmers Market Authority. The facility collects crustacean biomass waste from resident fishermen and biomass-handling plants that are then pressed for protein removal, passed through a fluidized bed dryer for dewatering, and shredded before bagging. At maximum capacity, the plant can process up to 10 tons of shrimp shells a day into sterilized, dry chitin-containing material. Finding uses for this waste material could save the seafood industry of Alabama hundreds of thousands of dollars a month in waste disposal costs as well as opportunities for new jobs as the shrimp shells can be converted into value-added products.

In 2012, an Alabama Innovation Fund (AIF) grant was given to The University of Alabama researchers and startup company 525 Solutions, Inc. to build a Laboratory Demonstration Pilot Unit (LDPU) that would enable scaling up the process of extracting chitin from the bench scale to a prototype scale needed for the manufacture of chitin-based products. The Department of Commerce award was part of a more complex project that included research and development, a business plan and market development, product demonstration, and prototype construction. The project was divided into several major phases: (1) design and development of a continuous process for chitin extraction, (2) scale-up of the process, (3) construction of a LDPU, (4) R&D of new chitin-based materials and new product-development technologies, and (5) demonstration of new products to provide higher value to the seafood wastes.

Under this Alabama Department of Commerce award, Rogers' group developed a prototype capable of the continuous processing of shellfish waste. The LDPU (Fig. 4.2) included a closed-loop where biomass and ionic liquid could be cycled through until the dissolution was complete. This provided excellent control over the heating rates of the solution. The unit consisted of a 3 L, glass-jacketed reactor (Fig. 4.2a) with an overhead mechanical stirrer (Fig. 4.2b) attached to a continuous-flow 2 kW microwave (Industrial Microwave Systems, Fig. 4.2c). The pumping was conducted using a peristaltic pump (Cole-Parmer, Fig. 4.2d).

Fig. 4.2 LDPU consisted of a 3 L glass-jacketed reactor (**a**), overhead mechanical stirrer (**b**), continuous-flow 2 kW microwave (**c**), and a peristaltic pump (**d**)

After an optimization study, continuous chitin dissolution was conducted using 3 L of [C_2C_1im][OAc] under cylindrical, 2 kW continuous microwave heating. The IL was fed into the unit and heated to 95–98 °C. Only 10% of the microwave capacity was necessary to achieve the dissolution of the biomass and not decompose the IL. The run suggested that the IL effectively absorbs the microwave energy and can be treated continuously in a microwave. To ensure that the recycled IL could be reclaimed and reused with no significant loss, the IL was circulated through the microwave for several cycles, and no obvious degradation was observed. This scaled dissolution of chitin in IL by microwave heating provided essential knowledge (thermal exchange data, microwave energy input/output, microwave energy efficiency, and cooling rate) for scaling up to a *pilot* system.

In addition, as a part of the project to provide higher value to the seafood wastes, we also focused on R&D of new chitin-based materials and new product development technologies. Chitin–IL solutions were shown to be suitable for the preparation of spun fibers, films, hydrogels, beads, and electrospun mats providing a route to a host of new materials (Fig. 4.3): chitin fibers (wet: a, dry: b), chitin electrospun nanomat (c, d), chitin beads (e, f), chitin hydrogel (g, h), and chitin film (i, j). [54–68].

4.3.3 Bench to Pilot Scale Prototype: Leveraging Sorbent Production Technology

Using AIF funds, dissolution of chitin in an ionic liquid by microwave heating using LDPU provided essential knowledge and revealed useful data needed for the further scale-up to a 20 L *pilot* system. Based on our earlier results with cellulose [75], we planned to develop and optimize the pilot scale-up process for this technology. However, from our earlier studies with cellulose we learned that there were few examples of successful academia-to-industry technology transfers. Successful transfer would require ensuring that both the technology and processes were scalable and, more

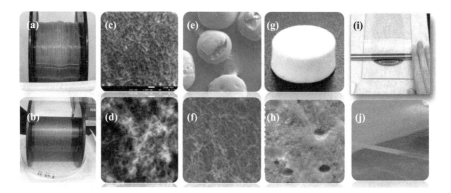

Fig. 4.3 Chitin products: chitin fibers (wet: **a**, dry: **b**), chitin electrospun nanomat (SEM: **c**, AFM: **d**), chitin beads (**e**), and chitin bead interior (**f**), chitin hydrogel (**g**) and hydrogel interior (**h**), wet chitin film cast on a glass plate (**i**) and wet chitin film in water (**j**). Images support the work described in our publications [54–68]

importantly, that the development of high-value end products was possible with a successful business plan rather than a scale-up with no purpose.

Considering that the extraction process could be key to a "chitin economy" and would, upon its success, provide a revenue stream for shrimpers to utilize their waste while helping to develop sustainable/green products, we started looking into ways to leverage our chitin extraction technology. Our attention turned to a U.S. Department of Energy (DOE) Nuclear Energy Program (uranium recovery from seawater), a part of U.S. DOE efforts that included collaborative efforts of several universities and small businesses.

The University of Alabama researchers and startup company 525 Solutions, Inc. combined efforts to develop highly economical and biodegradable uranium-selective sorbents, specifically for the U.S. DOE Nuclear Energy Program aimed at the extraction of uranium from seawater. The concept of sorbents was based on our previous work, where we investigated the electrospinability of IL-extracted chitin solutions from [C_2C_1im][OAc] [76]. The project focused on the delivery of the product to government-designated mining companies and at the same time proposed leveraging the U.S. DOE resources to generate a sustainable chitin products business. Such leveraging would allow both economic development and creation of jobs in R&D of chitin products and fishing industries. This way, the chitin nanomaterials would serve as a platform for the delivery of chitin to U.S. markets, as well as providing a range of medium- to high-value applications in medicine, energy, and environmental restoration sectors.

Because there was no existing industrial base for the extraction of uranium from seawater, the entry decision thus revolved more around how to build a successful business for this opportunity. If the commercialization strategy was built only around the sale of the sorbent, the company would cease to exist if government support of the program ended before an industry emerged. However, as mentioned earlier in this chapter, *numerous* market opportunities were emerging based on the underlying technology, chitin extraction. The ability to produce not only products from chitin, but chitin itself, provided a competitive advantage to diversify the range of products and enter several profitable specialized markets, while at the same time developing the sorbents for the DOE Nuclear Energy Program. Such leveraging would lower the bulk cost of the sorbent by building high-end markets that would help pay for the process development and economy of scale.

For 2014–2015, a $1.5 million grant was funded by the U.S. DOE Small Business Innovation Research Program, "Bench to pilot scale prototype for electrospinning biorenewable chitin sorbents for uranium from seawater: Process development, cost, and environmental analysis" (*DOE-SBIR Grant No. DE-SC0010152, Phase I/II*). The ultimate goal of this project was to collect the industrial process parameters, to conduct reliable economic estimates, and ultimately to generate data for the full-scale operating plant design.

To address the scalability of the IL technology platform for biomass processing, Rogers' group (together with 525 Solutions, Inc.) refined the pilot plant operating conditions and plant design and prepared input-output diagrams of the process. This provided the relationships between the major equipment of a pilot plant facility and the piping of the process flow together with all required equipment and instrumentation. Next, a scaled, highly automated customized 20-L early pilot stage system amenable for chitin extraction was built. It contained a custom-design, continuous-flow stirred-tank reactor (CSTR) and a 2 kW microwave (Fig. 4.4).

(a)   **(b)**

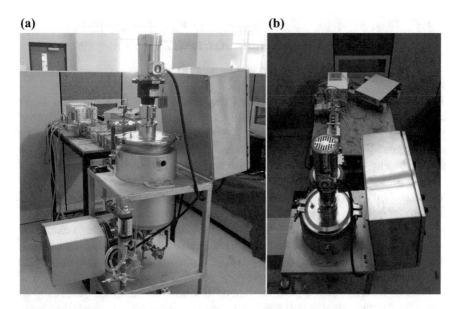

Fig. 4.4 A continuous scaled, automated, and customized 20-L early pilot stage system amenable for chitin extraction. (**a**): side view, (**b**): top view

After multiple, unsuccessful pilot trials, the optimal conditions (biomass load, temperature, flow rate through the microwave, residual time in the microwave, and process time) for the dissolution of biomass in the reactor were determined. Optimal conditions resulted in a high yield of approximately 95% (as percent of available chitin in biomass). A mass balance was completed using the optimal conditions of biomass dissolution based on different reagent streams, reagent consumption, and product recovery (i.e., using multi-parametric input). This permitted multi-step evaluations for the determination of the process inputs and outputs based on production cycles.

An energy balance was completed using the overall energy consumption in each energy-demanding step based on production. Cost analyses were conducted where scale-up of some 30 equipment items was handled either through increasing the size or capacity of the equipment (so-called economies of scale) or by increasing throughput by projecting the purchase of additional units of defined capacity. To date, no other entity has scaled this IL process to this size.

This project resulted in a fully engineered system, developed key engineering data and diagrams, as well as determined the equipment needed in a full-scale operating plant. Processing performance testing was conducted while manufacturing chitin on a pilot scale. The results of these studies were used to establish manufacturing capability and process robustness and to mitigate the risks before committing to a full-scale production process.

4.3.4   Mari Signum, Mid-Atlantic: The First Facility to Use Ionic Liquid-Based Chitin Extraction at a Production Scale

The knowledge obtained in these efforts was used to raise capital investment for building and operating a biomass/IL facility. Mari Signum, Mid-Atlantic, LLC (Mari Signum) [77] was formed as a chitin and chitin materials production company. The ultimate goal of Mari Signum, Mid-Atlantic, LLC is to become a sustainable source of high-quality chitin as well as chitin-based products developed in-house.

Mari Signum acquired the worldwide exclusive license for the portfolio of intellectual property (IP) that protects the manufacture of chitin. This IP will allow Mari Signum to maintain its position in the market and protect its competitive advantages. The licensing granted Mari Signum not only the rights to the chitin-extraction patents, but also all patents associated with high-value products. These products from chitin, and not chitin itself, would, indeed, be key components in the development of a "chitin economy".

Mari Signum's facility will be the first of its kind to use IL-based processing on a manufacturing scale that will allow the generation of sufficient supplies of high-quality chitin, which is unobtainable by any known chemical pulping processes. Mari Signum is currently building a processing plant for chitin isolation from crustacean biomass. The ability of Mari Signum to produce not only chitin itself but also products from chitin will give Mari Signum a competitive advantage to diversify the range of its products and to enter several profitable specialized markets.

4.4 Conclusion and Outcome

Chitin represents a billion-dollar industry worldwide, but despite the promising properties of chitin (and the large amounts of available shellfish waste in the United States), its potential production at an industrial level has been scarcely explored. There is no chitin producing plant in the United States mainly because of the current environmentally unfriendly chitin isolation process. At the same time, chitin isolation extends far beyond producing chitin itself. There are many opportunities for the manufacture of novel chitin products from shrimp shell waste that have not yet been tapped. Such high-value products will cause high industrial growth and have a positive environmental impact. A large enough chitin supply will be needed for the chitin industry to become a game-changer in a sustainable society.

Even with a billion-dollar industry opportunity, our own experience demonstrates that the transition of technology from academia to industry is rarely a straightforward process. As the technology progressed, it required proceeding through several time- and effort-consuming stages with each one successively larger in scale—(bench, pilot, demonstration, and production scale). Each stage used the knowledge accumulated from the previous round of scale-up. With the benefit of hindsight, this story might be useful for others who are ready to take this journey.

Hopefully, chitin products will make existing plastics obsolete.

Acknowledgements The authors would like to thank 525 Solutions, Inc., U.S. Department of Energy Small Business Innovation Research Program (DOE-SBIR Grant No. DE-SC0010152, Phase I/II), and DOE Office of Nuclear Energy, Nuclear Energy University Programs (DOE NEUP Grant No. DE-NE0000672) for financial support. We would also like to express our sincere gratitude to Dr. Eric Schneider (University of Texas at Austin, Mechanical Engineering in Cockrell School of Engineering) for the help with technology economic assessment, and Mr. Jonathan Bonner (Poly Engineering, Tuscaloosa, AL) for the help with scaling up the equipment to the production scale.

References

1. Plastics market analysis by product (PE, PP, PVC, PET, polystyrene, engineering thermoplastics), by application (film & sheet, injection molding, textiles, packaging, transportation, construction) and segment forecasts to 2020. Grand View Research, Inc., 2015. Available at: http://www.grandviewresearch.com/industry-analysis/global-plastics-market. Last accessed 04-02-18
2. University of Georgia (2017) More than 8.3 billion tons of plastics made: most has now been discarded. Science Daily, 19 July 2017. Available at: https://www.sciencedaily.com/releases/2017/07/170719140939.htm. Last accessed 12-10-19

3. Geyer R, Jambeck JR, Law KL (2017) Production, use, and fate of all plastics ever made. Sci Adv 3:e1700782. https://doi.org/10.1126/sciadv.1700782
4. Ellen MacArthur Foundation 2016 Report. Rethinking the future of plastics: https://www.ellenmacarthurfoundation.org/our-work/activities/new-plastics-economy/2016-report. Last accessed 07-28-19
5. Tunnicliffe H (2017) Turning ocean trash into cash. TCE: the chemical engineer 913/914:36–38. Available at: https://www.thechemicalengineer.com/features/turning-ocean-trash-into-cash/ Last accessed 07-16-19
6. North EJ, Halden RU (2013) Plastics and environmental health: the road ahead. Rev Environ Health 28:1–8. https://doi.org/10.1515/reveh-2012-0030
7. Popa V (2018) Biomass for fuels and biomaterials. In: Popa VI, Volf I (eds) Biomass as renewable raw material to obtain bioproducts of high-tech value, Elsevier, pp 1–37. https://doi.org/10.1016/B978-0-444-63774-1.00001-6
8. Intro to plantbottle packaging. Available at: https://www.coca-colacompany.com/plantbottle-technology. Last accessed 04-19-18
9. Edge M, Hayes M, Mohammadian M, Allen NS, Jewitt TS, Brems K, Jones K (1991) Aspects of poly(ethylene terephthalate) degradation for archival life and environmental degradation. Polym Degrad Stab 32:131–153. https://doi.org/10.1016/0141-3910(91)90047-U
10. Allen NS, Edge M, Mohammadian M, Jones K (1994) Physicochemical aspects of the environmental degradation of poly(ethylene terephthalate). Polym Degrad Stab 43:229–237. https://doi.org/10.1016/0141-3910(94)90074-4
11. Rogers RD (2015) Eliminating the need for chemistry. C&EN 93(48):42–43. https://cen.acs.org/articles/93/i48/Eliminating-Need-Chemistry.html
12. Gao X, Chen X, Zhang J, Guo W, Jin F, Yan N (2016) Transformation of chitin and waste shrimp shells into acetic acid and pyrrole. ACS Sustain Chem Eng 4:3912–3920. https://doi.org/10.1021/acssuschemeng.6b00767
13. Prudden JF, Migel P, Hanson P, Friedrich L, Balassa L (1970) The discovery of a potent pure chemical wound-healing accelerator. Am J Surg 119:560–564. https://doi.org/10.1016/0002-9610(70)90175-3
14. Jayakumar R, Prabaharan M, Kumar PTS, Sudheesh Kumar PT, Nair SV, Furnike T, Tamura H (2011) Novel chitin and chitosan materials in wound dressing. In: Laskovski AN (ed) Biomedical engineering trends in materials science, InTech 3–24. https://doi.org/10.5772/13509
15. Vázquez JA, Rodríguez-Amado I, Montemayor MI, Fraguas J, González M del P, Murado MA (2013) Chondroitin sulfate, hyaluronic acid and chitin/chitosan production using marine waste sources: characteristics, applications and eco-friendly processes: a review. Mar Drugs 11:747–774. https://doi.org/10.3390/md11030747
16. Mori T, Okumura M, Matsuura M, Ueno K, Tokura S, Okamoto Y, Minami S, Fujinaga T (1997) Effects of chitin and its derivatives on the proliferation and cytokine production of fibroblasts in vitro. Biomaterials 18:947–951. https://doi.org/10.1016/S0142-9612(97)00017-3
17. Hirano S, Nakahira T, Nakagawa M, Kim SK (1999) The preparation and application of functional fibres from crab shell chitin. J Biotechnol 70:373–377. https://doi.org/10.1016/S0079-6352(99)80130-1
18. Wan ACA, Tai BCU (2013) Chitin—a promising biomaterial for tissue engineering and stem cell technologies. Biotech Adv 31:1776–1785. https://doi.org/10.1016/j.biotechadv.2013.09.007
19. Barber PS, Kelley SP, Griggs CS, Wallace S, Rogers RD (2014) Surface modification of ionic liquid-spun chitin fibers for the extraction of uranium from seawater: seeking the strength of chitin and the chemical functionality of chitosan. Green Chem 16:1828–1836. https://doi.org/10.1039/C4GC00092G
20. Muzarelli RAA, Boudrant J, Meyer D, Manno N, DeMarchis M, Paoletti MG (2012) Current views on fungal chitin/chitosan, human chitinases, food preservation, glucans, pectins and inulin: a tribute to Henri Braconnot, precursor of the carbohydrate polymers science, on the chitin bicentennial. Carbohydr Polym 87:995–1012. https://doi.org/10.1016/j.carbpol.2011.09.063

21. Dutta PK, Dutta J, Tripathi VS (2004) Chitin and chitosan: chemistry, properties and applications. J Sci Ind Res 63:20–31.
22. Domard A (2011) A perspective on 30 years research on chitin and chitosan. Carbohydr Polym 84:696–703. https://doi.org/10.1016/j.carbpol.2010.04.083
23. Tharanathan RN, Kittur FS (2003) Chitin—the undisputed biomolecule of great potential. Critical Rev Food Sci Nutr 43:61–87. https://doi.org/10.1080/10408690390826455
24. Synowiecki J, Al-Khateeb NA (2003) Production, properties, and some new applications of chitin and its derivatives. Critical Rev Food Sci Nutr 2:145–171. https://doi.org/10.1080/10408690390826473
25. Ravi Kumar MNV (2000) A review of chitin and chitosan applications. React Funct Polym 46:1–27. https://doi.org/10.1016/S1381-5148(00)00038-9
26. Chitin market: agrochemical end use industry segment inclined towards high growth—moderate value during the forecast period: global industry analysis (2012–2016) and opportunity assessment (2017–2027). https://www.futuremarketinsights.com/reports/chitin-market. Last accessed 04-02-18
27. Yang T-L (2011) Chitin-based materials in tissue engineering: applications in soft tissue and epithelial organ. Int J Mol Sci 12:1936–1963. https://doi.org/10.3390/ijms12031936
28. Khor E, Lim LY (2003) Implantable applications of chitin and chitosan. Biomaterials 24:2339–2349. https://doi.org/10.1016/S0142-9612(03)00026-7
29. Mi F-L, Shyu S-S, Lin Y-M, Wu Y-B, Peng C-K, Tsai Y-H (2003) Chitin/PLGA blend microspheres as a biodegradable drug delivery system: a new delivery system for protein. Biomaterials 24:5023–5036. https://doi.org/10.1016/S0142-9612(03)00413-7
30. Rejinold NS, Chennazhi KP, Tamura H, Nair SV, Jayakumar R (2011) Multifunctional chitin nanogels for simultaneous drug delivery, bioimaging, and biosensing. ACS Appl Mater Interfaces 3:3654–3665; 11.13832. https://doi.org/10.1021/am200844m
31. Ding F, Deng H, Du Y, Shi X, Wang Q (2014) Emerging chitin and chitosan nanofibrous materials for biomedical applications. Nanoscale 6(16):9477–9493. https://doi.org/10.1039/C4NR02814G
32. Singh R, Shitiz K, Singh A (2017) Chitin and chitosan: biopolymers for wound management. Int Wound J 14:1276–1289. https://doi.org/10.1111/iwj.12797
33. Jayakumar R, Prabaharan M, Sudheesh Kumar PT, Nair SV, Tamura H (2011) Biomaterials based on chitin and chitosan in wound dressing applications. Biotech Adv 29:322–337. https://doi.org/10.1016/j.biotechadv.2011.01.005
34. Shigemasa Y, Minami S (1996) Applications of chitin and chitosan for biomaterials. Biotech Genetic Eng Rev 13:383–420. https://doi.org/10.1080/02648725.1996.10647935
35. Trutnau M, Bley T, Ondruschka J (2011) Chapter 1: Chitosan from fungi. In: Davis SP (ed) Chitosan: manufacture, properties, and usage, Nova Science Publishers, Inc., New York, NY, pp. 1–70.
36. Tidal Vision, USA. https://tidalvisionusa.com/chitosan/. Last accessed 04-03-18
37. CarboMer. https://www.carbomer.com/biopolymers. Last accessed 04-03-18
38. Shiau S-Y, Yu Y-P (1998) Chitin but not chitosan supplementation enhances growth of grass shrimp, *Penaeus monodon*. J Nutr 128:908–912. https://doi.org/10.1093/jn/128.5.908
39. Sharp RG (2013) A review of the applications of chitin and its derivatives in agriculture to modify plant-microbial interactions and improve crop yields. Agronomy 3:757–793. https://doi.org/10.3390/agronomy3040757
40. Bhatnagar A, Sillanpää M (2009) Applications of chitin- and chitosan-derivatives for the detoxification of water and wastewater—a short review. Adv Colloid Interface Sci 152:26–38. https://doi.org/10.1016/j.cis.2009.09.003
41. Araki J, Yamanaka Y, Ohkawa K (2012) Chitin-chitosan nanocomposite gels: reinforcement of chitosan hydrogels with rod-like chitin nanowhiskers. Polymer J 44:713–717. https://doi.org/10.1038/pj.2012.11
42. Unitika history creates its next history. https://www.unitika.co.jp/e/company/history/. Last accessed 04-04-18
43. Eisai Co. https://www.eisai.com/index.html, last accessed 04-04-18

44. Minami S, Okamoto Y, Miyatake A, Matsuhashi A, Kitamura Y, Tanigawa T, Tanaka Y, Shigemasa Y (1996) Chitin induces type IV collagen and elastic fiber in implanted non-woven fabric of polyester. Carbohydrate Polym 29:295–299. https://doi.org/10.1016/S0144-8617(96)00078-1
45. SyvekExcel. http://syvek.com/. Last accessed 04-04-18
46. Technical Information: ExcelArrest® XT. http://www.hemostasisllc.com/excelarrest-techinfo. html. Last accessed 04-04-18
47. Poeloengasih CD, Hernawan, Angwar M (2008) Isolation and characterization of chitin and chitosan prepared under various processing times. Indo J Chem 8:189–192. https://doi.org/10. 22146/ijc.21635
48. Beaney P, Lizardi-Mendoza J, Healy M (2005) Comparison of chitins produced by chemical and bioprocessing methods. J Chem Tech Biotech 80:145–150. https://doi.org/10.1002/jctb. 1164
49. Rinaudo M (2006) Chitin and chitosan: properties and applications. Prog Polym Sci 31:603–632. https://doi.org/10.1016/j.progpolymsci.2006.06.001
50. Khoushab F, Yamabhai M (2010) Chitin research revisited. Mar Drugs 8:1988–2012. https://doi.org/10.3390/md8071988
51. Shamshina JL, Barber PS, Gurau G, Griggs CS, Rogers RD (2017) Pulping of crustacean waste using ionic liquids: to extract or not to extract. ACS Sust Chem Eng 4:6072–6081. https://doi.org/10.1021/acssuschemeng.6b01434
52. Younes I, Rinaudo M (2015) Chitin and chitosan preparation from marine sources. Structure, properties and applications. Mar Drugs 13:1133–1174. https://doi.org/10.3390/md13031133
53. Silva SS, Mano JF, Reis RL (2017) Ionic liquids in the processing and chemical modification of chitin and chitosan for biomedical applications. Green Chem 19:1208–1220. https://doi.org/10.1039/C6GC02827F
54. King C, Shamshina JL, Gurau G, Berton P, Khan NFAF, Rogers RD (2017) A platform for more sustainable chitin films from an ionic liquid process. Green Chem 19:117–126. https://doi.org/10.1039/C6GC02201D
55. Shen X, Shamshina JL, Berton P, Bandomir J, Wang H, Gurau G, Rogers RD (2016) Comparison of hydrogels prepared with ionic-liquid-isolated vs commercial chitin and cellulose. ACS Sustainable Chem Eng 4:471–480. https://doi.org/10.1021/acssuschemeng.5b01400
56. Kadokawa J (2016) Dissolution, gelation, functionalization, and material preparation of chitin using ionic liquids. Pure Appl Chem 88:621–629. https://doi.org/10.1515/pac-2016-0503
57. Shamshina JL, Zavgorodnya O, Bonner JR, Gurau G, Di Nardo T, Rogers RD (2017) "Practical" electrospinning of biopolymers in ionic liquids. ChemSusChem 10:106–111. https://doi.org/10.1002/cssc.201601372
58. Zavgorodnya O, Shamshina JL, Bonner JR, Rogers RD (2017) Electrospinning biopolymers from ionic liquids requires control of different solution properties than volatile organic solvents. ACS Sustain Chem Eng 5:5512–5519. https://doi.org/10.1021/acssuschemeng.7b00863
59. Turner MB, Spear SK, Holbrey JD, Rogers RD (2004) Production of bioactive cellulose films reconstituted from ionic liquids. Biomacromol 5:1379–1384. https://doi.org/10.1021/bm049748q
60. Turner MB, Spear SK, Holbrey JD, Daly DT, Rogers RD (2005) Ionic liquid-reconstituted cellulose composites as solid support matrices for biocatalyst immobilization. Biomacromol 6:2497–2502. https://doi.org/10.1021/bm050199d
61. Sun N, Swatloski RP, Maxim ML, Rahman M, Harland AG, Haque A, Spear SK, Daly DT, Rogers RD (2008) Magnetite-embedded cellulose fibers prepared from ionic liquid. J Mater Chem 18:283–290. https://doi.org/10.1039/B713194A
62. Bagheri M, Rodríguez H, Swatloski RP, Spear SK, Daly DT, Rogers RD (2008) Ionic liquid-based preparation of cellulose–dendrimer films as solid supports for enzyme immobilization. Biomacromol 9:381–387. https://doi.org/10.1021/bm701023w
63. Shamshina JL, Gurau G, Block LE, Hansen LK, Dingee C, Walters A, Rogers RD (2014) Chitin–calcium alginate composite fibers for wound care dressings spun from ionic liquid solution. J Mater Chem B 2:3924–3936. https://doi.org/10.1039/C4TB00329B

64. Maxim ML, White JF, Block LE, Gurau G, Rogers RD (2012) Advanced biopolymer composite materials from ionic liquid solutions in ionic liquids: science and applications. In: Visser AE, Bridges NJ, Rogers RD (eds) Ionic liquids: science and applications, ACS Symp Ser 1117:167–187. https://pubs.acs.org/doi/10.1021/bk-2012-1117.ch007
65. Takegawa A, Murakami M, Kaneko Y, Kadokawa J (2010) Preparation of chitin/cellulose composite gels and films with ionic liquids. Carbohydr Polym 79:85–90. https://doi.org/10.1016/j.carbpol.2009.07.030
66. Singh N, Koziol KKK, Chen J, Patil AJ, Gilman JW, Trulove PC, Kafienah W, Rahatekar SS (2013) Ionic liquids-based processing of electrically conducting chitin nanocomposite scaffolds for stem cell growth. Green Chem 15:1192–1202. https://doi.org/10.1039/C3GC37087A
67. Mundsinger K, Müller A, Beyer R, Hermanutz F, Buchmeiser MR (2015) Multifilament cellulose/chitin blend yarn spun from ionic liquids. Carbohydr Polym 131:34–40. https://doi.org/10.1016/j.carbpol.2015.05.065
68. Sun N, Li W, Stoner B, Jiang X, Lu X, Rogers RD (2011) Composite fibers spun directly from solutions of raw lignocellulosic biomass dissolved in ionic liquids. Green Chem 13:1158–1161. https://doi.org/10.1039/C1GC15033B
69. Shamshina JL, Berton P, Rogers RD (2019) Advances in functional chitin materials: a review. ACS Sustain Chem Eng 7:6444–6457. https://doi.org/10.1021/acssuschemeng.8b06372
70. Qin Y, Lu X, Sun N, Rogers RD (2010) Dissolution or extraction of crustacean shells using ionic liquids to obtain high molecular weight purified chitin and direct production of chitin films and fibers. Green Chem 12:968–971. https://doi.org/10.1039/C003583A
71. Wang H, Gurau G, Rogers RD (2014) Dissolution of biomass using ionic liquids. In: Zhang S, Wang J, Lu X, Zhou Q (eds) Structures and interactions of ionic liquids 151. Springer, Berlin, Heidelberg, pp 79–105. https://doi.org/10.1007/978-3-642-38619-0_3
72. Shamshina JL, Zavgorodnya O, Rogers RD (2018) Advances in processing chitin as promising biomaterial from ionic liquids. In: Itoh T, Koo Y-M (eds) Application of ionic liquids in biotechnology. Advances in biochemical engineering/biotechnology. Springer, Cham, pp. 177–198. https://doi.org/10.1007/10_2018_63
73. Alabama Innovation Fund (AIF): Alabama EPSCOR. https://alepscor.org/alabama-innovation-fund/. Last accessed 4-03-18
74. Gulf Coast Agricultural and Seafood COOP. https://www.amcref.com/impact/gulf-coast-agriculture-and-seafood-co-op/. Last accessed 07-28-19
75. Zavgorodnya O, Shamshina JL, Berton P, Rogers RD (2017) Translational research from academia to industry: following the pathway of George Washington Carver. In: Shiflett MB, Scurto AM (eds) Ionic liquids: current state and future directions, ACS Symp Ser 1250:17–33. https://doi.org/10.1021/bk-2017-1250.ch002
76. Barber PS, Griggs CS, Bonner JR, Rogers RD (2013) Electrospinning of chitin nanofibers directly from an ionic liquid extract of shrimp shells. Green Chem 15:601–607. https://doi.org/10.1039/C2GC36582K
77. Mari Signum, Mid-Atlantic. http://www.marisignum.com/. Last Accessed 04-19-18

Chapter 5
Commercial Aspects of Biomass Deconstruction with Ionic Liquids

Aida R. Abouelela, Florence V. Gschwend, Francisco Malaret and Jason P. Hallett

Abstract The adaption of any new process from academic research to industrial-scale requires a robust economic profile demonstrating it can compete with peer technologies. IonoSolv process is a recently developed ionic liquid-based pretreatment that pioneers the use of low-cost ionic liquids for biomass fractionation. The use of low-cost protic ionic liquids elevated the techno-economic profile of the process, making its potential commercialization highly viable. In this chapter, the authors give an overview of key process-related aspects that underpinned this transformation, with special highlights on the progressive milestones achieved in developing a promising commercial ionic liquid-based pretreatment process. We also highlight the current challenges and knowledge gaps that need to be tackled to further elevate the technology readiness level.

Keywords Biomass deconstruction · Biorefinery · Cellulose · Delignification · ionoSolv

5.1 Introduction

Energy and chemical production in our society is currently experiencing a transition to replace the current fossil fuel-based economy with a more sustainable, renewable-based economy. This changing landscape requires designing cutting-edge technologies to make these new processes as efficient and reliable as conventional fossil fuels-based technologies [1]. While all renewable energy resources (e.g., solar, wind, nuclear) are inherently designed and implemented for energy generation, biomass is the sole renewable source for biofuels, chemicals, and materials production. For the past two decades, lignocellulosic biomass has attracted special attention as a low-cost feedstock for cellulosic ethanol production at the scale anticipated in the low-carbon energy scenarios [2, 3]. Benefits of using the abundant inedible ligno-cellulose biomass (grass, wood, and agricultural waste) as a biorefinery feedstock

A. R. Abouelela · F. V. Gschwend · F. Malaret · J. P. Hallett (✉)
Department of Chemical Engineering, Imperial College London, London SW7 2AZ, UK
e-mail: j.hallett@imperial.ac.uk

© Springer Nature Switzerland AG 2020
M. B. Shiflett (ed.), *Commercial Applications of Ionic Liquids*, Green Chemistry and Sustainable Technology, https://doi.org/10.1007/978-3-030-35245-5_5

include mitigating land-use changes associated with use of edible biomass and contributing to rural economic development [4]. In addition, the use of lignocellulose biomass as a platform for renewable chemicals and materials production is experiencing a renaissance due to the increasing demand to decarbonize the chemical industry [5].

Lignocellulose is the material that makes up the cell walls of woody plants and it is composed of three biopolymers: cellulose (35–50%), hemicellulose (30–40%), and lignin (10–30%). These three polymers are arranged in a three-dimensional structure that has naturally evolved to resist degradation and deconstruction, which is essential to maintain the plant's structural integrity. Unfortunately for us, the increased structural complexity of lignocellulose makes it highly recalcitrant to mild chemical or biological modification [6]. Therefore, the use of lignocellulose biomass as a viable feedstock in a biorefinery requires integrating an additional processing step called "pretreatment" [7]. The main objective of a pretreatment process is to effectively deconstruct the rigid lignocellulose structure to facilitate its subsequent processing of fuels, chemicals, and materials. For bioethanol production, the deconstruction effect improves the enzymatic hydrolysis of the carbohydrate polymers to their sugar monomers. For biochemicals and materials production, pretreatment is fundamentally important to effectively isolate each biopolymer from the lignocellulose matrix for subsequent upgrading and transformation to value-added products [8].

Although lignocellulosic materials are abundant and relatively cheap, the addition of a pretreatment process was projected to be the most capital-intensive step in a biorefinery facility. In fact, for bioethanol production, the cost of pretreatment was estimated to be higher than the enzymatic hydrolysis and fermentation cost combined [9]. Evaluating various pretreatment technologies has, therefore, become a major research field with an objective to find a low-cost technology that can penetrate the market and expedite industry learning. The mission became far more challenging recently due to the significant drop in oil prices, reduced energy security concerns, as well as the wide implementation of new shale oil and gas extraction technologies (e.g., hydraulic fracturing and horizontal drilling) [10].

Several lignocellulose pretreatment methods are under various degrees of development today, including steam explosion [11], ammonia fiber expansion (AFEX) [12], dilute acid [13], hot water [14], and Organosolv [15]. Ionic liquid (IL)-based pretreatment is one of the recent approaches introduced in the biomass pretreatment field. ILs have shown exciting potential to be used as solvents for the lignocellulosic biomass processing due to their outstanding capability to dissolve [16], fractionate [17], or convert lignocellulose biopolymers to higher value-added chemicals [18]. As these fields continue to grow, questions related to the economic justification and technical challenges of using ILs on a large-scale become more and more pressing. This is especially true in spite of the overwhelming popularity of ILs in academia; ILs-based processes are still very limited with only a few current industrial applications [19]. The most successful example of an industrial process using ILs is the BASIL™ process introduced by BASF in 2002. In this process, the ionic liquid 1-methylimidazolium chloride ($[C_1im][Cl]$) is produced in situ by using 1-methylimidazole as an HCl scavenger during the production of diethylphenylphosphonite. The use of an ionic liquid instead of the conventional amine process provided key advantages due to the nucleophilic catalytic activity of 1-alkylimidazole that increased the space-time yield (STY) by a factor of 86,000 as well as the economic reclamation of the 1-alkylimidazole by IL deprotonation [20, 21]. Such a remarkable improvement in process performance was crucial to accelerate the industrial deployment of new solvents, such as ILs. Likewise, the application of an IL-based pretreatment process will have to feature exclusive selling points and key processing advantages for it to compete with processes of similar or higher technology readiness level. From a high-level perspective, the key advantages of IL-based pretreatment compared to peer technologies include: (1) low operating pressure and reduced solvent losses as ILs are non-volatile solvents, (2) potentially lower capital cost of a pretreatment reactor compared to aqueous or organic solvent due to ILs' non-volatile nature, [22] (3) lower enzymes loading and smaller operation units due to lignin pre-extraction during pretreatment, [23] (4) production of high-quality lignin of high valorization potential, [24] and (5) robust process performance that is highly independent of feedstock type and composition [25–28].

Despite these appealing advantages, the very high cost of ILs compared to aqueous or organic solvents has been always highlighted as the key barrier for their large-scale application [29].

However, it was only recently demonstrated that the high cost associated with ILs is not inherent, but it is a consequence of choosing historically dominant ILs (dialkylimidazolium cations with poly-fluorinated anions) [30]. Overcoming the IL cost barrier was the key turning point that enabled the discussion of ionoSolv's potential commercialization.

5.1.1 Technology Readiness Level

The path of translating academic research to a large-scale industrial process follows several validation and development steps. Technology readiness level (TRL) is a popular concept adapted in several sectors (e.g., space, defence, and oil and gas operations) to measure technology maturity level and its readiness for large-scale deployment. The scale and definition are slightly different to suit the requirement of the different sectors; however, they all have the same general structure and pathway starting from proof of concept and lab experiments to pilot testing and demonstration and eventually to a mature industrially proven technology. Figure 5.1 shows the TRL scale and definition according to the 2017 European Commission, which can be used for renewable energy technologies [31]. In the context of this chapter, TRLs from 1 to 7 were used to describe the development pathway of using ILs for biomass pretreatment with special focus on our own experience in the development of the ionoSolv process.

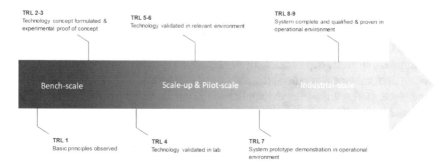

Fig. 5.1 Technology readiness level based on European Commission scale and definition

5.2 Basic Principles Observed (TRL 1)

When the basic principles of an innovative process are observed, this serves as a trigger that catalyses the rapid development of a new research field with a high level of excitement and expectations. The famous 2002 paper by Rogers and co-workers that re-discovered the capability of ILs to dissolve cellulose was the catalyst that sparked a surge of interest for using ILs for cellulose processing applications [32, 33].

In 2006, the interest spread to include using ILs for biomass processing upon the realization that common imidazolium-based ILs can dissolve the entire lignocellulose biomass constituents [16]. This timing was parallel to a renewed interest in second generation biofuel production technologies from lignocellulosic biomass [34].

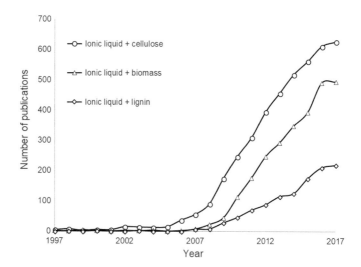

Fig. 5.2 Publications of ionic liquids use for cellulose, biomass, and lignin processing based on Web of Science (data taken on April 2018)

Figure 5.2 shows the rapid publication growth in using ILs for cellulose, biomass, and lignin processing and transformation.

5.2.1   Ionic Liquids and Biomass

Ionic liquids are a group of salts that are liquid at temperatures of <100 °C. ILs are known to be designer solvents as each anion–cation pair has its own unique properties which make them easy to be tailored to fit task-specific applications. Ionic liquids have been used in many fields of academic research, such as chemical synthesis [35], catalysis [36], electrochemistry [37], and biomass processing [38]. Today, ILs represent a new paradigm in the processing of biomass and biomass main platform macromolecules: cellulose and lignin [39]. This was especially exciting for cellulose as its highly crystalline compact structure makes it insoluble in most conventional solvents. Traditional cellulose dissolution processes for industrial fiber production involve the use of toxic carbon disulphide (CS_2) solvent to chemically dissolve cellulose (viscose process) or the application of harsher processing conditions to achieve direct solubilization using N-methylmorpholine-N-oxide (NMMO) as a solvent (Lyocell process) [40]. Concentrated phosphoric acid is also used as a solvent for cellulose dissolution [41]. All these solvents possess major drawbacks, such as toxicity (CS_2), instability at elevated temperatures (NMMO), or high corrosivity (phosphoric acid). The discovery (or re-discovery) that some ionic liquids can be used as solvents for cellulose dissolution and processing opened the opportunity to develop a more stable and environmentally friendly process for biorefinery applications.

The first patent on cellulose dissolution using an IL was issued in 1933 where N-ethylpyridinium chloride, in the presence of nitrogen-containing bases, was shown to be able to dissolve cellulose [42]. The innovation didn't make any major impact on the scientific community since very little was known about ionic liquids at the time. In 2002, Swatloski et al. revisited the innovation to re-discover the capability of ionic liquids to solubilize cellulose [33, 42]. The study investigated cellulose solubility using common ILs with 1-butyl-3-methylimidazolium $[C_4C_1im]^+$ cation with several anions, such as chloride Cl^-, bromide Br^-, thiocyanate $[SCN]^-$, tetrafluoroborate $[BF_4]^-$, and hexafluorophosphate $[PF_6]^-$. It was found that ILs with Cl^- and Br^- had greater cellulose solubilities of 14 and 7%, respectively, while the large non-coordinating anions $[BF_4]^-$ and $[PF_6]^-$ were non-solvents. The proposed mechanism of cellulose dissolution in $[C_4C_1im][Cl]$ was attributed to the ability of the chloride anions to break the compact intermolecular hydrogen-bond network and form new bonds with the hydroxyl groups in cellulose. These findings paved the way for using ILs as novel solvents for cellulose, potentially improving the efficiency and flexibility of cellulose processing and transformation to chemicals and materials [43].

Ioncell-F is an example of an IL-based process that pioneers the use of acetate-based ionic liquid to produce cellulosic fibers with attractive properties at lower processing temperatures compared to the Lyocell process [44].

Another important application for ILs in biomass processing is their use as a reaction medium for sugar conversion to 5-hydroxymethyl furfural (HMF) [45]. In the U.S. Department of Energy report on *Top Value-Added Chemicals from Biomass*, HMF was specified as a versatile platform chemical that can be used as a building block to synthesize numerous polymers and chemicals, such as 2,5-furandicarboxylic acid, 1,6-hexanediol, adipic acid, levulinic acid, 2-methylfuran, and caprolactone [46, 47]. The formation reaction of HMF requires the dehydration of hexose sugars (fructose or glucose) in the presence of a metal salt catalyst. In general, the IL used as well as the selected catalyst have important roles in the selectivity and yield of the reactions since the reaction is promoted via the formation of an IL-catalyst complex.

Synthesis of HMF from glucose is more challenging compared to synthesis from fructose as the reaction proceeds in two steps that require two different catalytic environments: (1) the isomerization of fructose to glucose which is catalyzed by a Lewis base followed by (2) dehydration of glucose to 5-HMF which is speculated to be promoted by Brønsted acids [48]. Conducting the reaction in an aqueous environment usually results in poor yield and selectively due to further dehydration of the HMF product to levulinic acid with a substantial formation of humins through side reactions [49]. On the other hand, superior product yield and selectivity were achieved when common imidazolium-based IL was used as the reaction environment, with 90% conversion achieved when glucose was used as a substrate [48]. The research field is still in its early stages and several challenges need to be resolved: (1) HMF product separation from the IL medium, (2) recycling of the IL and catalyst, (3) increasing glucose substrate loading, and (4) making the system robust enough to use cellulose as a direct substrate [50, 51]. Biomass pretreatment is one of the major applications of ionic liquids in biomass processing. In fact, processing and transformation of cellulose and lignin for fuel, chemical, or material production would first require their effective separation from the biomass polymer matrix, which is the objective of a pretreatment process. The use of ionic liquids in biomass pretreatment is discussed in Sect. 5.3.

5.2.2 Protic Ionic Liquids

Protic ionic liquids (PILs), a subclass of ILs, are prepared by a simple stoichiometric neutralization reaction between a Brønsted acid and a Brønsted base (Scheme 5.1) that results in an acid–base complex (i.e., the protic IL) [52].

The formation reaction is based on the proton transfer from the acid to the base which gives a PIL a distinguishable characteristic of having an available proton to form hydrogen bonds with an anion, other dissolved solutes, and solvents [53].

$$HA + B \leftrightarrow A^- + HB^+$$

Scheme 5.1 Protic IL formation reaction from a Brønsted acid and a Brønsted base

The simple acid–base neutralization reaction gives PILs a great economic advantage compared to all other ILs, which are by convention referred to as aprotic ILs (APILs). As mentioned earlier, the key barrier towards industrial large-scale use of commonly used ILs is their high production cost, which is usually 5–20 times more than conventional organic solvents [20]. The high production cost of APILs mainly stems from: (1) high cost of the starting materials, (2) low atom efficiency during their synthesis via alkylation and salt metathesis, and (3) lengthy extensive purification steps which lead to generation of salty wastewater that needs further treatment steps [54]. The significant reduction in processing steps and the use of inexpensive starting materials (e.g., mineral acids and amines) make PILs more attractive for industrial applications. This is especially true as historically many common industrial solvents were adapted for multi-ton industrial scale because they are generated as inexpensive by-products in the production line of other higher value products (e.g., acetone is a by-product of phenol production and toluene is a by-product of gasoline production).

APILs are usually hailed for their non-volatile nature as they have extremely low vapor pressures. However, this is not always the case for PILs as many PILs have a measurable boiling point and can be distilled [55, 56]. This feature is directly related to a PIL's formation mechanism. PILs exist in equilibrium with their parent acid and amine. When heated a PIL boils rather than decomposes as the equilibrium reaction is reversed to form the more volatile neutral acid or base species [52]. In fact, the volatility of a PIL is highly dependent on its ionicity and degree of proton transfer. Ideally, the proton transfer from the acid to the base must be completed such that only ionic species exist to form a true PIL that displays ideal Walden behavior [57]. However, this is not always the case depending on the parent acid and base; the proton transfer can be only partial [58]. Yoshizawa et al. proposed the use of ΔpK_a as a practical qualitative indication of the proton transfer completeness of PILs (or degree of ionicity) without conducting any direct measurement [57]. They defined $\Delta pK_a = pK_a(HB^+) - pK_a(HA)$ where $pK_a(HB^+)$ and $pK_a(HA)$ represent acid dissociation constants of HB^+ and HA in an aqueous environment. It was demonstrated that ΔpK_a has a direct impact on the boiling point and volatility of the prospective PIL, with greater ΔpK_a suggesting more complete proton transfer and less presence of molecular species, and hence a higher product boiling point [59, 60].

For example, comparing two PILs that share the same α-picoline parent base and two acids of different strengths showed that the PIL with trifluoroacetic acid had a ΔpK_a value of 7.3 and a measurable boiling point of 175 °C. On the other hand, a PIL synthesized with the much stronger triflic acid had a ΔpK_a value of 20 with no observed boiling point.

One of the key properties of PILs is their Brønsted acidity which makes them attractive media for acid-catalyzed reactions and for applications that require proton-conducting media, such as fuel cells [61], batteries [61], and capacitors [62]. A PIL's acidity can be measured using acidity formulation methods for non-aqueous media scales, such as potentiometric titration, calorimetric titration, or acidity functions coupled with spectroscopy (e.g., Hammett function) [63–65].

5.3 Technology Concept Formulated (TRL 2-3)

The natural high recalcitrance of wood and its insolubility in common solvents has severely hindered the development and the utilization of lignocellulosic biomass as a feedstock for biofuels and biochemicals production. The ability to solubilize cellulose in common imidazolium-based ILs was the gateway towards investigating the use of ILs as a pretreatment medium to deconstruct the biopolymers of lignocellulosic biomass [66]. Ionic liquids' structural variability and designer properties inspired scientists to use ILs to overcome the challenges faced in conventional pretreatment processes, such as solvent losses, high-pressure operation, waste generation, and sugar degradation [67–69].

5.3.1 Biomass Pretreatment Using Ionic Liquids

In 2007, Rogers et al. explored the use of ILs to extract cellulose from a variety of hardwood and softwood lignocellulosic feedstocks. It was found that $[C_4C_1im][Cl]$ was able to partially dissolve the woody material [16]. Cellulose was recovered by adding an antisolvent (1:1 acetone–water, dichloromethane, or acetonitrile) and the properties of the reconstituted cellulose were found to be very similar to the properties of pure cellulose subjected to the same process. In the same year, Kilpeläinen et al. [70] published a study to investigate the use of several imidazolium-based ILs with chloride anions in dissolving hardwood and softwood feedstocks. They found that the solubilization efficiency and dissolution profiles are highly affected by the wood particle size. The addition of water as an antisolvent recovered an amorphous material that was a mixture of the wood components. The intriguing nature of the amorphous material led to investigating its enzymatic digestibility, and they found that 60% of the theoretical amount of glucose was released versus only 12% released from the untreated control. Even though the solubility of the lignocellulosic materials did not exceed 10 wt%, these early studies established the foundation of a new research field using ionic liquids as a novel platform for lignocellulosic biomass processing.

In the past decade, two IL-based pretreatment approaches were developed: (1) biomass dissolution processes (non-selective solubilization of the biomass components) and (2) fractionation processes (selective solubilization of lignin and hemicellulose). Each approach uses ionic liquids that interact differently with lignocellulosic biomass biopolymers [71].

5.3.2 Biomass Dissolution

The biomass dissolution process uses ILs that are capable of dissolving the entire lignocellulosic biomass components, specifically cellulose, by disturbing its inter-

and intra-hydrogen-bond network and creating new ones with the IL [43]. The disturbance of the hydrogen-bond network in cellulose reduces cellulose crystallinity significantly [25]. Cellulose was recovered using a mixture of acetone and water (1:1 v/v), which is capable of dissolving lignin while simultaneously acting as an antisolvent for the cellulose-rich material. Following these findings, Rogers and co-workers [72] in 2009 reported another major milestone in the field when they reported the use of 1-ethyl-3-methylimidazolium acetate $[C_2C_1im][Ace]$ as a better solvent to dissolve wood than $[C_4C_1im][Cl]$. The complete dissolution of pine wood was successful at mild conditions and only required mild grinding of the wood chips, which translates to a lower energy requirement. Also, the use of $[C_2C_1im][Ace]$ offered several advantages due to its attractive properties, such as lower toxicity, lower corrosivity, lower melting point, and potentially higher biodegradability, compared to chloride-based ILs. The study also revealed the significant crystallinity reduction of the regenerated cellulose material compared to the original feedstock. By comparing XRD patterns, it was evident that the regenerated cellulose was transformed from its natural state in the untreated biomass (cellulose I) to a revised structure of reduced inter-chain interactions and more thermodynamically stable (cellulose II) [73] Since then, many studies investigating the effect of ionic liquid pretreatment on cellulose crystallinity have been published [26, 73–75].

The amorphous regenerated cellulose provides more surface area making enzymatic hydrolysis easier and significantly faster, which is a key success factor for a pretreatment process that targets bioethanol production [27, 73]. The IL anion plays the main role in determining cellulose dissolution ability with the cation having a minor effect [48, 49]. ILs with hydrogen-bonding basic anions, such as chloride [16, 76] and carboxylate (e.g., formate, acetate, phosphate, and lactate) [72, 77, 78], were shown to form strong hydrogen bonds with cellulose. Among several tested ILs, 1-ethyl-3-methylimidazolium acetate $[C_2C_1im][Ace]$ has been the IL predominantly used in dissolution biomass pretreatment due to its high cellulose-dissolving capability, non-toxic nature, as well as its ability to dissolve a wide variety of biomass feedstocks (e.g., *Miscanthus giganteus*, switchgrass, willow, oak, and pine). Singh et al. [27] studied the dynamic solubilization mechanism of switchgrass subjected to pretreatment with $[C_2C_1im][Ace]$ using the auto-fluorescence of plant cell walls.

They showed that, at first, the IL swelled the secondary cell walls, which resulted in a complete disruption of the cell wall structure after a treatment time of 2 h at 120 °C. Fig. 5.3 shows the dissolution phases of switchgrass stem using $[C_2C_1im][Ace]$ starting from plant cell disintegration in the first phase and separation to complete dissolution in the last phase [27].

However, facile extraction of the lignin was achieved. The close association of the anion hydrogen-bond basicity and cellulose dissolution was confirmed by NMR studies [76] and molecular dynamics studies [79]. The empirical solvatochromic Kamlet-Taft parameters were also used to correlate the anion hydrogen-bond basicity with the IL cellulose-dissolution ability. Kamlet-Taft parameters include: solvent hydrogen-bond acidity denoted as α, hydrogen-bond basicity denoted as β, and π^* for solvent polarizability. For ILs, β was found to be mainly influenced by the anion [80]. ILs that are capable of dissolving cellulose exhibited high hydrogen-bond

(a)   **(d)**

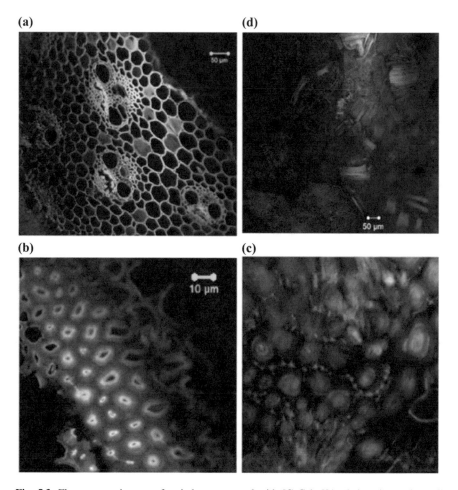

Fig. 5.3 Fluorescence images of switchgrass treated with [C$_2$C$_1$im][Ace] that shows the wall **a** before treatment, **b** 20 min pretreatment, **c** 50 min pretreatment, and **d** 120 min pretreatment where complete breakdown of cell walls occurred. Ref [27]

basicity ($\beta > 0.7$) [71]. A comprehensive review on using ILs for biomass dissolution was recently published by Rosatella and Afonso [81].

5.3.3 Biomass Fractionation

The first patent using ionic liquids to selectively extract lignin from the biomass matrix was published by MacFarlane et al. in 2007 [82]. They indicated that ILs containing an organic or inorganic cation along with an organic acid-based anion (mainly sulfonate) can fractionate lignocellulosic material, where lignin is solubilized in the ionic liquid and a cellulose-rich pulp remains highly intact during the process. The selective dissolution of lignin and hemicellulose in the IL medium gives the process the "fractionation" feature, with lignin removal (or delignification) being the main deconstruction mechanism. Lignin was subsequently recovered by adding an antisolvent (e.g., water), changing the solution pH or temperature. Another patent was filed in 2008 by Varanasi et al. describing the same biomass fractionation effect using ionic liquids and the subsequent successful enzymatic hydrolysis of cellulose to glucose [83]. Interestingly, the ionic liquids used for the fractionation were $[C_2C_1im][Ace]$ and $[C_4C_1im][Cl]$, which are the same ionic liquids reported earlier for their ability to dissolve all the biomass components, including cellulose. The major difference was the use of a much higher biomass loading (i.e., the weight percent of biomass in the reactor relative to IL; 33 wt% versus <5 wt% for dissolution [16]) that weakens the ability of the ionic liquid to dissolve the entire biomass components. The crystallinity of the cellulose-rich pulp recovered from the fractionation process is usually intact or is slightly increased due to the lignin and hemicellulose removal [29]. The extraction of lignin (i.e., delignification) increases the exposed surface of cellulose making it easier for the enzymes to access the cellulose fibers, thereby significantly increasing its hydrolysis to glucose. In general, ILs used in fractionation processes do not need to have high hydrogen-bond basicity to solubilize lignin [71]. Thus, a larger number of ILs are capable of dissolving lignin than cellulose. This includes ILs with moderate to strong hydrogen-bonding anions, such as halides (chlorides and bromides) [84, 85], acetate $[Ace]^-$, triflate $[OTf]^-$, methyl sulfate $[C_1SO_4]^-$ [85], xylenesulfonate $[XSO_3]^-$ [86], and hydrogen sulfate $[HSO_4]^-$ [87].

The ionoSolv process is a biomass fractionation process that uses ionic liquid and water mixtures to selectively dissolve lignin, producing a cellulose-rich pulp [22, 88]. The extracted lignin is recovered via precipitation with the addition of an anti-solvent, usually water, and the IL is recycled back for another pretreatment. The process has mainly focused on using protic ammonium-based ILs with a $[HSO_4]^-$ anion because they provide key technical and economic benefits, which will be discussed in detail in Sect. 5.4. Figure 5.4 shows the main steps involved in the lab-scale version of the ionoSolv process [22].

5.4 Early Technical Challenges

Historically, the biomass dissolution process has been the most prevalent ionic liquid pretreatment approach studied since the process was developed as a direct consequence of discovering cellulose-dissolving ILs [16]. As understanding the process continued to grow, technical challenges were associated with the practical implementation of biomass dissolution using ILs. These challenges were mainly related to the hyper-sensitivity of water in the process as well as to the thermal stability of the ILs used [71]. While thermal stability might be overcome by a more careful design of the IL, the water sensitivity problem cannot be solved the same way because cellulose

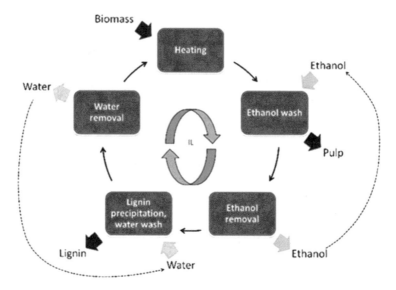

Fig. 5.4 Bench-scale workflow used in ionoSolv process. Ref [22]

solubility is inherently inhibited by water, regardless of what IL is used. The following sections discuss these two process aspects and how the ionoSolv pretreatment was designed to meet these challenges.

5.4.1 Process Water Tolerance

The first major technical challenge associated with the dissolution process was the very strict anhydrous conditions required for cellulose solubilization. The extremely low cellulose solubility in water makes the presence of <1% water in the IL highly inhibitory for cellulose dissolution [33]. As ILs used in the dissolution process (e.g., acetate, or chloride-based) are highly hygroscopic in nature, an IL-drying step will have to be integrated into the process scheme, which will be highly cost prohibitive [71]. In addition, the biomass feedstock will also need extensive drying as a newly harvested biomass contains large amounts of water, typically up to 50 wt%. Even after air drying, biomass will still contain moisture of 5–10 wt%, which is high enough to inhibit cellulose dissolution. Lab experiments use cellulose-dissolving ILs to dry their biomass feedstock at 90 °C overnight for moisture removal; a practice that will not be economically possible on an industrial scale.

In contrast, a key advantage of the ionoSolv process is the high performance when the IL is mixed with water. In fact, the initial identification of the fractionation ability of the ILs used in the ionoSolv process started with the observation that *M. giganteus* pretreatment with dry 1-butyl-3-methylimidazolium methyl sulfate

[C_4C_1im][C_1SO_4] (total water 1.1 wt% including biomass moisture) produced a very poorly enzymatically hydrolyzed pulp. On the contrary, mixing [C_4C_1im][C_1SO_4] with 20 wt% water produced a pulp that can be enzymatically hydrolyzed. This led to a more elaborative and comprehensive investigation of the importance of the presence of water during the pretreatment with this specific IL. Therefore, the first ionoSolv study done by Brandt et al. in 2011 investigated the performance of IL/water mixtures, mainly 1-butyl-3-methylimidazolium methyl sulfate [C_4C_1im][C_1SO_4] and 1-butylimidazolium hydrogen sulfate [C_4im][HSO_4] for lignocellulosic biomass pretreatment [87]. Glucose yields obtained from *M. giganteus* pulp after pretreatment were the highest when the ILs were mixed with 10–40 wt% water. This was the first demonstration that an IL-based pretreatment can be conducted with the presence of substantial amounts of water, eliminating the strict anhydrous conditions needed in an IL-dissolution process. The presence of water was found to be essential for the hydrolysis of the ether linkages in the lignin structure. Having the flexibility of operating the process with a substantial amount of water also offers several key processing advantages, including (1) reduced overall solvent mixture cost, as water is a cheap solvent, and (2) reduced solvent viscosity, which reduces the pumping power and makes filtration operations easier.

5.4.2 Ionic Liquid Thermal Stability

Solvent thermal stability is one of the key parameters that impact process performance in several ways. If a solvent has low thermal stability under process conditions, this translates to poor solvent recovery, higher solvent make-up, higher operating cost, accumulation of degradation products, changes in solvent properties, and, eventually, the uncertainty of process performance. In addition, solvent degradation products may also cause safety and process integrity issues, such as unpredictable corrosion behavior [89, 90]. Therefore, throughout the development of a new process, such as IL-based pretreatment, it is crucial to establish a very good understanding of the long-term thermal stability of the IL used as well as the degradation products formed in the process. IL thermal stability will have a direct impact on its recyclability efficiency, which is a major economic driver in the process [91].

ILs have been long known to be thermally stable solvents. However, such a general statement isn't possible for every anion/cation combination, especially given the sheer number of possible ILs [92]. There are some commonly used ILs that are highly thermally stable, but these usually contain highly non-coordinating fluorinated anions that are not used for biomass deconstruction [93]. IL-based pretreatment is usually operated at temperatures of 100–190 °C with retention times of 30 min up to 24 h [94]. The operating temperature range might be high enough to cause solvent degradation, especially in a closed-loop solvent system.

The thermal stability of ILs is influenced by both the anion and the cation, with the anion playing a more significant role [93]. Thermal gravimetric analysis (TGA) is a common analytical technique used to characterize solvent thermal stability by correlating the solvent mass loss with temperature. Using the step-tangent method, the TGA scan of different solvents can be compared based on their onset decomposition temperature, T_{onset}. The study of George et al. showed that all [HSO$_4$]-based ILs with different ammonium cations have higher thermal stabilities (T_{onset} of 290–320 °C) compared to the famously used [C$_2$C$_1$im][Ace] ($T_{onset} = 215$ °C) [29]. Another study evaluating the thermal stability of 66 ILs has also categorized acetate-based ILs among the least stable ILs ($T_{onset} < 250$ °C) while hydrogen sulfate-based ILs were classified to be more stable (250 °C $\leq T_{onset} < 300$ °C) [93]. It should be noted that T_{onset} represents the decomposition temperature; however, the significant mass loss still occurs at much lower temperatures [92]. For example, Rogers and co-workers have observed up to 15% weight loss within 10 min at 185 °C when treating bagasse and pine with [C$_2$C$_1$im][Ace] [16]. Brandt et al. have also reported that [C$_4$C$_1$im][Ace] lost 10% of its weight after 60 h at 120 °C, which is much lower than T_{onset} [71]. As a rough guide, process engineers typically expect a solvent lifetime of around 4 months or 2500 h of operation. As such, under process operating conditions, the combination of IL losses and degradation should not exceed 0.04% per hour of operation, which provides an upper bound for an acceptable solvent degradation rate.

In addition, since T_{onset} values are strongly subject to set-up (e.g., heating rate, gas flow, and pan material), some authors have proposed a more practical measurement parameter, $T_{x/z}$, where x represents a given decomposition extent (e.g., 1% weight loss) in z length of time (e.g., 10 h) [95]. Applying this to [C$_4$C$_1$im][Ace] IL, $T_{0.01/10h}$ was 102 °C and [C$_2$C$_1$im][Ace] is expected to have a very similar value. Although it was suggested that a maximum stable operating temperature for an ionic liquid-based process should be 10 °C lower than $T_{0.01/10h}$, a more realistic value would be at least 30 °C lower to maintain high solvent recovery and integrity. In both temperature limits, acetate ILs cannot be thermally stable, especially at the elevated temperatures required for biomass pretreatment.

The IL cation also has an important effect on its thermal stability. Achinivu et al. have recently reported the use of three acetate-based PILs with pyridinium, 1-methylimidazolium, and pyrrolidinium cations to selectively extract lignin from corn stover [96]. All three PILs suffered from low thermal stability with significant mass loss observed at temperatures <100 °C, making them easily susceptible to degradation at processing conditions. The poor thermal stability of acetate-based ILs is also largely attributed to the presence of dissociated molecular species due to incomplete protonation of the cation by the weak acetic acid [96].

The ionoSolv process typically uses ILs with ammonium-based cations, such as triethylammonium [22]. The most dominant degradation mechanism of an ammonium-based IL is the loss of the alkyl chain from the cation (dealkylation) before the breaking of C–H and C–C bonds at higher temperatures [97]. For [N$_{0\,2\,2\,2}$][HSO$_4$], diethylammonium or monoethylammonium are the expected dealkylation products (Scheme 5.2). A recent ionoSolv study recorded the ^{1}H-NMR spectra for recycled triethylammonium hydrogen sulfate [N$_{0\,2\,2\,2}$][HSO$_4$] (4th cycle at 120 °C and 9 h) to investigate the extent of IL degradation during pretreatment.

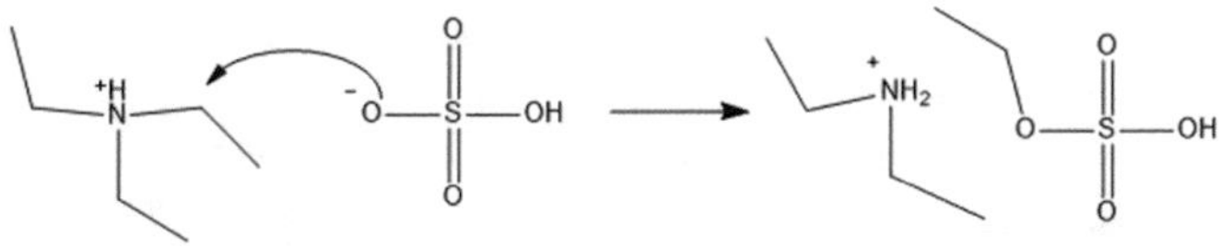

Scheme 5.2 Dealkylation of triethylammonium cation to form diethylammonium

However, no signal was detected for the expected dealkylation products, which demonstrates the robust thermal stability of the IL during the course of ionoSolv pretreatment. Even though these results are encouraging, there is still a need to have a more in-depth understanding and evaluation of the long-term thermal stability of the ILs used in the ionoSolv process.

5.4.3 IonoSolv Shift to Protic Ionic Liquids

The first IL used for biomass delignification in the ionoSolv process was $[C_4C_1im][C_1SO_4]$ upon the realization that the IL can be mixed with water giving highly digestible pulps [87]. It was soon realized that the presence of water caused the $[C_1SO_4]^-$ anion to hydrolyze forming $[HSO_4]^-$ and methanol. This led to investigating $[HSO_4]$-based ILs directly as they seemed to be the active species in the $[C_1SO_4]/H_2O$ system. It was shown that 1-butyl-3-methylimidazolium hydrogen sulfate $[C_4C_1im][HSO_4]$ is very effective in delignifying *M. giganteus* and, to a lesser extent, willow and pine (conditions at 120 °C and 22 h).

In an attempt to study the effect of the cation on pretreatment efficiency, a monoalkylated imidazolium cation, namely, 1-butylimidazolium $[C_4im]^+$ was used instead of the dialkylimidazolium $[C_4C_1im]^+$. Combining a $[HSO_4]^-$ anion with a monoalkylated imidazolium cation $[HC_4im]^+$ results in a Brønsted-acidic PIL that can be synthesized via a simple acid–base reaction between the parent acid and amine (in this case sulfuric acid and 1-butylimidazolium). In contrast, the synthesis of $[C_4C_1im][HSO_4]$ requires the reaction of 1-butyl-3-methylimidazole with dimethyl sulfate followed by hydrolysis at elevated temperature and then evaporation of the water and the methanol formed during the reaction (Fig. 5.5).

This synthesis route of $[C_4C_1im][HSO_4]$ has a near-quantitative atom efficiency providing an advantage compared to the poor atom efficiency in the synthesis of more common ILs [98]; however, dimethyl sulfate is a highly poisonous and dangerous chemical of carcinogenic, mutagenic, and corrosive nature. The viability of using the PIL analogue was well established when $[C_4im][HSO_4]$ was shown to be as effective for *M. giganteus* deconstruction and delignification, with 90% subsequent glucose yield during enzymatic hydrolysis.

Even though the $[C_4im][HSO_4]$ PIL was found to be effective for biomass delignification and a cheaper alternative compared to an APIL, it was intriguing to explore

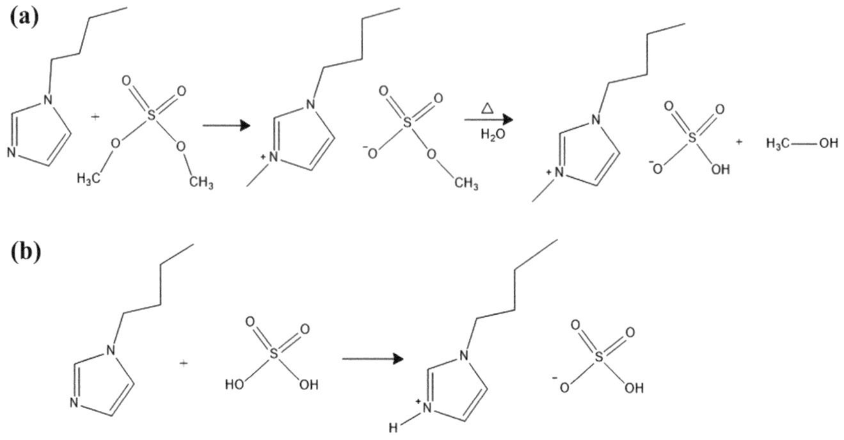

(a)

(b)

Fig. 5.5 **a** Synthesis route of $[C_4C_1im][HSO_4]$ aprotic ionic liquid using dimethyl sulfate and 1-butyl-3-methylimidazole **b** Synthesis of $[C_4im][HSO_4]$ protic ionic liquid using sulfuric acid and 1-butylimidazole

additional routes that might further reduce IL cost. This was accomplished by capitalizing on the structural variability of ILs and the fact that one can fine-tune their design to seek less costly ion combinations without impacting process performance. Since the anion determines the chemistry and solvation capability needed for lignin removal, keeping the $[HSO_4]^-$ anion was necessary. In addition, sulfuric acid (the precursor of $[HSO_4]^-$ anion) is inexpensive. So, there isn't an economical need to seek another alternative. This led to investigating different cations of cheaper and more massively produced alkylamines precursors. Therefore, a series of protic $[HSO_4]$-based ILs made from several low-cost alkylamines were investigated in terms of their effectiveness for biomass delignification. The degree of the ammonium cation substitution was shown to have a significant impact on pretreatment efficiency. Among the tested PILs, $[N_{0\,2\,2\,2}][HSO_4]$ IL was found to be the most effective and 75% as efficient as $[C_2C_1im][Ace]$ at a fraction of the projected large-scale production cost [29].

5.5 Technology Bench-Scale Validation (TRL 4)

Bench-scale validation of a new process usually focusses on studying and optimizing the key process variables that can be monitored in a lab environment. These developmental objectives can be perceived as micro-objectives with the aim to make the process more compact, energy-efficient, and environmentally sustainable [99]. On the other hand, macro-objectives aim to use engineering methods to evaluate the techno-economic viability and environmental sustainability of a process. These methods are important assessment tools to pin-point key process weaknesses and, therefore, steer research towards solving these weaknesses. In the case of the ionoSolv process, the effective application of [HSO$_4$]-based PILs for biomass fractionation held high potential for the development of an economical IL-based pretreatment process. However, it was important to understand the main factors that govern the process economics and resolve any problems. It was also of key importance to build an in-depth understanding of the process fundamentals and evaluate the process performance at intensified conditions that might potentially be translated to industry.

5.5.1 Resolving the Key Cost Driver

The first techno-economic assessment of IL-based pretreatment was conducted in 2011 using [C$_2$C$_1$im][Ace] as a cellulose-dissolving IL [100]. The motivation was to identify the main operational targets that are needed to be improved for an IL-based pretreatment to be economically competitive with peer technologies. The key variables driving the process economics were identified to include: (1) IL cost, (2) IL: solid loading, and (3) IL recovery. The authors emphasized that reducing the IL price is the sole factor that needs to be targeted to achieve an economically viable process. This is indeed true as the IL make-up amount (and, thereby, the associated IL make-up cost) is linked to the solid loading and recycle rate. For example, lowering the IL use in the process by increasing the biomass to ionic liquid ratio from 1:10 to 1:1 will increase IL make-up cost by an order of magnitude. The study took into consideration an IL cost range of $2.5–$50 kg^{-1}, which were considered reasonable given the scale of operation, yet they were still 2–50 times higher than organic solvent costs, such as toluene ($1.03 kg^{-1}) and acetone ($1.30 kg^{-1}). Another techno-economic study considered more intensified biomass dissolution pretreatment schemes that included a "one-pot" process and a conventional pretreatment dissolution route [101]. It was highlighted that even at the most economically optimistic conditions (which might be very challenging to achieve in reality) of 99.6% IL recovery and 50% biomass, the process minimum ethanol selling price (MESP) for both schemes was >$6 gal^{-1}, which is very high compared to a MESP of $2.15 gal^{-1} that was estimated for dilute acid pretreatment [102].

Even though the techno-economic assessments were done based on the IL biomass dissolution process, it is very likely that the same parameters (IL cost, IL: biomass loading, and IL recovery) will drive the process economics in IL fractionation processes. As stated earlier, the shift from using an APIL to a PIL for biomass delignification in the ionoSolv process was a key turning point that held so much potential in resolving the IL high-cost issue. To quantitively assess the anticipated cost reduction, Chen et al. performed the first economical evaluation that estimated the large-scale production cost of PILs used in the ionoSolv process [30]. Conducting such an economic evaluation was highly valuable not only to provide the IL production cost estimate but also to highlight the key aspects that contribute to the IL production cost, which can serve as a guideline for designing and estimating the cost of other PILs. In addition, although for the past two decades the cost of ILs has been always considered to be the bottleneck for their large-scale implementation, there were no previous reports assessing the manufacturing of ILs at scale.

Two PILs were taken as models for the process simulation and economic evaluation: triethylammonium hydrogen sulfate $[N_{0\,2\,2\,2}][HSO_4]$ and 1-methylimidazolium hydrogen sulfate $[C_1im][HSO_4]$. The study highlighted that the cost of raw materials accounts for 82% of the operating cost of the plant, which is mainly due to the simple equipment and cheap utilities (cooling water and electricity) involved in the manufacturing process of the PILs studied. Such an important outcome allowed the development of a simple formula that estimates the price of any other PILs where raw materials are expected to dominate the manufacturing price. The formula for the IL price is:

$$\text{IL price} = \frac{M_1 P_1 + M_2 P_2}{M_1 + M_2} \times 1.25$$

where M_1 and M_2 are the molecular weights of the two starting materials and P_1 and P_2 are the prices of the two starting materials. The formula indicates that the production cost of a PIL is dictated by the molecular weight of the more expensive parent chemical. In the case of $[HSO_4]$-based IL, the amine is the more expensive parent chemical and thus choosing a low-molecular weight amine will reduce the cost of the prospective IL. The impact of the cost of the parent amine is clear when comparing the estimated prices of $[N_{0\,2\,2\,2}][HSO_4]$ and $[C_4im][HSO_4]$, which were $1.24 and $2.96–5.88 kg^{-1}, respectively (based on 2014 chemical prices). The 2.5–5 increase in the IL cost is exclusively related to the higher cost of methylimidazole, which is a specialty chemical, versus the cost of triethylamine, a chemical that is produced in bulk scale [103]. The cost of these PILs is significantly lower than the estimated cost of $[C_2C_1im][Ace]$, which ranges between $20 and $101 kg^{-1} [100]. The low price of $[N_{0\,2\,2\,2}][HSO_4]$ made it particularly attractive as it is midway between a low-cost organic solvent, such as toluene ($1.03 kg^{-1}) and higher cost organic solvent, such as acetonitrile ($1.54 kg^{-1}). It is also important to note that the cost estimates were reported based on 2014 commodity prices that dropped dramatically after the decline in oil prices leading to a current IL price of $0.5 kg^{-1} [22].

In addition, the use of PILs also offers environmental advantage in addition to the economic advantage. Jessop proposed a simple approach to assess the environmental sustainability of solvents by creating a solvent synthesis tree (like a family tree) to count the synthesis steps and the nature of all parent chemicals associated with its synthesis [104]. Applying this approach to one of the commonly used ILs (e.g., $[C_4C_1im][Cl]$) will require 22 synthesis steps; on the other hand, the synthesis of $[N_{0\,2\,2\,2}][HSO_4]$ requires 7 steps [30, 104]. The reduced complexity in the chemical synthesis route translates to a tangible reduction in the environmental impact associated with IL production; simpler and more direct synthesis means less waste generated, lower energy and chemical use, and less solvent losses [105].

To conclude, when considering which IL to use for any specific application, researchers should be able to make an informed estimate about the cost of the IL.

This can be accomplished by conducting a techno-economic analysis, or it can be something as simple as making a back-of-the-envelope cost estimate based on some general considerations, such as the price of the starting materials and the level of synthesis complexity (degree of substitution and number of synthesis steps) [104, 105]. Working out the cost estimation for a new solvent, in this case, the IL, will highlight the hot-spots that can be altered to reduce the cost. It can also reveal whether using this new solvent for a specific process offers an economic advantage compared to alternative solvents [95]. For the ionoSolv process, switching from an APIL to a PIL when the process chemistry allows and then evaluating the economic viability of using a PIL at scale sets a good example of accelerating large-scale deployment of an IL-based process.

5.5.2 Improving Space-Time Yield

Lowering the cost of the pretreatment reactor is important to reduce the overall process investment cost, as it is often the largest piece of capital investment in a cellulosic ethanol plant [106]. The pretreatment reactor volume is mainly dictated by the pretreatment time and biomass loading where higher biomass loading and shorter time will result in a larger reactor volume [107]. Despite this, the majority of IL-based pretreatment studies have been conducted at low biomass loading (3–10 wt%) and long pretreatment times (8–72 h). While the knowledge obtained from these studies is important to develop the process chemistry, such conditions cannot be translated to an industrially relevant scale because they are uneconomical.

For example, Tao et al. estimated the capital cost of lime pretreatment to be eight times higher than dilute-acid pretreatment due to the significant number of reactors required (residence time of 4 h) and the need for a lime recovery unit [106].

To improve the space-time yield of a reactor, it is important to optimize the reaction kinetics, which is usually strongly dependent on temperature. Therefore, the ionoSolv pretreatment performance for *M. giganteus* feedstock [22, 108] was investigated at different temperatures and residence times for using $[N_{0\,2\,2\,2}][HSO_4]$ IL with 20 wt% H_2O. Enzymatic hydrolysis of cellulose to glucose changed dramatically as the recovered cellulose pulp composition varied depending on the pretreatment severity. Reaching the optimum glucose yield is significantly accelerated when the pretreatment is conducted at higher temperatures (Fig. 5.6).

For example, to achieve 75% glucose yield, the pretreatment should be conducted for 15 min at 180 °C or for 8 h at 120 °C. The ability to achieve high glucose yield while reducing the reactor volume by a factor of 32 offers a substantial economic advantage as it translates to an optimized reactor volume and higher throughput processing times in an industrial-scale process. This is true even if the process will be operated at higher temperatures as the reactor energy can be supplied through process heat integration.

As stated earlier, the ionoSolv pretreatment mechanism is based on lignin extraction from the biomass. Therefore, optimum glucose yield in enzymatic hydrolysis should be obtained when the biomass is highly delignified [22]. Understanding the

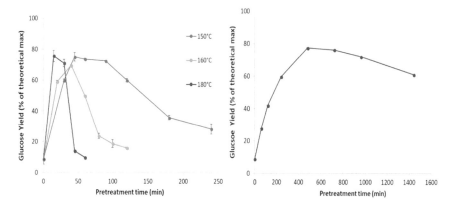

Fig. 5.6 Glucose release after 7 days of enzymatic saccharification from *M. giganteus* pulp after pretreatment at several temperatures and times with [N$_{0\,2\,2\,2}$][HSO$_4$]. Refs [22 and 108]

kinetics of lignin extraction is a challenging task as it involves a complex multi-step process of lignin fragmentation and depolymerization as well as the redeposition of condensed lignin and humins onto the pulp surface [109].

Increasing the pretreatment temperature from 120 to 150 °C had a dramatic impact on accelerating lignin extraction from the biomass, thereby increasing subsequent glucose yields. Similarly, operating at temperatures higher than 150 °C also improved and accelerated lignin removal to a lesser extent. In either case, approaching or exceeding the glass transition temperature (T_g) of lignin (estimated to be around 130–150 °C) was important to improve the kinetics of lignin removal. Li et al. reported accelerated dissolution of bagasse and pine using [C$_2$C$_1$im][Ace] (10 min when operating at temperatures above 170 °C compared to 16 h at 110 °C). This highlights the importance of operating at temperatures that exceed the T_g [110]. Arora et al. also observed a remarkable increase in switchgrass delignification using [C$_2$C$_1$im][Ace] at temperatures $\geq$150 °C, which was correlated to the increase in cellulose enzymatic hydrolysis to glucose [111].

Another important process intensification factor is biomass loading. As mentioned earlier, biomass loading is a key cost driver in an IL-based biomass pretreatment process as it reduces the amount of the IL required and, therefore, the cost associated with IL use in the process. Higher solid loading ($\geq$15 wt%) also brings several other advantages, such as a more concentrated product stream, lower energy input, lower water consumption, smaller reactor size, and a lower MESP [100, 112]. At the same time, increasing solid loading also brings several processing challenges, such as the difficulty of mixing and handling, poor heat and mass transfer, and increased inhibitor concentration [112, 113].

The use of high solid loading also has implications for the density and viscosity of the reaction product slurry, which would require higher power to overcome the high yield stresses at high concentrations [114]. Cruz et al. studied the impact of solid loading on the pretreatment efficiency using [C$_2$C$_1$im][Ace] and the changes in the rheological properties of the slurries produced [115]. It was found that the shear viscosity of the produced slurry increased by three orders of magnitude upon increasing the solid loading from 3 to 50 wt%. However, interestingly, increasing the solid loading of the biomass improved the slurry's shear thinning behavior, thereby significantly reducing its complex viscosity to values similar to low solid loading. The authors saw in this behavior an opportunity to design a high solid-loading reactor that induced high shear stress to overcome the poor mixing limitation. Wu et al. also investigated the role of solid loading on corn stover pretreatment efficiency using [C$_2$C$_1$im][Ace] aiming to find the minimum amount of IL required for an efficient pretreatment [116]. High glucose yields of ~80% were obtained at a wide range of solid loading (4.8–33 wt%), while at 50 wt% solid loading the glucose yield dropped to ~60%.

The decrease in glucose yield was attributed to a combined effect of lower lignin extraction efficiency and less disruption of the cellulose hydrogen bonds as less IL is present (mole ratio of [C$_2$C$_1$im][Ace] to glucose in cellulose). In ionoSolv processing, doubling the biomass loading from 10 to 20 wt% for pretreatment of *M. giganteus* was effective with no negative impact on process performance [108]. The combined effect of doubling biomass loading along with shortening the pretreatment time resulted in reducing the reactor volume by a factor of 64 which translates to a 90–95% reduction in reactor cost. A study on the impact of high solid loading on ionoSolv pretreatment efficiency is underway and will be published shortly.

5.5.3   Ionic Liquid Recovery and Recycle

High IL recovery and recycle rates are crucial for the economic success of the process as the amount of the IL make-up is one of the major contributors to the process operating expenditures (OPEX) [100]. The IL cost dictates the economic significance of the IL recycle rate and recovery; the higher the IL cost the narrower the tolerance and vice versa. Even in the case of using a low-cost IL, an IL recovery of >95% should be achieved for the solvent cost not to become limiting [22]. The technoeconomic model of the IL-dissolution process also highlighted the significant impact of the IL recovery on the process of capital expenditures (CAPEX). Low IL recovery reduces the capacity and the cost of the IL regeneration section. However, in the process configuration studied, because the non-recovered IL ends up flowing to the boiler, a larger and more expensive boiler will be needed to dispose of the IL by burning. Klein-Marcuschamer et al. showed that reducing the IL recovery from >99 to 94% at 10 wt% biomass loading decreases the CAPEX of the IL regeneration process from $96 to $93 M while at the same time tripling the boiler costs from ~$20 to ~$60 M [100].

Despite the technical and economic significance of IL recycle and recovery, few pretreatment studies have investigated this aspect of the process. Even the few studies that exist are based on IL-dissolution processes with low biomass loading (5 wt% or less) [117, 118]. Therefore, it was important to evaluate the ionoSolv process performance with recycled ionic liquid to quantify the recovery rate and its impact on the cellulose-pulp composition and enzymatic digestibility. Recycling experiments were conducted on *M. giganteus* using $[N_{0\,2\,2\,2}][HSO_4]$ at pretreatment conditions of 120 °C and 8 h. Over the course of four consecutive IL uses, the IL recovery achieved was 98–99% [22]. These recoveries are much higher than the reported $[C_2C_1im][Ace]$ recovery (85–90%) [119, 120] and also higher than organic solvent recovery in the organosolv process (96–98%) [67]. In addition, $[N_{0\,2\,2\,2}][HSO_4]$ recycling did not have a negative effect on process performance. In fact, the glucose yields gradually increased from 62 to 71% over three IL reuses, which was coupled with an increase of the glucan recovery in the pulp. Both effects were attributed to the hypothesis that the excess acid present in the first IL batch was consumed over the cycles causing more glucan to be retained and higher glucose to be released in enzymatic hydrolysis. The high IL recovery achieved in the ionoSolv process coupled with low IL cost shows high promise to be able to compete with established pretreatment technologies.

5.6 Technology Validation in a Relevant Environment (TRL 5-6)

A key step in the process development roadmap involves moving from the bench-scale to a larger scale. The "larger scale" can be either a laboratory scale-up study, where the process volume and equipment are scaled by several orders, or it can be a pilot-scale demonstration. Usually, both are necessary before taking the process to commercial operation. It is important at this stage for the process developers to draw a conceptual engineering design to think about the process flow and choose potential unit operations. This exercise coupled with the techno-economic evaluation are powerful tools to identify key process and scale-up challenges that may not have existed in a lab-scale environment.

5.6.1 Process Conceptual Design

Figure 5.7 shows a simplified process flow diagram of the ionoSolv process. The unit operations discussed here might be changed and adapted to meet the specific targets for a given project, such as cost cycle optimization, CAPEX minimization, and the scale of the plant. The main sections in the process can be divided into feed handling, pretreatment, lignin recovery, and IL regeneration. In feed handling, the supplied biomass is stored and prepared where a chipper or a mill is used to reduce the size of the biomass to facilitate pretreatment. In the pretreatment section, the chipped biomass is brought into contact with a mixture of recycled ionic liquid and water (typically 20% water content by weight) in a reactor and heated to temperatures

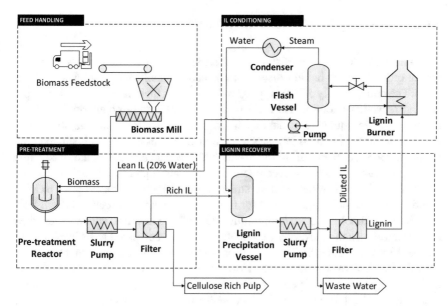

Fig. 5.7 Conceptual design of the ionoSolv process

ranging from 120 to 180 °C (depending on the biomass type) for a specific residence time (~15–60 min).

The temperature and time can be chosen based on the desired product specifications. For example, for bioethanol production obtaining highly delignified cellulose-rich pulp is key to ensure high sugar yield in the downstream enzymatic hydrolysis unit. For cellulose material applications, conditions should be optimized to have the required properties, such as fiber length, density, and tensile strength.

Once the pretreatment reaction is completed, the slurry is sent to a filter to separate the cellulose-rich pulp (solid) from the IL/water mixture containing the dissolved lignin (lignin-rich IL). The pulp is then washed thoroughly to ensure the complete removal of residual IL. Pulp washing on a large-scale system was highlighted as one of the potential process challenges and will be discussed in more detail in Sect. 5.6.3. The lignin-rich IL is then sent to the lignin recovery unit where lignin is precipitated by adding water as an antisolvent. The recovered lignin is filtered and washed with water to ensure minimal losses of the IL. Lignin is then sent to a boiler where heat is recovered and used to supply the process energy requirements, as no lignin valorization is considered at this stage of the process development. The diluted IL flowing from the lignin recovery unit needs to be re-concentrated before it is recycled back to the pretreatment reactor. The re-concentration or regeneration of the IL is expected to be a process-limiting step as water removal is naturally a highly energy-intensive process.

5.6.2 Ionic Liquid Re-Concentration

It is essential to develop robust and cost-effective IL recycle and recovery technologies in order for ionic liquid processes to be economically and ecologically viable [104]. The non-volatility of ionic liquids and, therefore, their near-quantitative recovery and recycling is one of the main reasons for their early popularity. However, studies on ionic liquids' recyclability in operational environments are still scarce [118]. In ionic liquid-based biomass pretreatment processes, the IL is diluted with a large amount of water during the product recovery step (ionic liquid dissolution process) or during the lignin precipitation step (ionic liquid fractionation process) [121]. The diluted IL-water mixture stream must be re-concentrated prior to ionic liquid recycling. Ionic liquids used for biomass pretreatment processes (both fractionation and dissolution) are hydrophilic and miscible in water, which makes their re-concentration and recovery a very challenging task compared to hydrophobic IL recovery and recycles [118]. Thermal evaporation of water is one of the first options to consider for IL recovery and recycle. However, thermal evaporation technologies are highly energy-intensive, especially when the solvent to be removed is non-volatile, such as water [122]. In addition, the IL–water mixture represents a real challenge due to the presence of strong hydrogen-bond interactions between the water molecules and the IL ions, especially at low water concentrations [123, 124]. In addition, boiling point elevation in the IL–water binary mixture due to the strong water–ions interaction makes thermal distillation more challenging and energy-intensive [121, 125].

Membrane-based technologies are considered as the most cost-effective and energy-efficient choice, and they are also industrially mature. Membrane-based technologies, such as nanofiltration (NF), reverse osmosis (RO), pervaporation and electrodialysis, were investigated for IL re-concentration and dehydration application. Haerens et al. have studied the use of pressure-driven membrane technologies, such as NF and RO, to re-concentrate 5 vol.% Ethaline200 aqueous solution [126]. Only a five-fold concentration of the IL was possible, reaching a maximum concentration of 20–25 vol.%. The intrinsically high osmotic pressure of ILs and low ILs rejection, especially at high IL concentrations, limit the application of membrane-pressure processes for IL dehydration. Therefore, the authors suggested using RO and NF to re-concentrate dilute IL solutions by a factor of four to five followed by another re-concentration process to achieve higher water removal. Lynam et al. studied a vacuum membrane distillation process to concentrate $[C_2C_1im][Cl]$ solution at high feed concentration (>20 wt%) using a hydrophobic polyacrylonitrile (PAN)-based membrane [127]. At optimal operating conditions, the processes showed good performance achieving a 65% final concentration and 99.5% IL recovery. However, the loss of performance with time due to membrane pore wetting was highlighted as a potential operational challenge.

Pervaporation technology was also recently investigated for $[C_2C_1im][Ace]$ re-concentration in a biomass pretreatment process [121]. The process was very effective in re-concentrating the IL from 20 wt% initial concentration up to 80 and 99 wt% at feed temperatures of 80 and 100 °C, respectively. In addition, a very high IL recovery of 99.9 wt% was able to be achieved in the process, which is of key importance for an economically viable process. Despite the encouraging results, more efforts are still needed to improve the permeate flux, selectivity, and membrane stability.

It is important to note that in IL biomass fractionation processes, such as iono-Solv where IL and water mixtures are used, only a certain amount of water needs to be removed during the re-concentration process. In contrast, very high dehydration levels (ideally ~100%) would be needed for an IL biomass dissolution process. In the ionoSolv process, the amount of water that needs to be removed during IL re-concentration is equivalent to the amount of water added to precipitate lignin from the lignin-rich IL stream. Therefore, the first attempt to optimize the process should focus on minimizing the amount of water needed for lignin precipitation to consequently reduce the energy requirement for IL re-concentration. In the current lab-scale protocol, the water used for lignin precipitation is fixed at 3 g per g of IL solution (3 water equivalents) while preliminary data suggest that 0.5 water equivalents are sufficient to precipitate >90% of the lignin [22]. A more detailed study on the water requirement for lignin precipitation and its impact on the pretreatment performance in the ionoSolv process is underway.

To illustrate the energy intensity of the IL re-concentration step, Fig. 5.8 compares the energy requirement needed for the pretreatment reactor and IL re-concentration step using a thermal evaporation unit in the ionoSolv process (based on a preliminary techno-economic evaluation—not published).

The energy needed to increase the temperature of the reaction mixture (20 wt% biomass loading in 20 wt% H_2O in $[N_{0\,2\,2\,2}][HSO_4]$) to 150 °C and reaction time of 15 min is about 5 MJ kg^{-1} of biomass. On the other hand, re-concentrating the IL solution from 60 wt% H_2O (after lignin precipitation, assuming 1.5 water equivalents) to initial water concentration of 20 wt% requires 15 MJ kg^{-1} of biomass. The reactor portion can be heat-integrated from available process energy reducing the energy

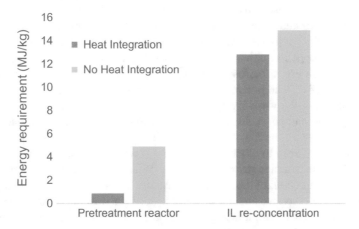

Fig. 5.8 Comparison of the energy requirement of the pretreatment reactor and IL re-concentration (thermal evaporation) in ionoSolv process with and without heat integration

requirement by ~80% (0.9 MJ kg^{-1}). However, IL re-concentration is more difficult to heat-integrate with only a 14% potential reduction in energy requirement after heat integration due to the substantial amount of energy needed during the phase change. The immense amount of energy required for IL re-concentration makes it imperative to dedicate more research efforts toward developing cost-effective and robust dehydration technologies to recover ILs from aqueous solutions with minimal IL losses.

5.6.3 Cellulose Pulp Washing

Washing the cellulose-rich pulp after the pretreatment is essential for three main reasons: (1) to reduce ionic liquid losses, (2) to remove residual inhibitory compounds, and (3) to maintain high pulp quality for downstream processing [128, 129]. For bioethanol production, the presence of small traces of an IL can inhibit commercial enzymes in the enzymatic hydrolysis unit or the microbes in the fermentation unit [129, 130]. Klein-Marcuschamer et al. tracked the residual amount of [C$_2$C$_1$im][Ace] in the recovered solid and liquid fractions to investigate the washing efficiency and its impact on the performance of commercial enzymes [100]. At high [C$_2$C$_1$im][Ace] concentrations of 16–46% (1st to 3rd water wash), the glucose yield of pretreated switchgrass did not exceed 60%. After the 5th water wash, the IL concentration dropped to 5.7% and the glucose yield increased to 80%. This highlights the importance of extensive washing to remove the residual IL. However, the extensive water washing step adds to the process of operating costs and the process capital cost for the wastewater treatment unit. The issue led to the exploration of options to reduce the water footprint, such as the development of the "single-pot" or "one-pot" wash-free process [101, 131]. The process combines the pretreatment and enzymatic saccharification in one step by using a specialty thermophilic enzyme cocktail, called JTherm, that can tolerate the presence of the ionic liquid [C$_2$C$_1$im][Ace] [132, 133]. Shi et al. showed that at 10 and 20% [C$_2$C$_1$im][Ace] concentrations and at 70 °C, JTherm retains 81 and 68% of its activity, respectively, while the activity of CTec2 commercial at 50 °C dropped significantly to 37 and 19%, respectively [134]. The authors found that the ability to perform enzymatic saccharification with the presence of 10% [C$_2$C$_1$im][Ace] can potentially reduce water use by 2–15 times compared to the conventional water wash configuration.

Although the one-pot configuration seems promising in terms of reducing the process of water consumption, the process is still in early stages of development, and the enzymes' tolerance to other ILs is still unknown. Recently, more ionic liquid-based pretreatment studies started to consider the use of IL–water mixtures instead of 100% neat ionic liquid [108, 135]. The use of IL–water mixtures should potentially reduce the excessive amount of water needed for washing prior to enzymatic saccharification. Therefore, it is important to target low initial IL concentration not only to reduce the IL use and cost in the process but also to achieve a less intensive and more cost-effective washing configuration. For example, in the ionoSolv process, it was shown that the *M. giganteus* pretreatment can be conducted with 10–40 wt% IL concentration [87], which gives larger flexibility to use a less intensive washing procedure. The lab-scale protocol of the ionoSolv process was optimized using ethanol as a washing solvent to give clean NMR spectra and to make sure that enzymatic saccharification is not inhibited by residual IL. Indeed, the practice is not representative of an industrial-scale process. Instead, a two-step washing procedure is suggested in the conceptual design of the process. The cellulose-rich pulp will be washed with a split stream of the lean hot recycled IL to remove the residual IL coating the pulp surface followed by a water wash to ensure efficient pulp washing and cleaning from inhibitory compounds. Washing with the lean hot IL stream will add three main benefits compared to a direct water wash configuration: (1) it will minimize IL losses as the residual IL will loop back to process, (2) it will minimize sending residual IL to the wastewater unit, and (3) it will potentially promote further lignin extraction from the pulp.

5.6.4 Materials of Construction

Protecting the production plant's integrity is a top priority of any industrial process as it directly impacts the safety and reliability of the operation. Selecting the appropriate materials of construction is, therefore, a key step in the process of scaling-up a technology. The selection process depends on several factors, mainly: (1) corrosion or degradation resistance of the material against processing fluids, (2) cost of the material selected and its contribution to the total capital cost, and (3) expected life of operation and type of process (i.e., continuous vs. batch) [43]. Metals and their alloys have been the most dominant construction materials in the industry. Over the decades, academic research and industrial experience have established an in-depth understanding of corrosion behavior and the mechanism and methods of protection of metals in conventional corrosive environments, such as saline water and acidic solutions. On the other hand, research on understanding the corrosion behavior of metals in ILs is scarce with existing studies focusing mainly on fluorine-containing ILs or functionalized ILs [128, 136–138]. However, the study of several classes of ILs as corrosion inhibitors has shown promising corrosion inhibition effects [139–141].

Uerdingen et al. conducted one of the first studies to investigate the corrosion behavior of common construction metals using several ILs [142]. The study highlighted how the chemical structure of the IL cation and the nature of the anion have a significant impact on the IL corrosive behavior.

In a pilot or industrial-scale operation of an IL-based pretreatment process, the pretreatment reactor's material compatibility with the hot ionic liquid needs to be known in advance. Corrosion rate is expressed in millimetres of metal lost per year (mm/y) and is used to indicate the corrosion severity during operation. It is a common industrial practice to use the corrosion rate value estimated in lab-scale or pilot-scale studies of an average cost material to permit some thickness allowance in the design specifications. As a general guideline in corrosion handbooks, corrosion rates

(CR) of metals in their operational environments are classified as follows: excellent (CR < 0.05 mm/y), very good (0.05 < CR < 0.5 mm/y), satisfactory (0.5 < CR < 1.27 mm/y), unsatisfactory (1.27 mm/y < CR) [143]. In the scale-up study of IL-based biomass pretreatment, Li et al. investigated the corrosion behavior and compatibility of Hastelloy C276 as the construction material of a Parr reactor in the scale-up of an acid-assisted IL pretreatment process [144]. The Hastelloy coupons were immersed in 1-butyl-3-methylimidazolium chloride $[C_4C_1im][Cl]$ and 1-ethyl-3-methylimidazolium chloride $[C_2C_1im][Cl]$ and biomass slurry for 144 h at 140 °C. At an HCl concentration of 0.6%, corrosion rates of all coupons were <0.025 mm/y. Increasing the HCl concentration to 1.8% increased corrosion rates by two to three times; however, corrosion rates were still <0.025 mm/y. Interestingly, at both HCl concentrations, $[C_4C_1im][Cl]$ was more corrosive than $[C_2C_1im][Cl]$. However, the use of expensive construction material, such as Hastelloy, is not favorable for the process economics, and the use of a less expensive material (e.g., stainless steel) is more economical for a large-scale operation [145].

Preliminary data on the corrosivity of the $[HSO_4]$-based ILs and water mixtures used in the ionoSolv process suggest that at room temperature, the corrosion resistance of stainless steel (SS-316 and SS-304) is excellent at water contents between 0–80 wt%. However, corrosion rates of both steel grades significantly increase by an order of magnitude at 70 °C. Since pretreatment takes place at higher temperatures ($\geq$150 °C), it is unlikely that conventional steels 304/316 could be used as construction materials for the pretreatment reactor. On the contrary, stainless steel showed very high corrosion resistance in ILs with fluorinated anions, such as bis(trifluoromethylsulfonyl)imide $[NTf_2]^-$ and triflate $[OTf]^-$ [137, 142, 146]. This highlights how the chemical structure differences of ILs (e.g., protic-aprotic, acidic-basic, and hydrophilic-hydrophobic) can have a significant impact on its interaction with the construction material.

The use of very highly corrosion-resistant metals and alloys, such as tantalum, titanium, and zirconium as construction materials, is often uneconomical as these metals are several times more expensive than steel (Fig. 5.9). One way to use these exotic corrosion-resistive metals is to apply them as a coating material for pipes and equipment. For example, Tantaline® is a product line of steel coated with tantalum, which provides superior corrosion resistance performance in very harsh environments, such as hot concentrated acid.

A more cost-effective way to reduce the material-of-construction cost is to use glass or fluoropolymer liners, such as polytetrafluoroethylene (PTFE), for pipes and equipment. However, the feasibility of implementing such liners is highly dependent on the process conditions. In addition, liners might pose some problems, such as permeation of the fluid, collapse or damage of the liners, joint creep or cold flow, and/or charge build-up. Because the interaction and the compatibility of ILs with these liners or exotic metal-coated steel are unknown, a dedicated study along with discussion with vendors would be crucial prior to testing. It is clear that there is a lack of information regarding the selection of construction materials for IL scale-up studies. However, the interaction of ILs with materials has gained much less attention compared to other process aspects. Since ILs present a new solvent paradigm that is

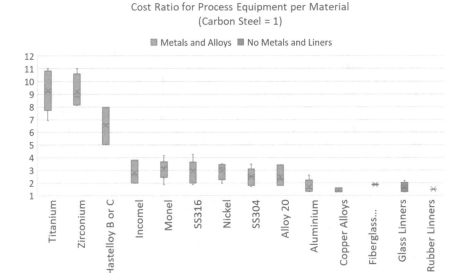

Fig. 5.9 Cost ratio of different construction material normalized to carbon steel

unknown to industry, it will be impossible to scale-up IL-based chemical processes without first establishing fundamentals of the chemical and electrochemical corrosion mechanisms in IL environments [147].

5.7   Prototype Demonstration in an Operational Environment (TRL 7)

The demonstration of a process in a real operational environment requires the consideration of several aspects beyond meeting the process technical objectives. Questions about where the technology will be applied, who the potential customers are, and how the technology will survive in the current environment and for the next 5–10 years are crucial to further process development. The answers to these questions require the development of a business model and market analysis that is also essential to evaluate competitive technologies, to estimate funding requirements, and to identify potential collaborators. The demonstration of a new chemical-based process is particularly challenging as chemical processes are inherently capital-intensive. It is, therefore, important to adapt strategies to reduce the risk associated with capital investment. Identifying an application where there is an urgent need for innovation is also a key to reduce investment risks and elevate the process market profile.

5.7.1 Reaching Economies of Scale

Most academic research is focussed on achieving high product yield as a key process success indicator. However, as we discussed earlier from different aspects, the translation of a process from a lab-scale to an industrially relevant scale depends on several economically and environmentally relevant metrices, such as solvent stability, process compactness, energy consumption, and material integrity. Development of a new process is generally capital-intensive, as piloting and demo facilities need to be built to go from the laboratory scale to an industrial scale. To reduce risk in at least the first few steps in the process, it is advantageous to consider a "bolt-on" configuration in a host established facility or alternatively conduct trials in open-access piloting facilities where critical process steps can be tested [148]. Typically, open-access facilities are run as a contracting service, where trained engineers are employed to run the tests for the customer further facilitating the scale-up process for an academic team with limited industrial experience. However, these facilities are often limited by the equipment available, which may not specifically be suited for the process purposes; this limits the possibilities of testing at open-access facilities. Also, as ILs are not yet commonly used in industry, the professionals working at these open-access facilities are usually unfamiliar with their handling, which can represent an additional challenge.

In addition, most equipment is typically made of stainless steel due to its good corrosion resistance. As discussed in Sect. 5.6, ILs have a diverse corrosion behavior based on their chemical structure and nature, and stainless steel may not be always the most compatible metal. This, in turn, requires evaluating the corrosion behavior of the IL understudy before starting the scale-up trials.

5.7.2 Finding the Right Niche

In the article "*The Grand Challenge of Cellulosic Biofuels*", Lynd diagnosed the reasons why investments in cellulosic biofuel had fallen way behind expectations [34]. Other renewable energy sectors were able to thrive even though all sectors faced the same economic conditions, especially the 2014 collapse in oil prices. A significant reason was the overestimation of the technology readiness level. For the past two decades, government, public, and private sectors were seeking big investments on mega-scale, stand-alone production facilities rather than niche applications. Raising large investments marked the beginning of divergence of reality from expectations. Meanwhile, other renewable sectors (e.g., solar and wind energy) invested in a stepwise manner, focusing on small projects where technological advancement was needed (i.e., finding a niche), the learning curve was rapid, and investment risk was minimal.

Since ILs are mostly unknown to industry, the use of ILs for cellulosic biofuel production or biomass pretreatment, in general, can be perceived as a (double) wild card.

Both the technology (the use of ILs) and intended application (biomass fractionation) have an embedded deployment risk. Although many process-related advantages have been claimed, none is yet proven at scale. Therefore, it is essential to find a currently under-served niche application where there is a clear unmet customer need that can be addressed. Niche applications should focus on using very low-cost feedstock and existing infrastructure and preferably provide an environmental and/or social solution to an existing problem. These considerations can expedite learning by doing at an industrially relevant level while still minimizing technological and investment risks. For example, if the process is first developed for the use of an unwanted waste, which is already collected at centralized locations and typically comes with a gate fee, this serves to de-risk the processing costs. Once successfully proven at a relatively small scale, which is viable for very low-cost feedstocks, the process can then be further scaled-up to a point where relatively more expensive raw materials become viable, such as sawdust or agricultural residues.

5.7.3 Toxicity and Safety Aspects

For any substance to be sold in the European Union in volumes exceeding one ton per annum (tpa), Registration, Evaluation, Authorization and Restriction of Chemicals (REACH) is required, and overseas markets will require similar registration. Therefore, careful design of the IL to reduce toxicity is paramount to facilitate the registration. However, more information on the health and environmental safety data of ILs is needed [149]. The sheer number of ionic liquids conceivable, as well as their very diverse structure, makes it impossible to give generally applicable statements about their health and safety implications. This is contrary to the popular history of ILs, as a huge part of their fame from the early 2000s until recently was hailed to the their "green credentials", which were largely attributed to their non-volatile nature. The non-volatility of many ILs can provide some processing advantages, such as relatively safer handling as no toxic fumes are produced and quantitative recovery without the need for a solvent condenser.

However, such an exaggerated and non-assessed claim was toned down in many recent IL studies for two main reasons. First is the increased awareness of the relativity concept as no solvent can be "green" in an absolute sense. The performance of the new solvent will always have to be assessed for the specifically intended application and then compared to other solvent options. The application of this relativity concept is imbedded in the life cycle assessment (LCA) approach, the standard method to evaluate the environmental impact and performance of technologies [150–152]. Currently, the lack of LCA studies to critically assess IL performance in many technologies was addressed as one of the main challenges. Second, there has been evidence that certain ILs have greater (eco)toxicity than molecular solvents [151], and there is, therefore, concern about the environmental benefits of using ILs over traditional solvents. The protic ILs used in the ionoSolv pretreatment, for example, have not been studied for their health and environmental impacts. However, they have been proven to be stable both in air and water, and their long-term stability has been confirmed well above the required process temperature range.

5.8 Conclusion and Lessons Learned

Ionic liquids have shown immense promise in academic research. They have at times been condemned for the relatively early stage of the technology and lack of past industrial application, but these facts are, as often, linked. Much development works remain to be done, but as the technical challenges shift from fundamental scientific understanding to process-oriented engineering exploitation, the research priorities will be expected to evolve in kind. For biomass pretreatment, ionic liquids have shown some unique *properties*; however, these should not be equated with *advantages*. As a simple example, cellulose decrystallization is an unusual feature of IL-based pretreatment but is not necessary for all applications and should not be treated as a pre-requisite. Selecting between amorphous cellulose and pure crystalline cellulose will depend on the application of interest and the cost of the respective process. Aspects, such as solvent cost, are multi-dimensional; the base cost of a solvent and its annual cost in a process are not identical—factors, such as stability, re-use, and loading, can be used to reduce base cost factors. A literal focus on "yield" often is counterproductive to a more realistic goal, such as unit cost.

However, a holistic view is needed to ensure a viable process option is achieved, and end applications will require such aspects to be fully considered as proof of principle to justify further process development.

Meanwhile, aspects, such as corrosion and energy usage on regeneration, have largely been ignored in the literature at the expense of a focus on the high activities that can be achieved by changing the solvent structure. Some balance is required to move the field toward implementation, but it would be remiss to ignore the simple advances in scientific understanding that can lead to step-changes in outlook. Ionic liquids are significantly diverse—there is, therefore, something to be tuned by academic synthetic chemists and industrial process engineers alike!

References

1. Beller M, Centi G, Sun L (2017) Chemistry future: priorities and opportunities from the sustainability perspective. ChemSusChem 10:6–13. https://doi.org/10.1002/cssc.201601739
2. Fulton LM, Lynd LR, Körner A, Greene N, Tonachel LR (2015) The need for biofuels as part of a low carbon energy future. Biofuels Bioprod Biorefining 9:476–483. https://doi.org/10.1002/bbb.1559
3. IPCC (2015) Foreword, preface, dedication and in memoriam. Climate change 2014: mitigation of climate change. Contribution of working group III to the fifth assessment report of the intergovernmental panel on climate change. pp. v–vi, Cambridge University Press, Cambridge. https://doi.org/10.1017/CBO9781107415416

4. Dale BE, Anderson JE, Brown RC, Csonka S, Dale VH, Herwick G, Jackson RD, Jordan N, Kaffka S, Kline KL, Lynd LR (2014) Take a closer look: biofuels can support environmental, economic and social goals. Environ Sci Technol 48:7200–7203. https://doi.org/10.1021/es5025433

5. Isikgor FH, Becer CR (2015) Lignocellulosic biomass: a sustainable platform for the production of bio-based chemicals and polymers. Polym Chem 6:4497–4559. https://doi.org/10.1039/C5PY00263J

6. Mosier N, Wyman C, Dale B, Elander R, Lee YY, Holtzapple M, Ladisch M (2005) Features of promising technologies for pretreatment of lignocellulosic biomass. Bioresour Technol 96:673–686. https://doi.org/10.1016/j.biortech.2004.06.025

7. Agbor VB, Cicek N, Sparling R, Berlin A, Levin DB (2011) Biomass pretreatment: fundamentals toward application. Biotechnol Adv 29:675–685. https://doi.org/10.1016/j.biotechadv.2011.05.005

8. Kumar AK, Sharma S (2017) Recent updates on different methods of pretreatment of lignocellulosic feedstocks: a review. Bioresour Bioprocess 4:7. https://doi.org/10.1186/s40643-017-0137-9

9. Alvira P, Tomás-Pejó E, Ballesteros M, Negro MJ (2010) Pretreatment technologies for an efficient bioethanol production process based on enzymatic hydrolysis: a review. Bioresour Technol 101:4851–4861. https://doi.org/10.1016/j.biortech.2009.11.093

10. Albers SC, Berklund AM, Graff GD (2016) The rise and fall of innovation in biofuels. Nat Biotechnol 34:814–821. https://doi.org/10.1038/nbt.3644

11. Oliveira FM, Pinheiro IO, Souto-Maior AM, Martin C, Gonçalves AR, Rocha GJ (2013) Industrial-scale steam explosion pretreatment of sugarcane straw for enzymatic hydrolysis of cellulose for production of second generation ethanol and value-added products. Bioresour Technol 130:168–173. https://doi.org/10.1016/j.biortech.2012.12.030

12. Bals B, Rogers C, Jin M, Balan V, Dale B (2010) Evaluation of ammonia fibre expansion (AFEX) pretreatment for enzymatic hydrolysis of switchgrass harvested in different seasons and locations. Biotechnol Biofuels 3:1. https://doi.org/10.1186/1754-6834-3-1

13. Saha BC, Iten LB, Cotta MA, Wu YV (2005) Dilute acid pretreatment, enzymatic saccharification and fermentation of wheat straw to ethanol. Process Biochem 40:3693–3700. https://doi.org/10.1016/j.procbio.2005.04.006

14. Zhuang X, Wang W, Yu Q, Qi W, Wang Q, Tan X, Zhou G, Yuan Z (2016) Liquid hot water pretreatment of lignocellulosic biomass for bioethanol production accompanying with high valuable products. Bioresour Technol 199:68–75. https://doi.org/10.1016/j.biortech.2015.08.051

15. Zhao X, Cheng K, Liu D (2009) Organosolv pretreatment of lignocellulosic biomass for enzymatic hydrolysis. Appl Microbiol Biotechnol 82:815–827. https://doi.org/10.1007/s00253-009-1883-1

16. Fort DA, Remsing RC, Swatloski RP, Moyna P, Moyna G, Rogers RD (2007) Can ionic liquids dissolve wood? Processing and analysis of lignocellulosic materials with 1-n-butyl-3-methylimidazolium chloride. Green Chem 9:63–69. https://doi.org/10.1039/B607614A

17. An Y-X, Zong M-H, Wu H, Li N (2015) Pretreatment of lignocellulosic biomass with renewable cholinium ionic liquids: Biomass fractionation, enzymatic digestion and ionic liquid reuse. Bioresour Technol 192:165–171. https://doi.org/10.1016/j.biortech.2015.05.064

18. Hu S, Zhang Z, Zhou Y, Han B, Fan H, Li W, Song J, Xie Y (2008) Conversion of fructose to 5-hydroxymethylfurfural using ionic liquids prepared from renewable materials. Green Chem 10:1280–1283. https://doi.org/10.1039/b810392e

19. Rogers RD, Seddon KR (2003) Ionic liquids–solvents of the future? Science 302:792–793. https://doi.org/10.1126/science.1090313

20. Plechkova NV, Seddon KR (2008) Applications of ionic liquids in the chemical industry. Chem Soc Rev 37:123–150. https://doi.org/10.1039/B006677J

21. García-Verdugo E, Altava B, Burguete MI, Lozano P, Luis SV (2015) Ionic liquids and continuous flow processes: a good marriage to design sustainable processes. Green Chem 17:2693–2713. https://doi.org/10.1039/C4GC02388A

22. Brandt-Talbot A, Gschwend FJV, Fennell PS, Lammens TM, Tan B, Weale J, Hallett JP (2017) An economically viable ionic liquid for the fractionation of lignocellulosic biomass. Green Chem 19:3078–3102. https://doi.org/10.1039/C7GC00705A

23. Jørgensen H, Pinelo M (2017) Enzyme recycling in lignocellulosic biorefineries. Biofuels Bioprod Biorefining 11:150–167. https://doi.org/10.1002/bbb.1724

24. Ragauskas AJ, Beckham GT, Biddy MJ, Chandra R, Chen F, Davis MF, Davison BH, Dixon RA, Gilna P, Keller M, Langan P, Naskar AK, Saddler JN, Tschaplinski TJ, Tuskan GA, Wyman CE (2014) Lignin valorization: improving lignin processing in the biorefinery. Science 344:1246843. https://doi.org/10.1126/science.1246843

25. Yuan T-Q, Wang W, Zhang L-M, Xu F, Sun R-C (2013) Reconstitution of cellulose and lignin after [C_2mim][OAc] pretreatment and its relation to enzymatic hydrolysis. Biotechnol Bioeng 110:729–736. https://doi.org/10.1002/bit.24743

26. Li C, Knierim B, Manisseri C, Arora R, Scheller HV, Auer M, Vogel KP, Simmons BA, Singh S (2010) Comparison of dilute acid and ionic liquid pretreatment of switchgrass: biomass recalcitrance, delignification and enzymatic saccharification. Bioresour Technol 101:4900–4906. https://doi.org/10.1016/j.biortech.2009.10.066

27. Singh S, Simmons BA, Vogel KP (2009) Visualization of biomass solubilization and cellulose regeneration during ionic liquid pretreatment of switchgrass. Biotechnol Bioeng 104:68–75. https://doi.org/10.1002/bit.22386

28. Shi J, Thompson VS, Yancey NA, Stavila V, Simmons BA, Singh S (2017) Impact of mixed feedstocks and feedstock densification on ionic liquid pretreatment efficiency. Biofuels 4:63–72. https://doi.org/10.4155/bfs.12.82

29. George A, Brandt A, Tran K, Zahari SMNS, Klein-Marcuschamer D, Sun N, Sathitsuksanoh N, Shi J, Stavila V, Parthasarathi R, Singh S, Holmes BM, Welton T, Simmons BA, Hallett JP (2015) Design of low-cost ionic liquids for lignocellulosic biomass pretreatment. Green Chem 17:1728–1734. https://doi.org/10.1039/C4GC01208A

30. Chen L, Sharifzadeh M, Mac Dowell N, Welton T, Shah N, Hallett JP (2014) Inexpensive ionic liquids: [HSO_4]$^-$-based solvent production at bulk scale. Green Chem 16:3098–3106. https://doi.org/10.1039/C4GC00016A

31. De Rose A, Buna M, Strazza C, Olivieri N, Stevens T, Peeters L, Tawil-Jamault D (2017) Technology readiness level: guidance principles for renewable energy technologies. Report EUR 27988 EN, European Commission, Brussels. https://doi.org:10.2777/577767

32. Gschwend FJV, Brandt A, Chambon CL, Tu W-C, Weigand L, Hallett JP (2016) Pretreatment of lignocellulosic biomass with low-cost ionic liquids. J Vis Exp 114:e54246. http://doi:10.3791/54246

33. Swatloski RP, Spear SK, Holbrey JD, Rogers RD (2002) Dissolution of cellulose with ionic liquids. J Am Chem Soc 124:4974–4975. https://doi.org/10.1021/ja025790m

34. Lynd LR (2017) The grand challenge of cellulosic biofuels. Nat Biotechnol 35:912–915. https://doi.org/10.1038/nbt.3976

35. Welton T (1999) Room-temperature ionic liquids. Solvents for synthesis and catalysis. Chem Rev 99:2071–2083. https://doi.org/10.1021/cr980032t

36. Dai C, Zhang J, Huang C, Lei Z (2017) Ionic liquids in selective oxidation: catalysts and solvents. Chem Rev 117:6929–6983. https://doi.org/10.1021/acs.chemrev.7b00030

37. Chen Y, Zhang X, Zhang D, Yu P, Ma Y (2011) High performance supercapacitors based on reduced graphene oxide in aqueous and ionic liquid electrolytes. Carbon 49:573–580. https://doi.org/10.1016/j.carbon.2010.09.060

38. Mood SH, Golfeshan AH, Tabatabaei M, Jouzani GS, Najafi GH, Gholami M, Ardjmand M (2013) Lignocellulosic biomass to bioethanol, a comprehensive review with a focus on pretreatment. Renew Sustain Energy Rev 27:77–93. https://doi.org/10.1016/j.rser.2013.06.033

39. Stark A (2011) Ionic liquids in the biorefinery: a critical assessment of their potential. Energy Environ Sci 4:19–32. https://doi.org/10.1039/C0EE00246A

40. Biganska O, Navard P (2009) Morphology of cellulose objects regenerated from cellulose-N-methylmorpholine N-oxide–water solutions. Cellulose 16:179–188. https://doi.org/10.1007/s10570-008-9256-y

41. Zhang Y-HP, Cui J, Lynd LR, Kuang LR (2006) A transition from cellulose swelling to cellulose dissolution by *o*-phosphoric acid: Evidence from enzymatic hydrolysis and supramolecular structure. Biomacromolecules 7: 644–648. https://doi.org/10.1021/bm050799c

42. Charles G (1933) Cellulose solution and cellulose derivative and process of making same, US patent no 1,924,238. US Patent and Trademark Office, Washington

43. Tadesse H, Luque R (2011) Advances on biomass pretreatment using ionic liquids: an overview. Energy Environ Sci 4:3913–3929. https://doi.org/10.1039/c0ee00667j

44. Michud A, Tanttu M, Asaadi S, Ma Y, Netti E, Määriainen P, Persson A, Berntsson A, Hummel M, Sixta H (2016) Ioncell-F: ionic liquid-based cellulosic textile fibers as an alternative to viscose and Lyocell. Text Res J 86:543–552. https://doi.org/10.1177/0040517515591774

45. Zhao H, Holladay JE, Brown H, Zhang ZC (2007) Metal chlorides in ionic liquid solvents convert sugar to 5-hydroxymethylfurfural. Science 316:1597–1600. https://doi.org/10.1126/science.1141199

46. van Putten RJ, van der Waal JC, de Jong ED, Rasrendra CB, Heeres HJ, de Vries JG (2013) Hydroxymethylfurfural, a versatile platform chemical made from renewable resources. Chem Rev 113:1499–1597. https://doi.org/10.1021/cr300182k

47. Werpy T, Petersen G (eds.) (2004) Top value added chemicals from biomass volume I—results of screening for potential candidates from sugars and synthesis gas. U.S. Department of Energy, National Renewable Energy Laboratory, Golden, CO. https://doi.org/10.2172/15008859

48. Eminov S, Brandt A, Wilton-Ely JDET, Hallett JP (2016) The highly selective and near-quantitative conversion of glucose to 5-hydroxymethylfurfural using ionic liquids. PLoS ONE 11:e0163835. https://doi.org/10.1371/journal.pone.0163835

49. Rosatella AA, Simeonov SP, Frade RFM, Afonso CAM (2011) 5-Hydroxymethylfurfural (HMF) as a building block platform: biological properties, synthesis and synthetic applications. Green Chem 13:754–793. https://doi.org/10.1039/c0gc00401d

50. Hou Q, Li W, Zhen M, Liu L, Chen Y, Yang Q, Huang F, Zhang S, Ju M (2017) An ionic liquid–organic solvent biphasic system for efficient production of 5-hydroxymethylfurfural from carbohydrates at high concentrations. RSC Adv 7:47288–47296. https://doi.org/10.1039/C7RA10237B

51. Lima S, Neves P, Antunes MM, Pillinger M, Ignatyev N, Valente AA (2009) Conversion of mono/di/polysaccharides into furan compounds using 1-alkyl-3-methylimidazolium ionic liquids. Appl Catal A Gen 363:93–99. https://doi.org/10.1016/j.apcata.2009.04.049

52. Greaves TL, Drummond CJ (2008) Protic ionic liquids: properties and applications. Chem Rev 108:206–237. https://doi.org/10.1021/cr068040u

53. Greaves TL, Kennedy DF, Mudie ST, Drummond CJ (2010) Diversity observed in the nanostructure of protic ionic liquids. J Phys Chem B 114:10022–10031. https://doi.org/10.1021/jp103863z

54. Deetlefs M, Seddon KR (2010) Assessing the greenness of some typical laboratory ionic liquid preparations. Green Chem 12:17–30. https://doi.org/10.1039/B915049H

55. Greaves TL, Weerawardena A, Krodkiewska I, Drummond CJ (2008) Protic ionic liquids: physicochemical properties and behavior as amphiphile self-assembly solvents. J Phys Chem B 112:896–905. https://doi.org/10.1021/jp0767819

56. Greaves TL, Weerawardena A, Fong C, Krodkiewska I, Drummond CJ (2006) Protic ionic liquids: Solvents with tunable phase behavior and physicochemical properties. J Phys Chem B Ibid. 110:26506. https://doi.org/10.1021/jp068102k

57. Yoshizawa M, Xu W, Angell CA (2003) Ionic liquids by proton transfer: vapor pressure, conductivity, and the relevance of ΔpK_a from aqueous solutions. J Am Chem Soc 125:15411–15419. https://doi.org/10.1021/ja035783d

58. Greaves TL, Drummond CJ (2015) Protic ionic liquids: evolving structure–property relationships and expanding applications. Chem Rev 115:11379–11448. https://doi.org/10.1021/acs.chemrev.5b00158

59. Penttilä A, Uusi-Kyyny P, Alopaeus V (2014) Distillable protic ionic liquid 2-(hydroxy)ethylammonium acetate (2-HEAA): density, vapor pressure, vapor–liquid equilibrium, and solid–liquid equilibrium. Ind Eng Chem Res 53:19322–19330. https://doi.org/10.1021/ie503823a
60. Earle MJ, Esperança JMSS, Gilea MA, Canongia Lopes JN, Rebelo LPN, Magee JW, Seddon KR, Widegren JA (2006) The distillation and volatility of ionic liquids. Nature 439:831–834. https://doi.org/10.1038/nature04451
61. Menne S, Vogl T, Balducci A (2014) Lithium coordination in protic ionic liquids. Phys Chem Chem Phys 16:5485–5489. https://doi.org/10.1039/c3cp55183k
62. Mayrand-Provencher L, Lin S, Lazzerini D, Rochefort D (2010) Pyridinium-based protic ionic liquids as electrolytes for RuO_2 electrochemical capacitors. J Power Sour 195:5114–5121. https://doi.org/10.1016/j.jpowsour.2010.02.073
63. Kanzaki R, Kodamatani H, Tomiyasu T, Watanabe H, Umebayashi Y (2016) A pH scale for the protic ionic liquid ethylammonium nitrate. Angew Chem Int Ed 55:6266–6269. https://doi.org/10.1002/anie.201511328
64. Kanzaki R, Doi H, Song X, Hara S, Ishiguro S, Umebayashi Y (2012) Acid–base property of N–methylimidazolium-based protic ionic. J Phys Chem B 116:14146−14152. https://doi.org/10.1021/jp308477p
65. Hashimoto K, Fujii K, Shibayama M (2013) Acid–base property of protic ionic liquid, 1-alkylimidazolium bis(trifluoromethanesulfonyl)amide studied by potentiometric titration. J Mol Liq 188:143–147. https://doi.org/10.1016/j.molliq.2013.08.023
66. da Costa Lopes AM, João KG, Morais ARC, Bogel-Łukasik E, Bogel-Łukasik R (2013) Ionic liquids as a tool for lignocellulosic biomass fractionation. Sustain Chem Process 1:3. https://doi.org/10.1186/2043-7129-1-3
67. Kautto J, Realff MJ, Ragauskas AJ (2013) Design and simulation of an organosolv process for bioethanol production. Biomass Convers Biorefinery 3:199–212. https://doi.org/10.1007/s13399-013-0074-6
68. Cherubini F (2010) The biorefinery concept: using biomass instead of oil for producing energy and chemicals. Energy Convers Manag 51:1412–1421. https://doi.org/10.1016/j.enconman.2010.01.015
69. Chen H, Liu J, Chang X, Chen D, Xue Y, Liu P, Lin H, Han S (2017) A review on the pretreatment of lignocellulose for high-value chemicals. Fuel Process Technol 160:196–206. https://doi.org/10.1016/j.fuproc.2016.12.007
70. Kilpelainen I, Xie H, King A, Granstrom M, Heikkinen S, Argyropoulos DS (2007) Dissolution of wood in ionic liquids. J Agric Food Chem 55:9142–9148. https://doi.org/10.1021/jf071692e
71. Brandt A, Gräsvik J, Hallett JP, Welton T (2013) Deconstruction of lignocellulosic biomass with ionic liquids. Green Chem 15:550–583. https://doi.org/10.1039/c2gc36364j
72. Sun N, Rahman M, Qin Y, Maxim ML, Rodríguez H, Rogers RD (2009) Complete dissolution and partial delignification of wood in the ionic liquid 1-ethyl-3-methylimidazolium acetate. Green Chem 11:646–655. https://doi.org/10.1039/b822702k
73. Cheng G, Varanasi P, Li C, Liu H, Melnichenko YuB, Simmons BA, Kent MS, Singh S (2011) Transition of cellulose crystalline structure and surface morphology of biomass as a function of ionic liquid pretreatment and its relation to enzymatic hydrolysis. Biomacromol 12:933–941. https://doi.org/10.1021/bm101240z
74. Wang Y, Zhang L, Zhang R, Liu G, Cheng G (2014) Understanding changes in cellulose crystalline structure of lignocellulosic biomass during ionic liquid pretreatment by XRD. Bioresour Technol 151:402–405. https://doi.org/10.1016/j.biortech.2013.10.009
75. Trinh LTP, Lee YJ, Lee J-W, Lee H-J (2015) Characterization of ionic liquid pretreatment and the bioconversion of pretreated mixed softwood biomass. Biomass Bioenerg 81:1–8. https://doi.org/10.1016/j.biombioe.2015.05.005
76. Remsing RC, Hernandez G, Swatloski RP, Massefski WW, Rogers RD, Moyna G (2008) Solvation of carbohydrates in N, N'-dialkylimidazolium ionic liquids: a multinuclear NMR spectroscopy study. J Phys Chem B 112:11071–11078. https://doi.org/10.1021/jp8042895

77. Fukaya Y, Hayashi K, Wada M, Ohno H (2008) Cellulose dissolution with polar ionic liquids under mild conditions: required factors for anions. Green Chem 10:44–46. https://doi.org/10.1039/B713289A

78. Fukaya Y, Sugimoto A, Ohno H (2006) Superior solubility of polysaccharides in low viscosity, polar and halogen-free 1,3-dialkylimidazolium formates. Biomacromol 7:3295–3297. https://doi.org/10.1021/bm060327d

79. Liu H, Sale KL, Holmes BM, Simmons BA, Singh S (2010) Understanding the interactions of cellulose with ionic liquids: a molecular dynamics study. J Phys Chem B 114:4293–4301. https://doi.org/10.1021/jp9117437

80. Ab Rani MA, Brant A, Crowhurst L, Dolan A, Lui M, Hassan NH, Hallett JP, Hunt PA, Niedermeyer H, Perez-Arlandis JM, Schrems M (2011) Understanding the polarity of ionic liquids. Phys Chem Chem Phys 13:16831–16840. https://doi.org/10.1039/c1cp21262a

81. Rosatella AA, Afonso CAM (2015) The dissolution of biomass in ionic liquids towards pretreatment approach. In: Bogel-Lukasik R (Ed.) Ionic liquids in the biorefinery concept, RSC Green Chemistry No. 36, The Royal Society of Chemistry, pp 38–64. https://doi.org/10.1039/9781782622598-00038

82. Upfal J, MacFarlane DR, Forsyth SA (2007) Solvents for use in the treatment of lignin-containing materials. US patent application no 2007/0215300A1

83. Varanasi S, Schall CA, Dadi AP, Anderson J, Rao K, Kumar G, Paripati P (2011) Biomass pretreatment. US patent no 8,030,030

84. Lee SH, Doherty TV, Linhardt RJ, Dordick JS (2009) Ionic liquid-mediated selective extraction of lignin from wood leading to enhanced enzymatic cellulose hydrolysis. Biotechnol Bioeng 102:1368–1376. https://doi.org/10.1002/bit.22179

85. Pu Y, Jiang N, Ragauskas AJ (2007) Ionic liquid as a green solvent for lignin. J Wood Chem Technol 27:23–33. https://doi.org/10.1080/02773810701282330

86. Tan SS, MacFarlane DR, Upfal J, Edye LA, Doherty WOS, Patti AF, Pringle JM, Scott JL (2009) Extraction of lignin from lignocellulose at atmospheric pressure using alkylbenzenesulfonate ionic liquid. Green Chem 11:339–345. https://doi.org/10.1039/b815310h

87. Brandt A, Ray MJ, To TQ, Leak DJ, Murphy RJ (2011) Ionic liquid pretreatment of lignocellulosic biomass with ionic liquid–water mixtures. Green Chem 13:2489–2499. https://doi.org/10.1039/c1gc15374a

88. Brandt-Talbot A, Murphy RJ, Leak DJ, Welton T, Hallett J (2017) Treatment of biomass to dissolve lignin with ionic liquid. US patent no 9,765,478

89. Sheldon R (2001) Catalytic reactions in ionic liquids. Chem Commun 2399–2407. https://doi.org/10.1039/B107270F

90. Supap T, Idem R, Tontiwachwuthikul P (2011) Mechanism of formation of heat stable salts (HSSs) and their roles in further degradation of monoethanolamine during CO_2 capture from flue gas streams. Energy Procedia 4:591–598. https://doi.org/10.1016/j.egypro.2011.01.093

91. Axelsson L, Franzén M, Ostwald M, Berndes G, Lakshmi G, Ravidranath NH (2012) Perspective: Jatropha cultivation in southern India: assessing farmers' experiences. Biofuels Bioprod Biorefining 6:246–256. https://doi.org/10.1002/bbb.1324

92. Fox DM, Awad WH, Gilman JW, Maupin PH, De Long HC, Trulove PC (2003) Flammability, thermal stability, and phase change characteristics of several trialkylimidazolium salts. Green Chem 5:724–727. https://doi.org/10.1039/B308444B

93. Cao Y, Mu T (2014) Comprehensive investigation on the thermal stability of 66 ionic liquids by thermogravimetric analysis. Ind Eng Chem Res 53:8651–8664. https://doi.org/10.1021/ie5009597

94. Chiaramonti D, Prussi M, Ferrero S, Oriani L, Ottonello P, Torre P, Cherchi F (2012) Review of pretreatment processes for lignocellulosic ethanol production, and development of an innovative method. Biomass Bioenerg 46:25–35. https://doi.org/10.1016/j.biombioe.2012.04.020

95. Subramaniam B, Helling RK, Bode CJ (2016) Quantitative sustainability analysis: a powerful tool to develop resource-efficient catalytic technologies. ACS Sustain Chem Eng 4:5859–5865. https://doi.org/10.1021/acssuschemeng.6b01571

96. Achinivu EC, Howard RM, Li G, Gracz H, Henderson WA (2014) Lignin extraction from biomass with protic ionic liquids. Green Chem 16:1114–1119. https://doi.org/10.1039/C3GC42306A

97. Maton C, De Vos N, Stevens CV (2013) Ionic liquid thermal stabilities: decomposition mechanisms and analysis tools. Chem Soc Rev 42:5963–5977. https://doi.org/10.1039/c3cs60071h

98. Hallett JP, Welton T (2011) Room-temperature ionic liquids: solvents for synthesis and catalysis. 2. Chem Rev 111:3508–3576. https://doi.org/10.1021/cr1003248

99. Sanders JPM, Clark JH, Harmsen GJ, Heeres HJ, Heijnen JJ, Kersten SR, van Swaaij WPM, Moulijn JA (2012) Process intensification in the future production of base chemicals from biomass. Chem Eng Process Process Intensif 51:117–136. https://doi.org/10.1016/j.cep.2011.08.007

100. Klein-Marcuschamer D, Simmons BA, Blanch HW (2011) Techno-economic analysis of a lignocellulosic ethanol biorefinery with ionic liquid pre-treatment. Biofuels Bioprod Biorefining 5:562–569. https://doi.org/10.1002/bbb.303

101. Murthy Konda NVSN, Shi J, Singh S, Blanch HW, Simmons BA, Klein-Marcuschamer D (2014) Understanding cost drivers and economic potential of two variants of ionic liquid pretreatment for cellulosic biofuel production. Biotechnol Biofuels 7:86. https://doi.org/10.1186/1754-6834-7-86

102. Humbird D, Davis R, Tao L, Kinchin C, Hsu D, Aden A, Schoen P, Lukas J, Olthof B, Worley M, Sexton S, Dudgeon D (2011) Process design and economics for biochemical conversion of lignocellulosic biomass to ethanol, NREL/TP-5100-4776 Technical Report

103. Ebel K, Koehler H, Gamer AO, Jäckh R (2011) Imidazole and derivatives. Ullmann's encyclopedia of industrial chemistry, vol 1. Wiley-VCH, New York, pp 131–139

104. Jessop PG (2011) Searching for green solvents. Green Chem 13:1391–1398. https://doi.org/10.1039/c0gc00797h

105. Clarke CJ, Tu W-C, Levers O, Bröhl A, Hallett JP (2018) Green and sustainable solvents in chemical processes. Chem Rev 118:747–800. https://doi.org/10.1021/acs.chemrev.7b00571

106. Tao L, Aden A, Elander RT, Pallapolu VR, Lee YY, Garlock RJ, Balan V, Dale BE, Kim Y, Mosier NS, Ladisch MR, Falls M, Holtzapple MT, Sierra R, Shi J, Ebrik MA, Redmont T, Yang B, Wyman CE, Hames B, Thomas S, Warner RE (2011) Process and technoeconomic analysis of leading pretreatment technologies for lignocellulosic ethanol production using switchgrass. Bioresour Technol 102:11105–11114. https://doi.org/10.1016/j.biortech.2011.07.051

107. Kazi FK, Fortman JA, Anex RP, Hsu DD, Aden A, Dutta A, Kothandaraman G (2010) Techno-economic comparison of process technologies for biochemical ethanol production from corn stover. Fuel 89:S20–S28. https://doi.org/10.1016/j.fuel.2010.01.001

108. Gschwend FJV, Malaret F, Shinde S, Brandt-Talbot A, Hallett JP (2016) Rapid pretreatment of *Miscanthus* using the low-cost ionic liquid triethylammonium hydrogen sulfate at elevated temperatures. Green Chem 20:3486–3498. https://doi.org/10.1039/C8GC00837J

109. Shinde SD, Meng X, Kumar R, Ragauskas AJ (2018) Recent advances in understanding the pseudo-lignin formation in a lignocellulosic biorefinery. Green Chem 20:2192–2205. https://doi.org/10.1039/C8GC00353J

110. Li W, Sun N, Stoner B, Jiang X, Lu X, Rogers RD (2011) Rapid dissolution of lignocellulosic biomass in ionic liquids using temperatures above the glass transition of lignin. Green Chem 13:2038–2047. https://doi.org/10.1039/c1gc15522a

111. Arora R, Manisseri C, Li C, Ong MD, Scheller HV, Vogel K, Simmons BA, Singh S (2010) Monitoring and analyzing process streams towards understanding ionic liquid pretreatment of switchgrass (*Panicum virgatum* L.). Bioenergy Res 3:134–145. https://doi.org/10.1007/s12155-010-9087-1

112. Modenbach AA, Nokes SE (2013) Enzymatic hydrolysis of biomass at high-solids loadings—a review. Biomass Bioenerg 56:526–544. https://doi.org/10.1016/j.biombioe.2013.05.031

113. Samaniuk JR, Scott CT, Root TW, Klingenberg DJ (2012) Rheological modification of corn stover biomass at high solids concentrations. J Rheol 56:649–665. https://doi.org/10.1122/1.3702101

114. Papa G, Feldman T, Sale KL, Adani F, Singh S, Simmons BA (2017) Parametric study for the optimization of ionic liquid pretreatment of corn stover. Bioresour Technol 241:627–637. https://doi.org/10.1016/j.biortech.2017.05.167

115. Cruz AG, Scullin C, Mu C, Cheng G, Stavila V, Varanasi P, Xu D, Mentel J, Chuang Y-D, Simmons BA, Singh S (2013) Impact of high biomass loading on ionic liquid pretreatment. Biotechnol Biofuels 6:52. https://doi.org/10.1186/1754-6834-6-52

116. Wu H, Mora-Pale M, Miao J, Doherty TV, Linhardt RJ, Dordick JS (2011) Facile pretreatment of lignocellulosic biomass at high loadings in room temperature ionic liquids. Biotechnol Bioeng 108:2865–2875. https://doi.org/10.1002/bit.23266

117. Auxenfans T, Buchoux S, Larcher D, Husson G, Husson E, Sarazin C (2014) Enzymatic saccharification and structural properties of industrial wood sawdust: Recycled ionic liquids pretreatments. Energy Convers Manag 88:1094–1103. https://doi.org/10.1016/j.enconman.2014.04.027

118. Qiu Z, Aita GM (2013) Pretreatment of energy cane bagasse with recycled ionic liquid for enzymatic hydrolysis. Bioresour Technol 129:532–537. https://doi.org/10.1016/j.biortech.2012.11.062

119. da Costa Lopes AM, João KG, Rubik DF, Bogel-Łukasik E, Duarte LC, Andreaus J, Bogel-Łukasik R (2013) Pre-treatment of lignocellulosic biomass using ionic liquids: wheat straw fractionation. Bioresour Technol 142:198–208. https://doi.org/10.1016/j.biortech.2013.05.032

120. Badgujar KC, Bhanage BM (2015) Factors governing dissolution process of lignocellulosic biomass in ionic liquid: current status, overview and challenges. Bioresour Technol 178:2–18. https://doi.org/10.1016/j.biortech.2014.09.138

121. Sun J, Shi J, Murthy Konda NVSN, Campos D, Liu D, Nemser S, Shamshina J, Dutta T, Berton P, Gurau G, Rogers RD, Simmons BA, Singh S (2017) Efficient dehydration and recovery of ionic liquid after lignocellulosic processing using pervaporation. Biotechnol Biofuels 1:154. https://doi.org/10.1186/s13068-017-0842-9

122. Mai NL, Ahn K, Koo Y-M (2014) Methods for recovery of ionic liquids—a review. Process Biochem 49:872–881. https://doi.org/10.1016/j.procbio.2014.01.016

123. Reid JESJ, Walker AJ, Shimizu S (2015) Residual water in ionic liquids: clustered or dissociated? Phys Chem Chem Phys 17:14710–14718. https://doi.org/10.1039/C5CP01854D

124. Mateyawa S, Xie DF, Truss RW, Halley PJ, Nicholson TM, Shamshina JL, Rogers RD, Boehm MW, McNally T (2013) Effect of the ionic liquid 1-ethyl-3-methylimidazolium acetate on the phase transition of starch: Dissolution or gelatinization? Carbohydr Polym 94:520–530. https://doi.org/10.1016/j.carbpol.2013.01.024

125. Hoerning A, Ribeiro FRG, Cardozo Filho L, Lião LM, Corazza M, Voll FAP (2016) Boiling point elevation of aqueous solutions of ionic liquids derived from diethanolamine base and carboxylic acids. J Chem Thermodyn 98:1–8. https://doi.org/10.1016/j.jct.2016.02.017

126. Haerens K, Van Deuren S, Matthijs E, Van der Bruggen B (2010) Challenges for recycling ionic liquids by using pressure driven membrane processes. Green Chem 12:2182–2188. https://doi.org/10.1039/c0gc00406e

127. Lynam JG, Chow GI, Coronella CJ, Hiibel SR (2016) Ionic liquid and water separation by membrane distillation. Chem Eng J 288:557–561. https://doi.org/10.1016/j.cej.2015.12.028

128. Li Y, Zhang S, Ding Q, Feng D, Qin B, Hu L (2017) The corrosion and lubrication properties of 2-mercaptobenzothiazole functionalized ionic liquids for bronze. Tribol Int 114:121–131. https://doi.org/10.1016/j.triboint.2017.04.022

129. Turner MB, Spear SK, Huddleston JG, Holbrey JD, Rogers RD (2003) Ionic liquid salt-induced inactivation and unfolding of cellulase from *Trichoderma reesei*. Green Chem 5:443–447. https://doi.org/10.1039/b302570e

130. Ganske F, Bornscheuer UT (2006) Growth of *Escherichia coli*, *Pichia pastoris* and *Bacillus cereus* in the presence of the ionic liquids [BMIM][BF$_4$] and [BMIM][PF$_6$] and organic solvents. Biotechnol Lett 28:465–469. https://doi.org/10.1007/s10529-006-0006-7

131. Xu F, Sun J, Murthy Konda NVSN, Shi J, Dutta T, Scown CD, Simmons BA, Singh S (2016) Transforming biomass conversion with ionic liquids: process intensification and the

development of a high-gravity, one-pot process for the production of cellulosic ethanol. Energy Environ Sci 9:1042–1049. https://doi.org/10.1039/C5EE02940F

132. Gladden JM, Allgaier M, Miller CS, Hazen TC, VanderGheynst JS, Hugenholtz P, Simmons BA, Singer SW (2011) Glycoside hydrolase activities of thermophilic bacterial consortia adapted to switchgrass. Appl Environ Microbiol 77:5804–5812. https://doi.org/10.1128/AEM.00032-11

133. Park JI, Steen EJ, Burd H, Evans SS, Redding-Johnson AM, Batth T, Benke PI, D'haeseleer P, Sun N, Sale KL, Keasling JD, Lee TS, Petzold CJ, Mukhopadhyay A, Singer SW, Simmons BA, Gladden JM (2012) A thermophilic ionic liquid-tolerant cellulase cocktail for the production of cellulosic biofuels. PLoS One 7:e37010. https://doi.org/10.1371/journal.pone.0037010

134. Shi J, Gladden JM, Sathitsuksanoh N, Kambam P, Sandoval L, Mitra D, Zhang S, George A, Singer SW, Simmons BA, Singh S (2013) One-pot ionic liquid pretreatment and saccharification of switchgrass. Green Chem 15:2579–2589. https://doi.org/10.1039/c3gc40545a

135. Shi J, Balamurugan K, Parthasarathi R, Sathitsuksanoh N, Zhang S, Stavila V, Subramanian V, Simmons BA, Singh S (2014) Understanding the role of water during ionic liquid pretreatment of lignocellulose: co-solvent or anti-solvent? Green Chem 16:3830–3840. https://doi.org/10.1039/C4GC00373J

136. Dilasari B, Jung Y, Sohn J, Kim S, Kwon K (2016) Review on corrosion behavior of metallic materials in room temperature ionic liquids. Int J Electrochem Sci 11:1482–1495. https://doi.org/10.1021/acssuschemeng.5b00974

137. Ma Y, Han F, Li Z, Xia C (2016) Corrosion behavior of metallic materials in acidic-functionalized ionic liquids. ACS Sustain Chem Eng 4:633–639. https://doi.org/10.1021/acssuschemeng.5b00974

138. Guo Y, Xu B, Liu Y, Yang W, Yin X, Chen Y, Le J, Chen Z (2017) Corrosion inhibition properties of two imidazolium ionic liquids with hydrophilic tetrafluoroborate and hydrophobic hexafluorophosphate anions in acid medium. J Ind Eng Chem 56:234–247. https://doi.org/10.1016/j.jiec.2017.07.016

139. Verma C, Ebenso EE, Quraishi MA (2017) Ionic liquids as green and sustainable corrosion inhibitors for metals and alloys: an overview. J Mol Liq 233:403–414. https://doi.org/10.1016/j.molliq.2017.02.111

140. Zhang QB, Hua YX (2009) Corrosion inhibition of mild steel by alkylimidazolium ionic liquids in hydrochloric acid. Electrochim Acta 54:1881–1887. https://doi.org/10.1016/j.electacta.2008.10.025

141. Kannan P, Karthikeyan J, Murugan P, Rao TS, Rajendran N (2016) Corrosion inhibition effect of novel methyl benzimidazolium ionic liquid for carbon steel in HCl medium. J Mol Liq 221:368–380. https://doi.org/10.1016/j.molliq.2016.04.130

142. Uerdingen M, Treber C, Balser M, Schmitt G, Werner C (2005) Corrosion behaviour of ionic liquids. Green Chem 7:321–325. https://doi.org/10.1039/b419320m

143. Schweitzer PA (2004) Corrosion resistance tables: metals, nonmetals, coatings, mortars, plastics, elastomers and linings, and fabrics. CRC Press, Boca Raton

144. Li C, Liang L, Sun N, Thompson VS, Xu F, Narani A, He Q, Tanjore D, Pray TR, Simmons BA, Singh S (2017) Scale-up and process integration of sugar production by acidolysis of municipal solid waste/corn stover blends in ionic liquids. Biotechnol Biofuels 10:13. https://doi.org/10.1186/s13068-016-0694-8

145. Shekiro J III, Kuhn EM, Nagle NJ, Tucker MP, Elander RT, Schell DJ (2014) Characterization of pilot-scale dilute acid pretreatment performance using deacetylated corn stover. Biotechnol Biofuels 7:23. https://doi.org/10.1186/1754-6834-7-23

146. Kermani NA, Petrushina I, Nikiforov A, Jensen JO, Rokni M (2016) Corrosion behavior of construction materials for ionic liquid hydrogen compressor. Int J Hydrogen Energy 41:16688–16695. https://doi.org/10.1016/j.ijhydene.2016.06.221

147. Dilasari B, Jung Y, Kwon K (2016) Comparative study of corrosion behavior of metals in protic and aprotic ionic liquids. Electrochem Commun 73:20–23. https://doi.org/10.1016/j.elecom.2016.10.009

148. Lynd LR, Liang X, Biddy MJ, Allee A, Cai H, Foust T, Himmel ME, Laser MS, Wang M, Wyman CE (2017) Cellulosic ethanol: status and innovation. Curr Opin Biotechnol 45:202–211. https://doi.org/10.1016/j.copbio.2017.03.008
149. Coleman D, Gathergood N (2010) Biodegradation studies of ionic liquids. Chem Soc Rev 39:600–637. https://doi.org/10.1039/b817717c
150. Righi S, Morfino A, Galletti P, Samorì C, Tugnoli A, Stramigioli C (2011) Comparative cradle-to-gate life cycle assessments of cellulose dissolution with 1-butyl-3-methylimidazolium chloride and N-methyl-morpholine-N-oxide. Green Chem 13:367–375. https://doi.org/10.1039/C0GC00647E
151. Zhang Y, Bakshi BR, Demessie ES (2008) Life cycle assessment of an ionic liquid versus molecular solvents and their applications. Environ Sci Technol 42:1724–1730. https://doi.org/10.1021/es0713983
152. Tao L, Tan ECD, Aden A, Elander RT (2014) Techno-economic analysis and life-cycle assessment of lignocellulosic biomass to sugars using various pretreatment technologies. In: Sun J, Ding S-Y, Doran-Peterson J (eds.) Biological conversion of biomass for fuels and chemicals: Exploration from natural utilization systems. RSC Energy and Environment Series 10, The Royal Society of Chemistry, pp. 358–380. https://doi.org/10.1039/9781849734738-00358

Part III
Ionic Liquid Products

Chapter 6
Gas Chromatography Columns Using Ionic Liquids as Stationary Phase

Mohsen Talebi, Rahul A. Patil and Daniel W. Armstrong

Abstract Ionic liquids satisfy the requirements of a gas chromatography stationary phase, among which characteristics include high viscosity, tunable selectivity through structural modifications, good wettability with respect to fused silica capillaries, and high thermal stability. When incorporated in either the first or second dimension of multidimensional gas chromatography, they offer unique selectivity compared to conventional gas chromatography stationary phases. The utilization of commercial ionic liquid columns for analysis of various analytes (i.e., fatty acids, flavors and fragrances, organic pollutants, and petrochemical samples) in complex matrices is described in this chapter. Thanks to their dual-nature behavior, IL-based stationary phases provide unique separation for both polar and nonpolar molecules in complex samples, such as essential oils. Moderately polar phosphonium-based ionic liquid columns (i.e., SLB-IL59, SLB-IL60, and SLB-IL61) with high operating temperature are suitable for analysis of complex petrochemical and environmental samples. The high polarity of ionic liquid columns allows analysis of positional isomers including those of unsaturated fatty acids. The other exceptional feature of ionic liquid columns is their stability in the presence of water and oxygen at high temperatures. Water-compatible ionic liquid-based gas chromatography capillary columns facilitate the direct injection of aqueous samples without the requirement of time-consuming sample pretreatment techniques.

Keywords Dicationic ionic liquids · Gas chromatography · Ionic liquid capillary columns · Polarity · Thermal stability

6.1 Introduction

Simple inorganic salts have very high melting points, and this limits their applications as solvents for chemical applications. Salts with bulky organic cations and inorganic or organic anions generally melt at lower temperatures, and those with melting points

M. Talebi · R. A. Patil · D. W. Armstrong (✉)
Department of Chemistry and Biochemistry, University of Texas at Arlington, Arlington, TX, USA
e-mail: sec4dwa@uta.edu

© Springer Nature Switzerland AG 2020   131
M. B. Shiflett (ed.), *Commercial Applications of Ionic Liquids*, Green Chemistry and Sustainable Technology, https://doi.org/10.1007/978-3-030-35245-5_6

less than 100 °C are known as ionic liquids (ILs) [1]. Salts which are liquid at room temperature (≤25 °C) are known as "room-temperature ionic liquids (RTILs)" [2]. ILs have "unique" properties, such as low volatility, non-flammability, wide liquid range, broad solubility properties, and high thermal stability, which make them versatile compounds that are useful in both academia and industry [3].

One of the crucial applications of ILs is their use as stationary phases for gas chromatography (GC) [4–9]. Molten salts which were stearate salts of bivalent metals were used as GC stationary phases for the first time by Barber et al. in 1959 [10]. Later, quaternary ammonium and phosphonium salts, such as ethylammonium nitrate, ethylpyridinium bromide, and tetraalkyl phosphonium salts, were used as GC stationary phases for packed and open tubular formats [11–15]. However, these ILs showed narrow liquid ranges, low column efficiencies, and poor thermal stabilities, which limited their practical applications as GC stationary phases. Toward the end of the last century, monocationic ILs with an imidazolium-based cation were developed and evaluated as GC stationary phases by Armstrong and coworkers [4, 5]. These stationary phases showed improved performance compared to the previously reported molten-salt stationary phases. These monocationic IL stationary phases were studied in detail, and some fundamental understanding was developed during these studies [16]. Further, a new class of dicationic ILs was synthesized and evaluated as GC stationary phases [6–8, 17]. This generation of ILs showed improved thermal stabilities, higher viscosities, and excellent performance as GC stationary phases compared to the monocationic ILs. The continued research led to the invention of tricationic IL stationary phases having performance akin to dicationic ILs [9]. This new generation of dicationic, tricationic, and even tetracationic ILs was shown to be excellent GC stationary phases, and many were commercialized by Supelco (now MilliporeSigma) beginning in 2008. These stationary phases showed very high thermal stabilities and more opportunities for structural modification [7–9]. The molecular structures of the commercial IL stationary phases in the columns along with their allowable operating temperatures are given in Table 6.1.

The dicationic and tricationic ILs have properties that make them unique and highly useful as GC stationary phases. These ILs show higher thermal stability than comparable traditional polarity columns containing polymeric stationary phases [18]. Low volatility helps in low column bleeding at high temperatures and makes ILs particularly useful for GC-MS applications [19]. ILs show good wettability, which makes it easier to coat them on the inner walls of fused silica capillary columns [4]. High viscosity is beneficial to keep the IL coated on the inner walls of the fused silica capillary column, especially at high temperatures [20, 21]. ILs have a wide liquid range, which makes them useful for performing gas–liquid chromatography over a wide temperature range [4, 17].

In addition, IL stationary phases show many advantages over traditional GC stationary phases. ILs show multiple solvation interactions, while traditional stationary phases often show one dominant type of interaction [16]. ILs show an unusual dual-nature behavior, separating both polar and nonpolar compounds through a wide range of interaction mechanisms [4, 22]. They perform like nonpolar stationary phases while separating nonpolar compounds and at the same time retain polar analytes

Table 6.1 Chemical structures of commercial ionic liquid columns

IL column	Temperature limits	IL molecular structure and IUPAC name
SLB-IL59	Subambient to 300 °C	
SLB-ILPAH	Subambient to 300 °C	
SLB-IL60	35–300 °C	1,12-Di(tripropylphosphonium)dodecane bis(trifluoromethanesulfonyl)imide
SLB-IL60i	35–280 °C	
SLB-IL61	40–290 °C	1,12-Di(tripropylphosphonium)dodecane bis(trifluoromethanesulfonyl)imide trifluoromethanesulfonate
SLB-IL76	Subambient to 270 °C	
SLB-IL76i	Subambient to 270 °C	Tri(tripropylphosphoniumhexanamido)triethylamine bis(trifluoromethanesulfonyl)imide
SLB-IL82	50–270 °C	1,12-Di(2,3-dimethylimidazolium)dodecane bis(trifluoromethanesulfonyl)imide
SLB-IL100	Subambient to 230 °C	1,9-Di(3-vinylimidazolium)nonane bis(trifluoromethanesulfonyl)imide
SLB-IL111	50–270 °C	
SLB-IL111i	50–260 °C	
SLB-ILD3606	50–260 °C	1,5-Di(2,3-dimethylimidazolium)pentane bis(trifluoromethanesulfonyl)imide
Watercol 1460	30–260 °C	Tri(tripropylphosphoniumhexanamido)triethylamine trifluoromethanesulfonate

(continued)

Table 6.1 (continued)

IL column	Temperature limits	IL molecular structure and IUPAC name
Watercol 1910	30–180 °C	 1,11-Di(3-hydroxyethylimidazolium)3,6,9-trioxaundecane trifluoromethanesulfonate
Watercol 1900	30–180 °C	 1,11-Di(3-methylimidazolium)3,6,9-trioxaundecane trifluoromethanesulfonate

showing polar behavior [18]. These properties provide unique selectivities for the separation of a wide variety of compounds. IL stationary phases show high thermal stability along with high polarity [23], while in case of traditional phases, polarity is achieved at the cost of thermal stability (e.g., the "highly polar" 1,2,3-tris(2-cyanoethoxy)propane (TCEP) stationary phase has an upper temperature limit of 145 °C). IL phases are air and moisture stable compared to the polymeric phases used in traditional GC columns. This gives them tremendous advantages in the analysis of water and water-based samples [24–26]. The structure of dicationic ILs can be extensively modified by structural variations in order to optimize and "fine-tune" their physicochemical properties, selectivities, and polarities [20, 21, 27]. Different structural variables that affect the behavior of ILs are discussed in the following sections.

Dicationic ILs can be considered as a combination of three structural moieties: (1) cationic head groups, (2) a linker or spacer chain, and (3) the associated anions [20]. The typical structure of dicationic ILs is shown in Fig. 6.1. The terminal cationic head group often consists of imidazolium, phosphonium, or pyrrolidinium groups. Also, the cationic head groups can have different side chain substituents (Fig. 6.1). Different cationic head groups can be used at the two ends of the linkage chains, and such ILs are known as "unsymmetrical" ILs [27]. Dicationic ILs with the same terminal cationic groups are known as "symmetrical" or "geminal" dicationic ILs [17, 20]. The two cationic head groups can be connected by linkage chains of different

Fig. 6.1 General structure of dicationic ionic liquids

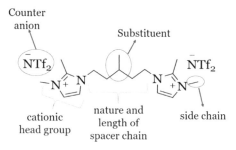

lengths [17, 27]. Different linkages, such as alkane, polyethylene glycol, and partially fluorinated carbon chains, can be used to connect cationic head groups [7, 28]. Instead of a straight chain linkage, branched linkage chains can be used as a "tether," and they can have different substituents [21]. Such ILs are known as branched-chain dicationic ILs [21, 29]. Hence, different cations can be synthesized by using different combinations of these structural moieties, and then, these cations can be paired with different anions: bis(trifluoromethanesulfonyl)imide ($[NTf_2]^-$), perfluorooctanesulfonate ($[PFOS]^-$), triflate ($[OTf]^-$), hexafluorophosphate ($[PF_6]^-$), among others [17, 23]. The structure–property relationship of ILs is studied by systematic variation of structural moieties and examining changes in their physicochemical properties [20, 21, 23]. Understanding the effects of different structural modifications on physicochemical properties is essential if one wants to introduce or vary any desired property of an IL.

6.2 Selectivity of Ionic Liquid Stationary Phases

Physicochemical characterization of ionic liquids is essential to understand the effects of different structural variations and their behavior when used as stationary phases. One type of characterization is known as "inverse GC" in which different analyte probes are used to understand the solvent properties of GC stationary phases at different temperatures [8, 16]. Different models and methods used for characterization of IL GC stationary phases are explained in the following three sections.

6.2.1 Rohrschneider–McReynolds Parameters

Early on, Rohrschneider and McReynolds developed a system to characterize and compare different stationary phases used in GC [30, 31]. For characterization, the retention behavior of five "informative" probe analytes is compared on different IL stationary phases. The probes are benzene (X) for π–π interactions, 1-butanol (Y) for hydrogen-bond donor and acceptor interactions, 2-pentanone (Z) for weak hydrogen-bond acceptor and dipole interactions, nitropropane (U) for polar interactions, and pyridine (S) for basic or strong hydrogen-bond acceptor interactions. Kovats retention indices of all five probe molecules are determined on the IL GC stationary phase to be characterized and a squalane stationary phase column of the same dimensions at the same isothermal temperature. The McReynolds constants for the probes represent the difference between the Kovats index of the probe on the IL stationary phase and the Kovats index of the same probe on the squalane stationary phase. The overall or average polarity (P) of the IL stationary phase is measured by taking the sum of five McReynolds constants. The polarity number (PN) of an IL stationary phase is calculated by Eq. 6.1, in which $P_{\text{SLB-IL100}}$ is the average polarity of

Table 6.2 Normalized McReynolds constants, overall polarity P, and polarity numbers PN, for commercial IL stationary phases

IL column	McReynolds constants					P	PN
	Benzene (X)	1-Butanol (Y)	2-Pentanone (Z)	Nitropropane (U)	Pyridine (S)		
SLB-IL59	338	505	549	649	583	2624	59
SLB-IL60	362	492	525	679	564	2622	59
SLB-IL61	371	551	516	624	648	2710	61
SLB-IL76	456	690	643	845	745	3379	76
SLB-IL82	532	676	701	921	808	3638	82
SLB-IL100	602	853	884	1017	1081	4437	100
SLB-IL111	766	930	957	1192	1093	4938	111

All the data is taken from manufacturer's website: Supelco Ionic Liquid GC Columns (2013) http://www.sigmaaldrich.com/content/dam/sigma-aldrich/countries/czech/gc_kolony.pdf

1,9-di(3-vinylimidazolium)nonane bis(trifluoromethylsulfonyl)imide (SLB-IL100) IL GC stationary phase [32].

$$PN = \left(\frac{P_x}{P_{SLB\text{-}IL100}} \right) \cdot 100 \qquad (6.1)$$

The McReynolds constants and polarity numbers for all the commercial IL phases are provided in Table 6.2.

The Rohrschneider–McReynolds approach uses only five analytes to discriminate five different types of interactions of ILs. The small number of analytes limits the ability of this scale to define each solvation parameter accurately.

6.2.2 Abraham Linear Solvation Energy Relationship

Abraham designed a solvation parameter model to characterize either liquid- or gas-phase interactions between solute molecules and solvents [33]. The model known as the linear solvation energy relationship (LSER) uses different analyte probes capable of numerous interactions to characterize the stationary phase by inverse GC [34, 35]. The use of multiple probe analytes in the LSER provides information on the interactions of solute molecules with the IL stationary phases. The relationship between the properties of solute and stationary phase (or solute–solvent) is given by Eq. 6.2.

$$\log K = c + r R_2 + s \pi_2^{H} + a \sum \alpha_2^{H} + b \sum \beta_2^{H} + l \log L^{16} \qquad (6.2)$$

Alternatively, the LSER model can also be represented by Eq. 6.3.

$$\log k = c + eE + sS + aA + bB + lL \tag{6.3}$$

Here, K and k are the partition coefficient and retention factor, respectively. The solute descriptors: E and R_2, S and π_2^H, A and α_2^H, B and β_2^H, and L and L^{16} represent the excess molar refraction, dipolarity, H-bond acidity, H-bond basicity, and gas–hexadecane partition coefficient at 25 °C, respectively. The solute descriptor values for more than 300 solutes are reported in the literature [33]. The phase constants e, s, a, b, and l are a measure of the ability for the stationary phase or solvent to interact with analyte or solute via π or nonbonding electrons, dipole–dipole interactions, H-bond basicity, H-bond acidity, or dispersion forces, respectively. The equation constant (c) does not define any fundamental property, but when the $\log k$ is used as the dependent variable, it is dominated by the phase ratio for the column. The retention factors (k) for every probe analyte/solute are measured experimentally, and the phase constants are calculated by subjecting the data set to multiple linear regression analysis (MLRA). Interaction parameters obtained from the LSER model for commercial IL stationary phases are given in Fig. 6.2.

The LSER evaluation of IL stationary phases is performed at different isothermal temperatures (60, 100 °C, etc.). IL stationary phases usually show dominant dipolarity (s), hydrogen-bond basicity (a), and dispersion (l) interactions. Dispersion forces are nearly constant for all the commercial IL stationary phases, and their values are lower than for the nonpolar polysiloxane-based stationary phases [36]. The most significant interaction of ILs, hydrogen-bond basicity (a), is mainly dominated by

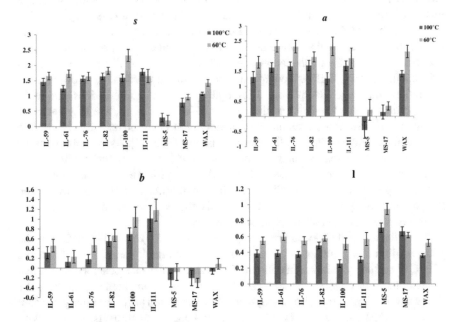

Fig. 6.2 Calculated values for phase parameters s, a, b, and l. MS-5, 5% phenyl PDMS; MS-17, 50% phenyl PDMS; WAX, cross-bonded polyethylene glycol. Reprinted with permission [36]

the anion part of the IL [16]. Anions with high nucleophilicity (e.g., halides) show high (a) values compared to the less nucleophilic anions (e.g., $[PF_6]^-$, $[NTf_2]^-$, [16]. All the commercial IL stationary phases (SLB-IL series) contain the $[NTf_2]^-$ anion, and hence, they show similar (a) values [36]. However, their (a) values are higher compared to the polysiloxane-based nonpolar GC stationary phases [36]. The dipolarity (s) is weakly dependent on the type of cation, linkage chain, and anion [16, 36]. All the SLB-IL series stationary phases showed similar (s) values, only IL-100 being slightly higher than others [36]. The hydrogen-bond acidity (b) of SLB-IL phases was observed to increase with an increase in polarity (except SLB-IL59) [36]. The polysiloxane-based and wax stationary phases do not possess significant hydrogen-bond acidity (b) values compared to IL phases of all polarities [36].

6.2.3   Characterization Using Test Mixtures

Retention behavior, separation efficiencies, and selectivities of IL stationary phases can be observed by analyzing different test mixtures composed of a variety of compounds having different functional groups and polarities. This procedure is mainly followed in the early development of new IL GC stationary phases, and it is a quick method to determine the effect of structural modifications on the selectivities and polarities of IL stationary phases. Different test mixtures, such as the Grob mixture, fatty acid methyl esters (FAME) isomer mixture, and polar compound/alkane mixtures, are used for stationary phase characterization studies [7–9, 18].

The Grob test mixture consists of 12 different components: dicyclohexylamine, methyl decanoate, methyl undecanoate, methyl laurate, decane, undecane, 1-nonanal, 1-octanol, 2-ethylhexanoic acid, 2,3-butanediol, 2,6-dimethylaniline, and 2,6-dimethylphenol. Different components in this mixture are useful to evaluate the separation efficiency, adsorptive activity, hydrogen bonding interactions, and inertness (acid/base characteristic) of the column [37, 38]. The mixture of polar compounds (alcohols, amines, ketone, etc.) and straight-chain alkanes can be used to evaluate the relative polarity of IL stationary phases [7, 9, 18]. As the polarity of the stationary phase increases, the relative retention of polar compounds with respect to the nonpolar alkanes increases. This quick method is useful in designing new stationary phases and understanding the influence of different structural variables on the polarities of IL phases. An example of this approach to examine the effects of cations on the polarities of ILs is shown in Fig. 6.3.

The polar test probe 2-octanone (compound 1 in Fig. 6.3) elutes before C15 on the phosphonium IL column, while it elutes after C16 on the imidazolium column and after C17 on the pyrrolidinium column. The elution order of 2-octanone shows that ILs with the tripropylphosphonium cation are less polar compared to analogous ILs with a dimethylimidazolium or methylpyrrolidinium cation [18]. The same procedure is used to investigate the effects of linkage chain lengths, substituents, and anions on the polarities of ILs.

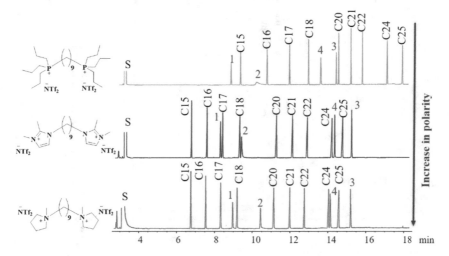

Fig. 6.3 Polarity comparison of ionic liquid stationary phases containing different cationic head groups. (1) 2-octanone, (2) 1-octanol, (3) 2,6-dimethylaniline, (4) 2,6-dimethylphenol

A simple eleven-component rapeseed fatty acid methyl esters (FAME) mixture also can be used to evaluate IL structural variation on the selectivities and polarities of these stationary phases [18, 23, 29]. In case of FAMEs, the relative retention of methyl stearate (C18:0), methyl oleate (C18:1n9), methyl linoleate (C18:2n6), methyl linolenate (C18:3n3), methyl arachidate (C20:0), and *cis*-11-eicosenoic acid methyl ester (C20:1) is mainly observed for comparison of stationary phases. The C18:3 FAME is more polarizable compared to the C20:1 and C20:0 FAMEs. In general, the relative retention of C20:0 with respect to C18:3 decreases with an increase in polarity of the stationary phase. An example showing the analysis of FAMEs on two columns with different linkage chain lengths is shown in Fig. 6.4. The relative retention of C20:0 with respect to C18:3 is decreased on the IL with the larger C12 linkage chain compared to the analogous IL with a C9 linkage. This shows that the polarity of an IL increases with a decrease in length of the linkage chain. The Kovats retention indices (KRIs) of the three C18 FAMEs are also monitored for the stationary phase polarities. The KRIs of the three C18 FAMEs are higher on the more polar stationary phases compared to the stationary phases of lower polarity. Another method for gaining useful information on the stationary phases by using FAMEs is known as the equivalent chain length (*ECL*), and this will be discussed in detail in Sect. 6.3.1.

6.3   Analysis of Fatty Acids

Fatty acids (FAs) are recognized as one of the most important classes of lipids because of their biofunctional significance in living organisms [39, 40]. Before analysis, fatty

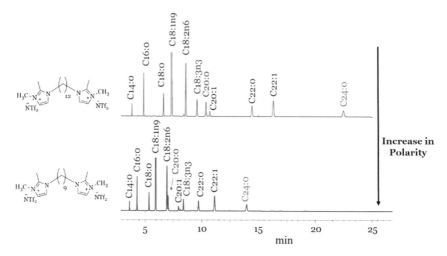

Fig. 6.4 Effect of linker chain length on the polarity and selectivity of ionic liquid stationary phases

acids are often converted to their methyl esters (FAMEs) through a transesterification reaction. This is done to reduce the polarity of carboxyl functional groups and to increase their volatility by decreasing strong intermolecular interactions (i.e., hydrogen-bonding interactions) between fatty acid molecules, which lowers their boiling points. The resulting methyl esters offer high stability that can be effectively and quantitatively analyzed by GC [41]. GC along with flame ionization detection (FID) or mass spectrometry (MS) is generally used for analysis of FAMEs [42, 43]. Although MS is a powerful tool for peak identification, it is somewhat difficult to adequately identify FA isomers that have identical molecular mass using MS alone. Of course, FID, a nonselective detector, is not capable of identifying unknown peaks in complex samples. Thus, high chromatographic resolution and selectivity is still required to interpret MS data reliably [44].

The natural occurrence of saturated and unsaturated fatty acids with many isomers means that a large variety of selected compounds have to be analyzed in biological and food samples. Also, possible *cis/trans* geometrical configurations of double bonds add further complexity to the separation and analysis of these samples. Considering the slight difference between the *cis* and *trans* FAME isomers of the same carbon length and degree of unsaturation, very efficient capillary GC columns composed of highly selective stationary phases are required to achieve adequate separation [42–47]. Accordingly, the official American Oil Chemists' Society (AOCS Method Ce 1 h-05) suggests the use of a 100 m poly(biscyanopropyl siloxane) column operating isothermally at 180 °C [48]. Additionally, a prior fractionation of fatty acids by silver-ion chromatography is required to achieve better identification of *cis* and *trans* isomers [43, 45, 49–52].

The invention of IL columns has provided a remarkable opportunity for gas chromatographic separation of FAMEs [53]. Utilization of the extremely polar SLB-IL111 column eliminates the need for complimentary silver-ion fractionation and allows faster separation of FAMEs [42, 43]. Herein, the applicability of IL-based columns for FAMEs separation and their unique selectivity to distinguish between positional FAME isomers are discussed.

6.3.1 Selectivity

Highly selective IL stationary phases are particularly adept in analyzing complete mixtures of fatty acids [18, 42–44, 46, 53–60]. Generally, the following behaviors can be observed when separating FAMEs on polar ionic liquid columns: (1) shorter analysis times than traditional GC stationary phases, (2) increased retention with greater degrees of unsaturation, (3) elution of *trans* FAME isomers before *cis* isomers, and (4) longer retention of unsaturated FAMEs with the double bond located closer to the hydrocarbon chain terminus (e.g., ω-3 fatty acids tend to retain longer than ω-6 isomers) [29, 57].

Retention patterns on IL columns are greatly dependent on chromatographic conditions. For example, the elution orders of fatty acids can sometimes be reversed with a slight change in column temperature. Therefore, the peak identification should be verified when working with IL columns at different temperatures, especially if a nonselective detector (e.g., FID) is utilized [47, 60].

Equivalent chain length (*ECL*) is the most established method to assess the selectivity of columns based on retention of unsaturated FAMEs. It corresponds to the theoretical fractional saturated chain length that would result in the same retention that is observed for the unsaturated FAME [47, 61, 62]. The *ECL* value can be calculated using the two saturated FAMEs that bracket the peak of the unsaturated FAME [29]. A simplified equation for calculation of *ECL* values is shown in Eq. 6.4.

$$ECL(C_n:x) = n + 2 \frac{\log(t'_R(C_n:x)) - \log(t'_R(C_n:0))}{\log(t'_R(C_{n+2}:0)) - \log(t'_R(C_n:0))} \tag{6.4}$$

Here, n represents the FAME chain length, x is the number of double bonds, and t'_R is the adjusted retention time. A significant increase in the *ECL* values with increasing column polarity can be expected for IL columns, specifically in case of polyunsaturated FAMEs [29, 44, 47]. Moreover, determination of *ECL* values is particularly important for the evaluation of overlap patterns on different stationary phases. For instance, the FAMEs' elution order is identical for the IL59, IL60, and IL61 columns [44]. Several overlaps between nutritionally important fatty acids occur on these columns limiting their application for clinical and nutritional studies of the fatty acids components [47]. On the other hand, the highly polar SLB-IL111 with the greatest *ECL* values among commercially available columns offers exceptional selectivity for the separation of unsaturated FAMEs.

6.3.2 Biodiesel

Biodiesel, an alternatively derived fuel, is obtained from different biological sources of fatty acids. Animal fats, vegetable oils, recyclable cooking oils, and plant and waste products are usually processed for extraction of fatty acid methyl esters [63]. A petroleum-based diesel is often blended with a biomass-based diesel before consumer use. Due to the overlap between the petroleum-based compounds of saturated and aromatic hydrocarbons in blended biodiesel, the FAME profile elucidation is difficult to see using traditional GC stationary phases [64–69]. The importance of fatty acid ester analysis in biofuel and fuel samples is twofold. First, the analysis is used to monitor cross-contamination in tankers and storage facilities used for multiple fuel types [64]. This is specifically important in case of fuels that are used in aviation and other specialized fuels employed by defense forces where small amounts of FAME contamination contribute to poor thermal and storage stability of the fuel [70]. Second, the analysis is used to investigate the influence of FAME chemical structure on exhaust emissions [68]. Studies have shown that the hydrocarbon chain length and degree of unsaturation can affect the carbon monoxide (CO) levels and nitrogen oxide (NO_x) emissions from combustion of fatty esters [71–73]. According to the UNI EN 14331 procedure, a hydrocarbon liquid chromatography (LC) preseparation step is required before gas chromatographic determination of the FAME profile [74]. This procedure can be time- and labor-intensive. Therefore, a simple, rapid, and less expensive approach to effectively separate the FAMEs from the petroleum-based portion appears to be necessary. Fast analysis of biodiesel can be achieved on SLB-IL100 and SLB-IL-111 micro-bore columns [65, 68]. The key feature of these stationary phases, namely extremely high polarity, allows excellent separation of the FAME components from the less-retained petroleum-based hydrocarbons. Figure 6.5 shows that the analysis of the soybean B20 biodiesel blend on the micro-bore SLB-IL100 (12 m × 0.1 mm × 0.08 μm d_f) column resulted in the separation of compounds before 2.5 min. Particularly, the methyl palmitate (C16:0) eluted after C26 (n-hexacosane), the last significant n-alkane in the sample [65]. Biodiesel samples obtained through the transesterification reaction of different vegetable oils were analyzed utilizing a customized narrow-bore SLB-IL111 (14 m × 0.1 mm × 0.08 μm d_f) column [68]. This fast method could be used to quantify the FAME content in basic and acidic transesterified vegetable oils and to control the quality of biodiesel using routine analyses.

6.3.3 Biological Samples

The *Rhodobacter sphaeroides* is a rod-shaped, gram negative, purple non-sulfur bacterium utilized as a model microorganism for anoxygenic photosynthesis and carbon fixation [46, 56, 75–80]. The use of a 100 m SLB-IL111 column allowed effective separation of geometrical isomers of monounsaturated C16:1 and C18:1

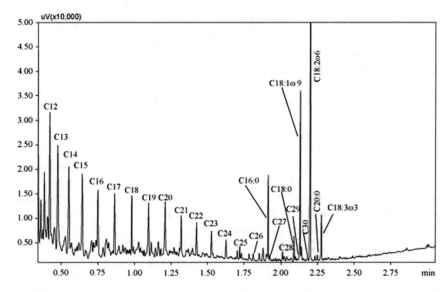

Fig. 6.5 Analysis of the soybean B20 biodiesel blend on the SLB-IL100 12 m column. Reprinted with permission from Ref. [65]

fatty acids extracted from *R. sphaeroides* 2.4.1 [56]. An unprecedented separation of oleic acid (C18:1 *cis-*Δ^9) from the most predominant *cis*-vaccenic acid (C18:1 *cis-*Δ^{11}) isomer was achieved using the highly polar IL column. In addition, the unusual occurrence of 11-methyl-*trans-*Δ^{12}-octadecanoic acid was demonstrated in the lepidic matrix of *R. sphaeroides*. A metabolic pathway was postulated to support these findings.

The 100 m SLB-IL111 could also be used to characterize the fatty acids found in human sebum, hair, and nail lipids [81]. A baseline separation between petroselinic acid (C18:1 *cis-*Δ^6) and C18:1 *cis-*Δ^8 was achieved in samples of human origin. Thanks to the incredible selectivity of the IL column, a series of Δ^6-monounsaturated fatty acids, namely C14:1 *cis-*Δ^6, C15:1 *cis-*Δ^6, i-C16:1 *cis-*Δ^6, C16:1 *cis-*Δ^6, a-C17:1 *cis-*Δ^6, C17:1 *cis-*Δ^6, and C18:1 *cis-*Δ^6 were identified in these samples. The occurrence of these Δ^6-monounsaturated fatty acids was then explained by a biosynthetic pathway.

6.3.4 Food Samples

Determining the fatty acid composition of a food product may be challenging because foods contain a complex mixture of saturated (SFA), monounsaturated (MUFA), and polyunsaturated (PUFA) fatty acids with a variety of hydrocarbon chain lengths. Furthermore, the presence of *trans* fatty acids in processed foods, containing partially hydrogenated oils, interferes with the natural metabolic process. This necessitates the determination by food manufacturers of the levels of each type of fatty acid, as each has different known or suspected health effects [42, 43, 46, 77, 82].

The SLB-IL111 column with a length of 100 m or 200 m is usually used for detailed analysis of fatty acids in food samples [42, 43, 58, 59]. The exceptionally polar characteristic of a SLB-IL111 column allows separation of key *cis/trans* FAME isomers that are hard to resolve on other columns. It also provides a complementary elution profile of FAMEs typically separated on the biscyanopropyl siloxane columns [42]. Excellent performance of the SLB-IL111 was shown in separation of *cis/trans* conjugated linoleic acid (CLA) isomers, where *c*9, *t*11-CLA, and *t*7, *c*9-CLA, the most abundant CLA isomers in ruminant fats, were completely resolved from each other [42]. Octadecenoic acid (C18:1) is the foremost constituent of dietary fats and oils with distributions of 26 *cis* and *trans* isomers having various double bond positions in the range of Δ4–16. Oleic acid (C18:1 *cis-*Δ^9), the most naturally abundant isomer, is quantifiable with the SLB-IL111 column separation and not with any other commercial column due to coelution with other C18:1 isomers [83]. The major *trans* C18:1 fatty acid in ruminant-derived foods, especially milk fat, is *trans* vaccenic acid (*t*VA, *t*11-C18:1). Studies have shown that *t*11-C18:1 is beneficial for the human body, and it is a precursor for conjugated linoleic acids (CLA). *Trans*-10 is another notable *trans* C18:1 isomer, which unlike *trans*-11 is not known to provide any useful health benefits in humans [84–87]. While using the official AOCS method results in coelution of *trans* 9–11 C18:1 isomers, most of the *trans* FAME geometrical isomers including *t*10-C18:1 and *t*11-C18:1 can be separated using the SLB-IL111 column [43, 46]. The SLB-IL111 is also proficient at separating odd and branched-chain fatty acids (OBCFAs) from other milk FAs eluting in the same chromatographic region [88]. These FAs are the chief lipids in bacterial membranes and are considered as biomarkers of ruminant fat intake [89]. By definition, the branching on the *iso*-methyl-branched fatty acids is positioned on the penultimate carbon, while the methyl substituent in *anteiso*-branched fatty acids is located on the *ante*-penultimate carbon atom. The elution behavior of *iso* fatty acids on the SLB-IL111 is very similar to the conventional polymer stationary phases. The *iso* FA with *n* C-atoms elutes at a carbon number (*CN*) of (n–1 + 0.5). Accordingly, *iso*-C13:0 elutes between C12:0 and C13:0 fatty acids (or $CN = 12.5$). On the other hand, the *aiso*, a FA containing *n* C-atoms, retains slightly longer than the *iso* isomer, at *CN* of (n–1 + 0.7). Furthermore, unlike *cis* and *trans* FA isomers, the relative retention times are not affected for *iso* and *aiso* isomers by changing the isothermal program on ionic liquid columns [88].

The 200 m SLB-IL111 column was further used for determination of total, *trans*, saturated, and *cis* unsaturated fats in 32 representative fast food samples [46]. The content of *trans* fat ranged from 0.1 to 3.1 g per serving, as determined according to American Oil Chemists' Society (AOCS) official method Ce 1j-07. The improved separation of *trans* C18:1 positional isomers and enhanced resolution of *trans* C18:2 and C18:3 FAME isomers are two major advantages of the SLB-IL111 column, relative to the biscyanopropyl siloxane stationary phases.

The fatty acid composition in marine oils and products is very complex with a wide variety of chain length and unsaturation [47]. The SLB-IL111 column has been successfully used for separation of extracted fats from marine products. Using this IL column along with time-of-flight (TOF) mass spectrometry, the separation and accurate identification of 125 fatty acids in menhaden oil, including two novel branched fats having a *trans* configuration, namely 7-methyl-6-hexadecenoic and 7-methyl-6-octadecenoic fatty acids, were achieved. The presence of small amounts of t6-17:1 and t8-18:1 suggested that t6-18:1 may be chain-elongated in fish [59]. In addition, analysis of fatty acids in marine oil omega-3 supplements on a 200 m SLB-IL111 column permitted quantification of rarely reported fatty acids, such as C21:5n3 and the n-4 and n-1 polyunsaturated fatty acid (PUFA). The presence of predominant mono-*trans* isomers of eicosapentaenoic acid (EPA; C20:5n3) and docosahexaenoic acid (DHA; C22:6n3) was confirmed in these samples [90]. The SLB-IL100 was tested for analysis of *cis* positional isomers of eicosenoic (C20:1) and docosenoic (C22:1) fatty acids of fish origin. The high polarity of SLB-IL100 allowed separation of several isomers including C20:1n11 from C20:1n13, which are unresolvable on conventional polar polymer phase columns [91].

Steryl esters are a minor lipid class in animal and plant lipids with high molecular mass (600–700 g/mol) that are usually analyzed by reverse-phase high-performance liquid chromatography (RP-HPLC) [92–94]. However, several coelutions may occur between steryl esters using C18 or hexyl-phenyl-modified HPLC stationary phases. Also, high elution temperatures required for gas chromatographic analysis of these compounds raise challenges to separate fatty acids present in the sterol backbones efficiently [95]. A 12 m SLB-IL59 column was employed to separate fatty acid esters of cholesterol and phytosterols. The high temperature limit (300 °C) and medium polarity of SLB-IL59 allow separation of steryl esters both by total carbon number and by the degree of unsaturation [92].

Application of the recently developed vacuum ultraviolet (VUV) detector along with a SLB-IL111 column offers additional selectivity in the characterization of fats and edible oils [76, 96]. The gas-phase VUV absorption spectra are significantly selective to the molecular structure of analytes. VUV provides completely different absorption profiles for unsaturated and saturated FAMEs. In addition, *cis* and *trans* isomers can be easily distinguished in VUV due to their distinct spectra.

6.3.5   Multidimensional GC Separations

Ionic liquid columns have been used both in the first and second dimension of two-dimensional gas chromatography (GC × GC) to enhance the separation of FAMEs in complex matrices [58, 64, 69, 70, 97–103]. The major advantage of using GC × GC with IL columns is to obtain higher peak capacity, hence enhancing the resolution between positional FAME isomers. The arrangement of highly polar SLB-IL100 in the first dimension along with an intermediate polar column in the second dimension resulted in the effective separation of C18:1 isomers [103]. The SLB-IL82 and SLB-IL100 were evaluated for one-dimensional and two-dimensional GC analyses of fatty acids in marine biota [99]. Conventional GC analysis showed that these columns provide similar selectivity to polar biscyanopropyl siloxane stationary phases. However, when employed in the second dimension, an incomparable group-type separation of FAMEs with different carbon numbers and degrees of unsaturation could be obtained using ionic liquid columns. The degree of orthogonality of ionic liquid columns in comprehensive two-dimensional gas chromatography was evaluated using a 1-phase GC × GC system. Among the commercial IL columns, SLB-IL111 showed the best results regarding thermal sensitivity and degree of difference over the tested temperature range [98]. The orthogonality between the two separations is usually achieved using GC columns of different polarities. The greatest spread of FAMEs in 2D space can be achieved using a combination of the highest (IL111) and the lowest polarity (IL59) columns. This configuration was found to be effective for separation and quantification of safflower oil FAs [100]. Using SLB-IL111 in both dimensions along with a capillary tube coated with palladium between ^{1}D and ^{2}D offered exceptional separation of poorly resolved FAMEs that could not be achieved by conventional GC [58]. Instead of employing two GC columns of different polarities or using different elution temperatures, separation could be performed by modification of FAMEs' chemical structure between the two dimensions. All the saturated FAMEs lie on a straight diagonal line bisecting the separation plane. The FAMEs with the same carbon skeleton but a different degree of unsaturation, the position of double bonds, or geometric configuration lie on lines parallel to the ^{1}D time axis. Thanks to their upper temperature limit of 300 °C, SLB-IL60 and SLB-IL61 were effectively applied in the second dimension of GC × GC to analyze biodiesel samples [64]. Using the IL phase along with a nonpolar phase provided adequate orthogonality to identify suspected low-level biodiesel contamination and resolved semi-polar FAME compounds from the nonpolar and aromatic hydrocarbons. Also, chemical markers of biodiesel sources were identified and characterized based on their FAME composition.

Furthermore, a recent generation of inert ionic liquid (iIL) columns was examined for separation of FAMEs [102]. A modified surface treatment of iIL phases results in improved surface interaction and reduced adsorption toward polar or active compounds. Although similar retention times with minor selectivity can be observed using iIL phases, the average tailing factors and peak widths were significantly reduced compared to the corresponding conventional IL columns. Also, a polar/nonpolar column set was found to provide the best separation of FAMEs.

6.4   Analysis of Essential Oils

Essential oils are major constituents of flavors and fragrances and are widely used in pharmaceutical, cosmetic, and food industries. The analysis of essential oils is essential for quality control (purity determination, adulteration, etc.) and the determination of natural or synthetic origin. Essential oils are complex mixtures consisting of a large number and variety of compounds with different physicochemical properties. The compounds include hydrocarbons (monoterpenes, sesquiterpenes, and aromatics), oxygenated compounds (aromatics, phenols, alcohols, sesquiterpene alcohols, aldehydes, ketones, esters, lactones, ethers, oxides, etc.), and other compounds. Most of the compounds are isomers or analogs of each other and generate the same mass spectra. Hence, chromatographic separation of these compounds is important. The separation demands stationary phases with broad selectivities. Traditional wax (polyethylene glycol-based) and cyanopropyl polysiloxane-based phases are used for the analysis of essential oils. However, these stationary phases produce coelutions of many of the compounds. Multidimensional GC can also be used for the analysis of such complex mixtures, but such approaches involve operational complexity and expensive instrumentation. However, the IL GC stationary phases show dual-nature behavior and broad selectivities, which provide them advantages in the separation of flavors and fragrances. Applications of different IL GC stationary phases in the analysis of essential oils are discussed in the following section. Benefits of new inert ionic liquid (iIL) GC stationary phases in the analysis of essential oils are also considered.

The advantages of IL stationary phases in the separation of flavor and fragrance compounds were observed during the initial development of dicationic IL GC stationary phases [7–9]. Furthermore, a dicationic IL 1,9-di(3-vinylimidazolium)nonane bis(trifluoromethanesulfonyl)imide $[C_9(vim)_2][NTf_2]_2$, now known as SLB-IL100 was used in the analysis of fennel, cinnamon, and nutmeg essential oils [104]. This study involved the use of four different IL columns: (1) a (1:1:1) mixture of dicationic IL $[C_9(vim)_2][NTf_2]_2$, a monocationic IL 1-vinyl-3-nonylimidazolium bis[(trifluoromethyl)sulfonyl]imide $[vC_9im][NTf_2]$, and OV-1701 polysiloxane, (2) the dicationic $[C_9(vim)_2][NTf_2]_2$ IL, (3) a 5% phenyl-methylpolysiloxane (HP-5 MS) column, and (4) a polyethylene glycol (HP-Innowax) column. Comparison of the analyses of three essential oils on the four columns showed that column 1 (a mixture of dicationic IL and monocationic IL) showed better efficiency and wider selectivity compared to the other three columns. Column 1 had better selectivity in the separation of both nonpolar and polar compounds compared to column 2, and not all the hydrocarbons in the essential oils were separated on column 2. The dicationic IL is polar (polarity number 100) and showed lower retention of nonpolar compounds and retained the polar compounds longer. However, the polysiloxane part of column 1 played a role in the better separation of hydrocarbons. Also, the dicationic IL column displayed an affinity for oxygen-containing compounds, while the polyethylene glycol stationary phase retained the "hydrogen-bond acidic" compounds (i.e., alcohols). This study used a polar IL $[C_9(vim)_2][NTf_2]_2$, $(PN = 100)$ and showed coelution for some hydrocarbons.

Later, a medium-polarity commercialized column SLB-IL59 $(PN = 59)$ was used in the analysis of essential oils and it showed a better separation of essential oils components (both polar and nonpolar compounds) [32]. Separations of four test compounds (p-cymene, 1,8-cineole, limonene, and (Z)-β-ocimene) were compared on three columns (5% phenyl PDMS, PEG, and SLB-IL59). The PEG column and

SLB-IL59 column have similar polarities, but they showed complimentary selectivities for (Z)-β-ocimene and 1,8-cineole. All four test compounds were separated with better resolution on the SLB-IL59 column compared to the 5% phenyl PDMS column. Also, a big difference in the elution order was observed on the two phases; on 5% phenyl PDMS, the elution order was p-cymene < limonene < (Z)-β-ocimene < 1,8-cineole, while on SLB-IL59 the order was limonene < (Z)-β-ocimene < 1,8-cineole < p-cymene. Peak symmetry and capacity factors (k) of the IL and wax columns were evaluated by using different test compounds representing a specific class of compounds: limonene for monoterpene hydrocarbons, linalool for monoterpene alcohols, (E)-caryophyllene for sesquiterpene hydrocarbons, neral and geranial for monoterpene aldehydes, and neryl acetate for monoterpene esters. Lower capacity factors (k) were observed for sesquiterpene hydrocarbons, monoterpene hydrocarbons, alcohols, and esters on the SLB-IL59 column compared to the WAX column. Comparable k values were observed for monoterpene aldehydes on both compared phases. When the SLB-IL59 column was evaluated for the analysis of lemon essential oil, all the 41 components of the oil were separated and could be identified. However, neryl acetate and β-bisabolene coeluted on the WAX column, and p-cymene, limonene, 1,8-cineole, and (Z)-β-ocimene coeluted on the 5% PDMS column. This shows the overall better performance of the SLB-IL59 column in the separation of lemon essential oil compared to polar WAX and nonpolar phases.

Furthermore, commercial IL columns were examined for the analyses of a 29 allergen mixture and the performance was compared to the OV-1701 column [105]. The SLB-IL76 column was not a good choice for the analysis because the separation resulted in broad peaks with low efficiency. The highly polar SLB-IL100 and 111 columns showed lower retention of volatile, nonpolar components. The SLB-IL59 column showed a higher average peak width compared to the other columns. SLB-IL61 column performed better in the separation of alcohol components. The separation capability of IL columns was determined by calculating the separation measure value (Δs) [105, 106].

$$\Delta s = \int_{t_A}^{t_B} \frac{dt}{\sigma} \qquad (6.5)$$

σ is the peak width and the term Δs defined as the number of consecutive non-overlapping σ intervals within an arbitrary time interval ($t_B - t_A$). The separation measure of a time interval is the number of adjacent non-overlapping σ-slots in it or the number of consecutive non-overlapping σ-intervals within an arbitrary time interval. A higher separation capacity was observed for the IL columns compared to conventional columns (OV-1701:1582) and the capacity increased from IL-59 (1516) to IL82 (1787). This highlighted the ability of IL columns to separate a greater number of compounds.

6.5   Analysis of Environmental Pollutants

Phthalates or phthalic acid esters (PAEs) and adipates or adipic acid esters are mainly used as plasticizers in the manufacturing of plastic products [107, 108]. Phthalates are not biodegradable, and they have adverse effects on humans as they lead to congenital disabilities, heart diseases, and respiratory problems and they are endocrine disruptors [109–112]. In addition to humans, they cause reproductive and fertility problems in wildlife [113]. The Environmental Protection Agency (EPA) demands strict control of the use and monitoring of toxic phthalates. GC-MS is a widely utilized technique for the analysis of PAEs as lower sensitivity is obtained with LC-MS [113]. Chromatographic separation of phthalates is important because identification and quantification by MS can be difficult due to the structural similarity of phthalates (most phthalates have a common base peak ion at m/z 149) [114].

IL stationary phases have broad selectivity toward a variety of compounds and were evaluated for the separation of phthalates. Three toxic adipates and eight phthalates were analyzed on the commercial SLB-IL59, SLB-IL76, SLB-IL82, SLB-IL100, and SLB-IL111 GC columns [115]. All plasticizers were completely separated on the SLB-IL59, SLB-IL76, and SLB-IL111 columns, while diethylbutyl phthalate and diethylhexyl adipate coeluted on SLB-IL82 and SLB-IL100. All the IL columns showed the same retention order except diethylhexyl adipate, which was less retained on the SLB-IL 111 column compared to other columns. Symmetric peak shapes and narrow peak widths were observed for plasticizers compared to the 5% phenyl-95% methylpolysiloxane column. When non-commercial phosphonium-based dicationic ILs with aromatic substituents were used to analyze twelve regulated phthalates, complimentary selectivities were observed for most of the phthalates with respect to the 5% phenyl polydimethylsiloxane (PDMS) phase [19]. Phthalates with benzene and cyclohexyl ring substituents were observed to retain longer on the studied phosphonium ILs compared to the 5% PDMS phase. Three pairs of compounds (bis(4-methyl-2-pentyl) phthalate/bis(2-ethoxyethyl) phthalate, di-n-hexyl phthalate/butyl benzyl phthalate, and dicyclohexyl phthalate/bis(2-ethylhexyl) phthalate) were difficult to separate on the 5% phenyl column. However, they were well separated on the phosphonium dicationic IL GC stationary phases studied [19].

Polychlorinated biphenyls (PCBs) are toxic environmental pollutants, and they have adverse effects on humans. Use of PCBs is currently banned, but due to their persistent nature and widespread use in the last century, they are present in the environment at low concentrations. There are 209 possible congeners of PCBs, and they cannot be separated chromatographically by using any single column [116, 117] (PCB nomenclature with numbers is reported in the literature [118]). Hence, research is targeted toward the development of new stationary phases for the effective separation of PCB congeners. The PCBs are classified into two main categories: planar and non-planar. Non- and mono-ortho-substituted PCBs are planar, while other ortho-substituted PCBs are non-planar. Planar PCBs are more toxic compared to non-planar PCBs and should be effectively separated from each other for better quantification.

Recently commercialized IL stationary phases were evaluated for the separation of PCB congeners, and their performance was compared to the 5%-phenyl/95%-methyl polysiloxane (DB-5MS) column [116]. The number of coelutions on the IL columns and DB-5MS were not much different, while the important difference observed was in the elution order of the PCBs. More polar mono-ortho-substituted PCBs were retained longer compared to the less polar di-ortho-substituted PCBs on the IL columns. The SLB-IL111 column showed the highest deviation in the elution order compared to the DB-5MS column. Trichlorobiphenyls (28 and 30) were retained longer than tetrachlorobiphenyls (47, 49, and 52) on the SLB-IL111 column, while the opposite elution order was observed on the DB-5MS column and all other IL columns. PCBs 28 and 30 were partially separated on the SLB-IL59, but they could not be separated on all the other columns evaluated in the study. This separation was important as PCB 28 is one of the so-called indicator PCBs. Seven indicator PCBs 28, 52, 101, 118, 138, 153, and 180 were best separated on the DB-5MS column compared to all the commercialized IL columns.

In a separate study, a total of 69 PCBs were investigated on the IL columns, and the separation performances were assessed based on the polarity of IL stationary phases [119]. The extremely polar stationary phase SLB-IL111 showed the highest number of coelutions, that is, 31 congeners. Among three IL columns (SLB-IL100, SLB-IL82, and SLB-IL76), the number of coelutions increased with the polarity of the IL stationary phase. Between SLB-IL59 and SLB-IL61, SLB-IL59 showed better performance in the separation of PCBs. Moreover, more symmetrical and narrower peak shapes were obtained with SLB-IL59. The phosphonium-based IL stationary phases SLB-IL59 and SLB-IL76 performed better compared to the imidazolium-based IL stationary phases in the separation of PCBs. SLB-IL61 containing a triflate anion and the same phosphonium dication as SLB-IL59 performed poorly compared to the SLB-IL59. Non-commercial phosphonium ILs with aromatic substituents were shown to be effective in the separation of toxic planar PCBs. A dicationic IL stationary phase, 1,9-bis(diphenyl-o-tolyl-phosphonium)nonane bis(trifluoromethanesulfonyl)imide, effectively separated twelve tetra- to hepta-chloro-substituted toxic PCB congeners [19]. The same IL stationary phase also showed a baseline separation of the most toxic 2,3,7,8-tetrachlorodibenzodioxin (TCDD) isomer from the other five TCDD isomers when compared to the commercial SLB-5MS and SP-2331 columns [19].

Polycyclic aromatic compounds (PACs) are common environmental pollutants comprised of condensed multiring benzenoid compounds [120]. Polycyclic aromatic hydrocarbons (PAHs) and polycyclic aromatic sulfur heterocycles (PASHs) are classes of PACs and have carcinogenic and mutagenic activities. Gas chromatography is the preferred technique for the separation and determination of PACs [121]. Many of the PAHs and PASHs are structural isomers, and their chromatographic separation is essential as they generate the same mass spectra. The structural isomers can have different toxicities, and their separation is important for quantification.

PASHs are a dominant part of fossil fuel-based pollutants and coal tar. Their analyses are mainly done by using moderate- to nonpolar columns in GC or GC × GC

(5%-diphenyl/95%-dimethyl polysiloxane, 50%-diphenyl/50%-dimethyl polysilox-ane, 14%-cyanopropyl/86%-polydimethylsiloxane, etc.). However, none of these polysiloxane-based columns provide broad selectivities for the separation of isomers within a homologous series [122]. Polysiloxane-based columns showed 38–56% coelution during analyses of 119 PASHs [123]. However, the IL-based GC station-ary phases performed better with 32–38% coelutions, and the SLB-IL111 column was the best with 32% coelutions [123]. Separation and determination of ratios of some target PASHs is important in environmental forensic studies to determine the source and the extent of fossil fuel weathering. For example, ratios of 1-, 2-, 3-, and 4-methyldibenzothiophene (MeDBT) vary with petroleum source. 2- and 3-MeDBT coelute on the polysiloxane-based stationary phases, while they can be baseline sep-arated on the SLB-IL111 column [123]. IL columns also have advantages in the separation of isomeric pairs of 1,3,7- and 3,4,7-trimethyldibenzothiophene, 6- and 8-methylbenzo[*b*]naphthol[1,2-*d*]thiophene, 2- and 9-methylbenzo[*b*]naphthol[1,2-*d*]thiophene, and 4- and 7-methylbenzo[*b*]naphthol[2,1-*d*]thiophene compared to siloxane columns [123].

The retention behavior of the PASHs is determined by the polarity and polarizabil-ity of the analytes [123]. Polarizability is mainly dependent on the molecular geom-etry. Linear PASHs (e.g., anthracene analogs) are more polarizable than condensed PASHs (e.g., pyrene analogs) with the same number of rings [123]. The polarity is dependent on the position of the sulfur atom. For example, PASHs with "protected" sulfur in a bay-region (benzo[*b*]naphthol[2,1-*d*]thiophene) or by an alkyl group on the adjacent carbon (6-methylbenzo[*b*]naphthol[1,2-*d*]thiophene) are less polar com-pared to PASHs with an exposed sulfur (benzo[*b*]naphthol[1,2-*d*]thiophene) [123]. PASHs with higher polarizability and polarity showed higher retention with respect to low polarizable and polar PASHs on all the columns. The relatively more polar PASHs (exposed sulfur) showed higher retention on the IL column compared to polysiloxane columns [123]. The IL column SLB-IL60 showed increased separation space and resolution for the separation of PASHs in the GC × GC applications compared to the polysiloxane columns. In GC × GC applications, IL columns provided increased sep-aration of more polarizable PASHs compared to the polydimethylsiloxane columns [123].

Polycyclic aromatic hydrocarbons (PAHs) are common environmental pollutants derived from incomplete combustion of organic material. PAHs are toxic pollutants, and they have mutagenic and carcinogenic activities. Most of the PAHs are structural isomers, and their chromatographic separation is important for the quantitation of individual isomers. A specially developed IL column SLB-ILPAH was recently intro-duced for the effective separation of 22 PAHs. The SLB-ILPAH column consists of 1,12-di(tripropylphosphonium) dodecane bis(trifluoromethanesulfonyl)imide IL as a stationary phase and the column dimensions are 20 m × 0.18 mm i.d., 0.05 μm d_f. Structural isomers of "benzo-fluoranthene" and "dibenzo-pyrene" are difficult to sep-arate on the other GC stationary phases but can be easily and rapidly separated using a SLB-ILPAH column. When a non-commercialized IL [2mC$_3$(mim)$_2$][CTf$_3$]$_2$ con-taining tris(trifluoromethanesulfonyl)methide [CTf$_3$]$^-$ anion was used during PAH analysis, an unusual selectivity in the separation of anthracene and phenanthrene was observed [23]. Anthracene eluted before phenanthrene on this IL column, while the reverse elution order was observed on all other commercial GC stationary phases.

6.6 Characterization of Petroleum Samples

Ionic liquid columns have been effectively utilized for gas chromatographic analysis of complex petroleum samples. The comparatively high thermal stability of IL-based columns allows convenient elution of the heavy aromatic compounds in low-boiling petrochemical samples. The very polar SLB-IL100 and SLB-IL111 columns were successfully used for one-dimensional and comprehensive two-dimensional flow-modulated GC analysis of aromatic hydrocarbons in gasoline, reformate, and fluid catalytic cracking samples [124]. The effective selectivity of these polar columns produced excellent fractionation of hydrocarbons from aromatic compounds in low-boiling petroleum products. The combination of a SLB-IL111 column in the first dimension and a 5% phenyl polydimethylsiloxane monolithic silica (PDMS) column in the second dimension offered the best results for the individual determination of C6–C11 aromatic hydrocarbons. As illustrated in Fig. 6.6, the C8–C11 aromatic compounds, indene, and the methyl-indenes, as well as naphthalene and methyl-naphthalenes, are group-separated employing the SLB-IL111 (^{1}D) + HP-5MS (^{2}D) column series.

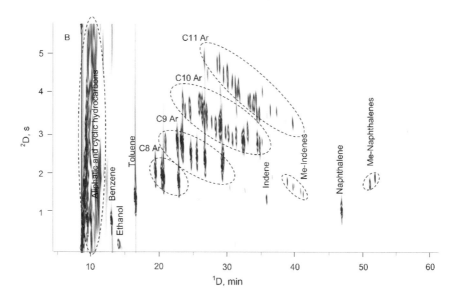

Fig. 6.6 Plot of flow-modulated GC × GC-FID analysis of gasoline sample on the SLB-IL 111 (1D) + HP-5MS (2D) column series. Reprinted with permission from Ref. [124]

Analysis of alkyl phosphates is of utmost importance due to the problems they pose in petroleum refining processes [125]. A relationship between molecular structure and GC retention of alkyl phosphates was established by employing a series of ionic liquid columns [126]. The dependence of elution order on separation temperature for a homologous series of alkyl phosphates was observed on the three ionic liquid columns with highest polarities (i.e., SLB-IL82, SLB-IL100, and SLB-IL111). It was found that at high temperatures, the elution order was reversed with trioctyl phosphate eluting before trihexyl phosphate. Moreover, a thermodynamic model was developed to predict the interactions between alkyl phosphates and IL stationary phases. Subsequently, the SLB-IL100 column was found to have the optimal selectivity for separation of short-chain alkyl phosphates from petroleum hydrocarbons.

The SLB-IL59 column was utilized for high-temperature GC × GC separation of high-boiling point species in heavy petroleum fractions [127, 128]. The IL59 column provided enhanced separation between neutral and basic nitrogen-containing PACs and provided quantitative distribution of acridines and carbazoles in heavy matrices [127]. In addition, an unprecedented group-type separation of heavy sulfur species including naphtheno-aromatic S-compounds family was achieved in vacuum gas oil (VGO) using a SLB-IL59 column in reversed mode 2D GC [128].

Ionic liquid columns were evaluated for characterization of sulfur and nitrogen compounds in Brazilian petroleum derivatives with comprehensive 2D GC [129]. The normal configuration comprised of the DB-5MS/IL59 provided the best orthogonality for separation of organic sulfur compounds (OSCs), while nitrogen compounds were better fractioned using IL59/DB-5MS reversed mode. The matrix interference that resulted from coelution of OSCs and PAHs was significantly eliminated by using ionic liquid columns. Consequently, a higher number of sulfur compounds was detected in comparison with the utilization of traditional stationary phases.

6.7   Water Analysis

The accurate quantification of water is one of the most ubiquitous yet important analytical measurements worldwide [130]. Water content is measured at a broad range of concentrations (from sub-ppm levels to above 99%) in a greater array of matrices than any other analytes [131]. Measuring water is usually mandated by regulatory agencies in certain products including foodstuffs, pharmaceuticals, and other consumer products [132]. Given the fact that water is an omnipresent interfering constituent in hydrophilic samples, further challenges may be met in accurately quantifying trace amounts of moisture [131]. Several analytical techniques have been developed depending on the nature of the sample and the amount of moisture present to effectively determine the water content. Accordingly, a widely used approach for most sample types is highly desirable. Refractive index (RI) measurement, gravimetric determination of water loss after drying (LOD), and Karl Fischer titration (KFT) are the most recurrent methods for determination of water [130]. However, there are certain challenges with each approach. RI measurement requires a thermal pretreatment step for the sample that leads to some loss of moisture content and therefore inaccurate results [133]. Employing the LOD technique usually generates numbers that are lower than the true water content of the samples [134]. While KFT has a broad dynamic range, it can be troublesome for samples with low levels of water [135, 136]. Expensive consumables, limited solubility of samples, reactive impurities, side reactions, and other matrix effects are frequent challenges associated with using the KFT approach [137–140].

Gas chromatography is another approach for the determination of water content in different matrices [24, 25, 130–132, 135, 141–145]. Traditionally, the combination of a packed column with a thermal conductivity detector (TCD) was used for GC analysis of moisture. However, there were several problems, such as non-ideal adsorption isotherms of water to the diatomaceous earth and various other supports (e.g., molecular sieves) resulting in unsymmetrical poor peak shapes, bad response, and even interference with other common analytes [146–150]. Also, employing capillary columns coated with conventional stationary phases leads to film degradation by repeated high temperature exposure to water [131]. The convenient quantitation of water can be performed using water-compatible columns that eliminate all of the concerns mentioned above.

The three triflate-based IL columns were recently made available commercially by MilliporeSigma, under the trade name Watercol™. The Watercol™ 1460 column is coated with tri-(tripropylphosphoniumhexanamido)trimethylamine trifluoromethanesulfonate, a trigonal phosphonium-based ionic liquid. The Watercol™ 1900 and Watercol™ 1910 columns are composed of dicationic imidazolium-based ILs, connected with a polyethylene glycol spacer chain [24, 131]. The chemical structures of the three IL stationary phases that were specifically designed for water analysis are shown in Table 6.1.

The Watercol columns provide optimal selectivity in differentiating water from a vast variety of polar and nonpolar compounds. They also produce water peaks of good efficiency and symmetry to allow for proper integration and subsequent quantitation. The retention of water is lowest on the tricationic Watercol 1460 column with the lowest polarity among water columns. This column is excellent for separation of water in matrices with high-boiling point components. On the other hand, while Watercol 1900 and Watercol 1910 columns have very close Kovats retention indices for water, they provide slightly different selectivities in separating other compounds. This selectivity difference primarily results from different substituents on the imidazolium ring of the dicationic IL moieties [24].

The Watercol 1910 column was used for determination of water in active pharmaceutical ingredients (APIs) as well as solid pharmaceutical products [25, 141]. Monitoring water content in pharmaceutical products is important due to microbial growth in the formulation of drugs. In addition, physical and chemical stability of APIs is significantly correlated to the water content present in these compounds. The 1-ethyl-3-methylimidazolium tris(pentafluoroethyl)trifluorophosphate ($[C_2C_1im][FAP]$) IL was used as headspace solvent for dissolving solid samples. A barrier discharge ionization detector (BID) was utilized for sensitive detection of trace water levels in pharmaceutical products.

In addition, a rapid analysis (under 4 min) with high selectivity and resolution of water and ethanol was obtained in consumer products using a Watercol 1910 column. The established method allowed measurement of both ethanol and water at all concentrations without using any internal standard [132].

The Watercol columns were also used for the determination of trace water content in petroleum products [24]. Due to the complex matrices of petroleum and petroleum-based products, it is usually difficult to analyze water content using traditional approaches. These samples contain many compounds tending to react with iodine, which leads to inaccurate results by KFT. A fully automated protocol using headspace GC along with the BID was developed to accurately quantify trace water levels in 12 petroleum products. Figure 6.7 shows chromatograms of a typical analysis of water in CLP oil on the three Watercol columns. Using the choice of three different IL columns enabled effective separation of the water peak from any interfering compound. Depending on matrix composition, one column could be superior to the others for specific samples.

Further applications of the IL-based Watercol 1910 column were investigated in food samples to determine water levels in liquid sweeteners and honey [142, 143]. The use of headspace GC in combination with BID facilitated the accurate water measurement in these samples. A similar method was utilized for examination of varied and changing ethanol content in kombucha products [26]. Kombucha is a traditional fermented drink prepared by mixing tea and sugar with bacteria and yeast [151]. Although kombucha producers claim that beverages are non-alcoholic [<0.5% alcohol by volume (ABV)], it was found that the range of ethanol in these products

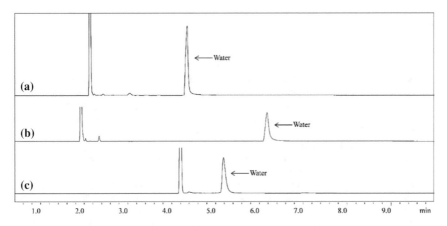

Fig. 6.7 a Analysis of CLP oil on the 30 m Watercol 1460 at 70 °C. **b** Chromatogram of CLP oil on the 30 m Watercol 1900 at 70 °C. **c** Chromatogram of CLP oil at 150 °C on the 60 m Watercol 1910. Reprinted with permission from Ref. [24]

could be between 1.12 and 2.00% (v/v) [26]. The ethanol and other volatile compounds were successfully separated using the Watercol 1910 IL column without any change in column performance after 1200 injections. In fact, the chemical stability of Watercol columns allows long-term exposure to water in aqueous samples without sacrificing column efficiency [132].

6.8   Conclusions

Ionic liquid columns have opened a new era in the field of GC separations. They are the first new class of stationary phases in the last 40 years. The commercial availability of IL-based columns with flexible selectivities and high thermal stabilities has enabled more scientists to use them for separating complex samples, such as fatty acid methyl esters, essential oils, chlorinated hydrocarbons, polycyclic aromatic compounds, and related compounds. Specific IL stationary phases significantly retain more-polar solutes compared with less-polar analytes. The highly polar SLB-IL111 column also provides improved selectivity for the separation of *cis* and *trans* FAME isomers and other geometrical isomers. Water-compatible GC stationary phases based on ILs can be applied routinely for direct analysis of samples containing water as the main solvent, eliminating time-consuming sample preparation steps. In addition, ionic liquid columns offer great orthogonality when incorporated in the second dimension of two-dimensional gas chromatography. Overall, it is expected that interest in ILs as GC stationary phases will continue to increase, thanks to the combination of their exceptional polarity and the capacity to separate a broad array of compounds.

References

1. Weingärtner H (2008) Understanding ionic liquids at the molecular level: facts, problems, and controversies. Angew Chem Int Ed 47:654–670. https://doi.org/10.1002/anie.200604951
2. Soukup-Hein RJ, Warnke MM, Armstrong DW (2009) Ionic liquids in analytical chemistry. Ann Rev Anal Chem 2:145–168. https://doi.org/10.1146/annurev-anchem-060908-155150
3. Welton T (1999) Room-temperature ionic liquids. Solvents for synthesis and catalysis. Chem Rev 99:2071–2083. https://doi.org/10.1021/cr980032t
4. Armstrong DW, He L, Liu Y-S (1999) Examination of ionic liquids and their interaction with molecules, when used as stationary phases in gas chromatography. Anal Chem 71:3873–3876. https://doi.org/10.1021/ac990443p
5. Anderson JL, Armstrong DW (2003) High-stability ionic liquids. A new class of stationary phases for gas chromatography. Anal Chem 75:4851–4858. https://doi.org/10.1021/ac0345749
6. Anderson JL, Armstrong DW (2005) Immobilized ionic liquids as high-selectivity/high-temperature/high-stability gas chromatography stationary phases. Anal Chem 77:6453–6462. https://doi.org/10.1021/ac051006f

7. Huang K, Han X, Zhang X, Armstrong DW (2007) PEG-linked geminal dicationic ionic liquids as selective, high-stability gas chromatographic stationary phases. Anal Bioanal Chem 389:2265–2275. https://doi.org/10.1007/s00216-007-1625-0

8. Breitbach ZS, Armstrong DW (2008) Characterization of phosphonium ionic liquids through a linear solvation energy relationship and their use as GLC stationary phases. Anal Bioanal Chem 390:1605–1617. https://doi.org/10.1007/s00216-008-1877-3

9. Payagala T, Zhang Y, Wanigasekara E, Huang K, Breitbach ZS, Sharma PS, Sidisky LM, Armstrong DW (2008) Trigonal tricationic ionic liquids: a generation of gas chromatographic stationary phases. Anal Chem 81:160–173. https://doi.org/10.1021/ac8016949

10. Barber DW, Phillips CSG, Tusa GF, Verdin A (1959) The chromatography of gases and vapours. Part VI. Use of the stearates of bivalent manganese, cobalt, nickel, copper, and zinc as column liquids in gas chromatography. J Chem Soc 18–24. https://doi.org/10.1039/JR9590000018

11. Pacholec F, Butler HT, Poole CF (1982) Molten organic salt phase for gas–liquid chromatography. Anal Chem 54:1938–1941. https://doi.org/10.1021/ac00249a006

12. Pacholec F, Poole C (1983) Stationary phase properties of the organic molten salt ethylpyridinium bromide in gas chromatography. Chromatographia 17:370–374. https://doi.org/10.1007/BF02262375

13. Pomaville RM, Poole SK, Davis LJ, Poole CF (1988) Solute—solvent interactions in tetra-n-butylphosphonium salts studied by gas chromatography. J Chromatogr A 438:1–14. https://doi.org/10.1016/S0021-9673(00)90227-9

14. Gordon JE, Selwyn JE, Thorne RL (1966) Molten quaternary ammonium salts as stationary liquid phases for gas–liquid partition chromatography. J Org Chem 31:1925–1930. https://doi.org/10.1021/jo01344a059

15. Poole CF, Butler HT, Coddens ME, Dhanesar SC, Pacholec F (1984) Survey of organic molten salt phases for gas chromatography. J Chromatogr A 289:299–320. https://doi.org/10.1016/S0021-9673(00)95096-9

16. Anderson JL, Ding J, Welton T, Armstrong DW (2002) Characterizing ionic liquids on the basis of multiple solvation interactions. J Am Chem Soc 124:14247–14254. https://doi.org/10.1021/ja028156h

17. Anderson JL, Ding R, Ellern A, Armstrong DW (2005) Structure and properties of high stability geminal dicationic ionic liquids. J Am Chem Soc 127:593–604. https://doi.org/10.1021/ja046521u

18. Patil RA, Talebi M, Sidisky LM, Armstrong DW (2017) Examination of selectivities of thermally stable geminal dicationic ionic liquids by structural modification. Chromatographia 80:1563–1574. https://doi.org/10.1007/s10337-017-3372-5

19. Patil RA, Talebi M, Sidisky LM, Berthod A, Armstrong DW (2018) Gas chromatography selectivity of new phosphonium-based dicationic ionic liquid stationary phases. J Sep Sci 41:4142–4148. https://doi.org/10.1002/jssc.201800695

20. Patil RA, Talebi M, Xu C, Bhawal SS, Armstrong DW (2016) Synthesis of thermally stable geminal dicationic ionic liquids and related ionic compounds: an examination of physicochemical properties by structural modification. Chem Mater 28:4315–4323. https://doi.org/10.1021/acs.chemmater.6b01247

21. Talebi M, Patil RA, Armstrong DW (2018) Physicochemical properties of branched-chain dicationic ionic liquids. J Mol Liq 256:247–255. https://doi.org/10.1016/j.molliq.2018.02.016

22. Sun P, Armstrong DW (2010) Ionic liquids in analytical chemistry. Anal Chim Acta 661:1–16. https://doi.org/10.1016/j.aca.2009.12.007

23. Talebi M, Patil RA, Sidisky LM, Berthod A, Armstrong DW (2018) Variation of anionic moieties of dicationic ionic liquid GC stationary phases: effect on stability and selectivity. Anal Chim Acta 1042:155–164. https://doi.org/10.1016/j.aca.2018.07.047

24. Frink LA, Armstrong DW (2016) Determination of trace water content in petroleum and petroleum products. Anal Chem 88:8194–8201. https://doi.org/10.1021/acs.analchem.6b02006

25. Frink LA, Armstrong DW (2016) Water determination in solid pharmaceutical products utilizing ionic liquids and headspace gas chromatography. J Pharm Sci 105:2288–2292. https://doi.org/10.1016/j.xphs.2016.05.014

26. Talebi M, Frink LA, Patil RA, Armstrong DW (2017) Examination of the varied and changing ethanol content of commercial Kombucha products. Food Anal Methods 10:4062–4067. https://doi.org/10.1007/s12161-017-0980-5

27. Payagala T, Huang J, Breitbach ZS, Sharma PS, Armstrong DW (2007) Unsymmetrical dicationic ionic liquids: manipulation of physicochemical properties using specific structural architectures. Chem Mater 19:5848–5850. https://doi.org/10.1021/cm702325a

28. Zeng Z, Phillips BS, Xiao J-C, Shreeve JM (2008) Polyfluoroalkyl, polyethylene glycol, 1,4-bismethylenebenzene, or 1,4-bismethylene-2,3,5,6-tetrafluorobenzene bridged functionalized dicationic ionic liquids: synthesis and properties as high temperature lubricants. Chem Mater 20:2719–2726. https://doi.org/10.1021/cm703693r

29. Talebi M, Patil RA, Sidisky LM, Berthod A, Armstrong DW (2018) Branched-chain dicationic ionic liquids for fatty acid methyl ester assessment by gas chromatography. Anal Bioanal Chem 410:4633–4643. https://doi.org/10.1007/s00216-017-0722-y

30. McReynolds WO (1970) Characterization of some liquid phases. J Chromatogr Sci 8:685–691. https://doi.org/10.1093/chromsci/8.12.685; ibid. (1971) 9:15A. https://doi.org/10.1093/chromsci/9.5.15A-a.

31. Petsch M, Mayer-Helm BX, Söllner V (2005) Preparation and characterization of fused-silica capillary columns coated with m-carborane–siloxane copolymers for gas chromatography. Anal Bioanal Chem 383:322–326. https://doi.org/10.1007/s00216-005-3399-6

32. Ragonese C, Sciarrone D, Tranchida PQ, Dugo P, Dugo G, Mondello L (2011) Evaluation of a medium-polarity ionic liquid stationary phase in the analysis of flavor and fragrance compounds. Anal Chem 83:7947–7954. https://doi.org/10.1021/ac202012u

33. Abraham MH (1993) Scales of solute hydrogen-bonding: their construction and application to physicochemical and biochemical processes. Chem Soc Rev 22:73–83. https://doi.org/10.1039/CS9932200073

34. Abraham MH, Ibrahim A, Zissimos AM (2004) Determination of sets of solute descriptors from chromatographic measurements. J Chromatogr A 1037:29–47. https://doi.org/10.1016/j.chroma.2003.12.004

35. Abraham MH, Poole CF, Poole SK (1999) Classification of stationary phases and other materials by gas chromatography. J Chromatogr A 842:79–114. https://doi.org/10.1016/S0021-9673(98)00930-3

36. Weber W, Andersson JT (2014) Ionic liquids as stationary phases in gas chromatography—an LSER investigation of six commercial phases and some applications. Anal Bioanal Chem 406:5347–5358. https://doi.org/10.1007/s00216-014-7972-8

37. Grob Jr K, Grob G, Grob K (1978) Comprehensive, standardized quality test for glass capillary columns. J Chromatogr A 156:1–20. https://doi.org/10.1016/S0021-9673(00)83120-9

38. Grob K, Grob G, Grob Jr K (1981) Testing capillary gas chromatographic columns. J Chromatogr A 219:13–20. https://doi.org/10.1016/S0021-9673(00)80568-3

39. Dawczynski C, Martin L, Wagner A, Jahreis G (2010) n − 3 LC-PUFA-enriched dairy products are able to reduce cardiovascular risk factors: a double-blind, cross-over study. Clin Nutr 29:592–599. https://doi.org/10.1016/j.clnu.2010.02.008

40. Dawczynski C, Massey KA, Ness C, Kiehntopf M, Stepanow S, Platzer M, Grün M, Nicolaou A, Jahreis G (2013) Randomized placebo-controlled intervention with n-3 LC-PUFA-supplemented yoghurt: effects on circulating eicosanoids and cardiovascular risk factors. Clin Nutr 32:686–696. https://doi.org/10.1016/j.clnu.2012.12.010

41. Christie WW (2003) Lipid analysis: isolation, separation, identification and structural analysis of lipids. Elsevier, Amsterdam

42. Delmonte P, Fardin Kia A-R, Kramer JKG, Mossoba MM, Sidisky L, Rader JI (2011) Separation characteristics of fatty acid methyl esters using SLB-IL111, a new ionic liquid coated capillary gas chromatographic column. J Chromatogr A 1218:545–554. https://doi.org/10.1016/j.chroma.2010.11.072

43. Delmonte P, Fardin-Kia AR, Kramer JKG, Mossoba MM, Sidisky L, Tyburczy C, Rader JI (2012) Evaluation of highly polar ionic liquid gas chromatographic column for the determination of the fatty acids in milk fat. J Chromatogr A 1233:137–146. https://doi.org/10.1016/j.chroma.2012.02.012

44. Zeng AX, Chin S-T, Nolvachai Y, Kulsing C, Sidisky LM, Marriott PJ (2013) Characterisation of capillary ionic liquid columns for gas chromatography–mass spectrometry analysis of fatty acid methyl esters. Anal Chim Acta 803:166–173. https://doi.org/10.1016/j.aca.2013.07.002

45. Delmonte P, Kia A-RF, Hu Q, Rader JI (2009) Review of methods for preparation and gas chromatographic separation of *trans* and *cis* reference fatty acids. J AOAC Int 92:1310–1326. https://www.ingentaconnect.com/content/aoac/jaoac/2009/00000092/00000005/art00007f

46. Tyburczy C, Delmonte P, Fardin-Kia AR, Mossoba MM, Kramer JKG, Rader JI (2012) Profile of *trans* fatty acids (FAs) including *trans* polyunsaturated FAs in representative fast food samples. J Agric Food Chem 60:4567–4577. https://doi.org/10.1021/jf300585s

47. Lin C-C, Wasta Z, Mjøs SA (2014) Evaluation of the retention pattern on ionic liquid columns for gas chromatographic analyses of fatty acid methyl esters. J Chromatogr A 1350:83–91. https://doi.org/10.1016/j.chroma.2014.05.023

48. Firestone D (2009) Official methods and recommended practices of the AOCS. In: American Oil Chemists' Society. http://agris.fao.org/agris-search/search.do?recordID=US201300136250

49. Precht D, Molkentin J (2000) Identification and quantitation of *cis/trans* C16:1 and C17:1 fatty acid positional isomers in German human milk lipids by thin-layer chromatography and gas chromatography/mass spectrometry. Eur J Lipid Sci Technol 102:102–113. https://doi.org/10.1002/(SICI)1438-9312(200002)102:2%3C102::AID-EJLT102%3E3.0.CO;2-C

50. Precht D, Molkentin J (1996) Rapid analysis of the isomers of *trans*-octadecenoic acid in milk fat. Int Dairy J 6:791–809. https://doi.org/10.1016/0958-6946(96)00004-0

51. Sehat N, Rickert R, Mossoba MM, Kramer JKG, Yurawecz MP, Roach JAG, Adlof RO, Morehouse KM, Fritsche J, Eulitz KD, Steinhart H, Ku Y (1999) Improved separation of conjugated fatty acid methyl esters by silver ion–high-performance liquid chromatography. Lipids 34:407–413. https://doi.org/10.1007/s11745-999-0379-3

52. Yurawecz MP, Roach JAG, Sehat N, Mossoba MM, Kramer JKG, Fritsche J, Steinhart H, Ku Y (1998) A new conjugated linoleic acid isomer, 7 *trans*, 9 *cis*-octadecadienoic acid, in cow milk, cheese, beef and human milk and adipose tissue. Lipids 33:803–809. https://doi.org/10.1007/s11745-998-0273-z

53. Fanali C, Micalizzi G, Dugo P, Mondello L (2017) Ionic liquids as stationary phases for fatty acid analysis by gas chromatography. Analyst 142:4601–4612. https://doi.org/10.1039/C7AN01338H

54. Shibamoto S, Gooley A, Yamamoto K (2015) Separation behavior of octadecadienoic acid isomers and identification of *cis*- and *trans*-isomers using gas chromatography. Lipids 50:85–100. https://doi.org/10.1007/s11745-014-3966-8

55. Delmonte P (2016) Evaluation of poly(90% biscyanopropyl/10% cyanopropylphenyl siloxane) capillary columns for the gas chromatographic quantification of *trans* fatty acids in non-hydrogenated vegetable oils. J Chromatogr A 1460:160–172. https://doi.org/10.1016/j.chroma.2016.07.019

56. Granafei S, Losito I, Salivo S, Tranchida PQ, Mondello L, Palmisano F, Cataldi TRI (2015) Occurrence of oleic and 18:1 methyl-branched acyl chains in lipids of *Rhodobacter sphaeroides* 2.4.1. Anal Chim Acta 885:191–198. https://doi.org/10.1016/j.aca.2015.05.028

57. Ragonese C, Tranchida PQ, Dugo P, Dugo G, Sidisky LM, Robillard MV, Mondello L (2009) Evaluation of use of a dicationic liquid stationary phase in the fast and conventional gas chromatographic analysis of health-hazardous C_{18} cis/trans fatty acids. Anal Chem 81:5561–5568. https://doi.org/10.1021/ac9007094

58. Delmonte P, Fardin-Kia AR, Rader JI (2013) Separation of fatty acid methyl esters by GC-online hydrogenation × GC. Anal Chem 85:1517–1524. https://doi.org/10.1021/ac302707z

59. Fardin-Kia AR, Delmonte P, Kramer JKG, Jahreis G, Kuhnt K, Santercole V, Rader JI (2013) Separation of the fatty acids in menhaden oil as methyl esters with a highly polar ionic liquid

gas chromatographic column and identification by time of flight mass spectrometry. Lipids 48:1279–1295. https://doi.org/10.1007/s11745-013-3830-2

60. Dettmer K (2014) Assessment of ionic liquid stationary phases for the GC analysis of fatty acid methyl esters. Anal Bioanal Chem 406:4931–4939. https://doi.org/10.1007/s00216-014-7919-0

61. Mjøs SA (2003) Identification of fatty acids in gas chromatography by application of different temperature and pressure programs on a single capillary column. J Chromatogr A 1015:151–161. https://doi.org/10.1016/S0021-9673(03)01240-8

62. Mjøs SA, Grahl-Nielsen O (2006) Prediction of gas chromatographic retention of polyunsaturated fatty acid methyl esters. J Chromatogr A 1110:171–180. https://doi.org/10.1016/j.chroma.2006.01.092

63. Pinto AC, Guarieiro LLN, Rezende MJC, Ribeiro NM, Torres EA, Lopes WA, de P Pereira PA, de Andrade JB (2005) Biodiesel: an overview. J Braz Chem Soc 16:1313–1330. http://dx.doi.org/10.1590/S0103-50532005000800003

64. Webster RL, Rawson PM, Evans DJ, Marriott PJ (2016) Quantification of trace fatty acid methyl esters in diesel fuel by using multidimensional gas chromatography with electron and chemical ionization mass spectrometry. J Sep Sci 39:2537–2543. https://doi.org/10.1002/jssc.201600307

65. Ragonese C, Tranchida PQ, Sciarrone D, Mondello L (2009) Conventional and fast gas chromatography analysis of biodiesel blends using an ionic liquid stationary phase. J Chromatogr A 1216:8992–8997. https://doi.org/10.1016/j.chroma.2009.10.066

66. Goding JC, Ragon DY, O'Connor JB, Boehm SJ, Hupp AM (2013) Comparison of GC stationary phases for the separation of fatty acid methyl esters in biodiesel fuels. Anal Bioanal Chem 405:6087–6094. https://doi.org/10.1007/s00216-013-7042-7

67. Mogollon NGS, de Lima Ribeiro FA, Lopez MM, Hantao LW, Poppi RJ, Augusto F (2013) Quantitative analysis of biodiesel in blends of biodiesel and conventional diesel by comprehensive two-dimensional gas chromatography and multivariate curve resolution. Anal Chim Acta 796:130–136. https://doi.org/10.1016/j.aca.2013.07.071

68. Takahashi Sato R, Stroppa PHF, da Silva AD, de Oliveira MAL (2016) Fast GC-FID method for monitoring acidic and basic catalytic transesterification reactions in vegetable oils to methyl ester biodiesel preparation. Quim Nova 39:352–355. http://dx.doi.org/10.5935/0100-4042.20160027

69. Mogollón NGS, Ribeiro FAL, Poppi RJ, Quintana AL, Chávez JAG, Agualongo DAP, Aleme HG, Augusto F (2017) Exploratory analysis of biodiesel by combining comprehensive two-dimensional gas chromatography and multiway principal component analysis. J Braz Chem Soc 28:740–746. http://dx.doi.org/10.21577/0103-5053.20160222

70. Webster RL, Evans DJ, Marriott PJ (2015) Detailed chemical analysis using multidimensional gas chromatography–mass spectrometry and bulk properties of low-temperature oxidized jet fuels. Energy Fuels 29:2059–2066. https://doi.org/10.1021/acs.energyfuels.5b00264

71. McCormick RL, Graboski MS, Alleman TL, Herring AM, Tyson KS (2001) Impact of biodiesel source material and chemical structure on emissions of criteria pollutants from a heavy-duty engine. Environ Sci Technol 35:1742–1747. https://doi.org/10.1021/es001636t

72. Sushchik NN, Kuchkina AYu, Gladyshev MI (2013) Fatty acid content and composition of sediments from Siberian eutrophic water bodies: Implications for biodiesel production. Water Res 47:3192–3200. https://doi.org/10.1016/j.watres.2013.03.031

73. Knothe G, Sharp CA, Ryan III TW (2006) Exhaust emissions of biodiesel, petrodiesel, neat methyl esters, and alkanes in a new technology engine. Energy Fuels 20:403–408. https://doi.org/10.1021/ef0502711

74. CSN EN 14331 (2004) Liquid petroleum products - Separation and characterization of fatty acid methyl esters (FAME) from middle distillates - Liquid chromatography (LC)/gas chromatography (GC). European Committee for Standardization, Brussels

75. Katona G, Andréasson U, Landau EM, Andréasson L-E, Neutze R (2003) Lipidic cubic phase crystal structure of the photosynthetic reaction centre from *Rhodobacter sphaeroides* at 2.35 Å resolution. J Mol Biol 331:681–692. https://doi.org/10.1016/S0022-2836(03)00751-4

76. Weatherly CA, Zhang Y, Smuts JP, Fan H, Xu C, Schug KA, Lang JC, Armstrong DW (2016) Analysis of long-chain unsaturated fatty acids by ionic liquid gas chromatography. J Agric Food Chem 64:1422–1432. https://doi.org/10.1021/acs.jafc.5b05988

77. Tyburczy C, Mossoba MM, Rader JI (2013) Determination of *trans* fat in edible oils: current official methods and overview of recent developments. Anal Bioanal Chem 405:5759–5772. https://doi.org/10.1007/s00216-013-7005-z

78. Verméglio A, Joliot P (1999) The photosynthetic apparatus of *Rhodobacter sphaeroides*. Trends Microbiol 7:435–440. https://doi.org/10.1016/S0966-842X(99)01625-X

79. Zeng X, Roh JH, Callister SJ, Tavano CL, Donohue TJ, Lipton MS, Kaplan S (2007) Proteomic characterization of the *Rhodobacter sphaeroides* 2.4.1 photosynthetic membrane: identification of new proteins. J Bacteriol 189:7464–7474. https://doi.org/10.1128/JB.00946-07

80. Tucker JD, Siebert CA, Escalante M, Adams PG, Olsen JD, Otto C, Stokes DL, Hunter CN (2010) Membrane invagination in *Rhodobacter sphaeroides* is initiated at curved regions of the cytoplasmic membrane, then forms both budded and fully detached spherical vesicles. Mol Microbiol 76:833–847. https://doi.org/10.1111/j.1365-2958.2010.07153.x

81. Destaillats F, Guitard M, Cruz-Hernandez C (2011) Identification of Δ6-monounsaturated fatty acids in human hair and nail samples by gas-chromatography–mass-spectrometry using ionic-liquid coated capillary column. J Chromatogr A 1218:9384–9389. https://doi.org/10.1016/j.chroma.2011.10.095

82. Inagaki S, Numata M (2015) Fast GC analysis of fatty acid methyl esters using a highly polar ionic liquid column and its application for the determination of *trans* fatty acid contents in edible oils. Chromatographia 78:291–295. https://doi.org/10.1007/s10337-014-2837-z

83. Yoshinaga K, Asanuma M, Mizobe H, Kojima K, Nagai T, Beppu F, Gotoh N (2014) Characterization of *cis*- and *trans*-octadecenoic acid positional isomers in edible fat and oil using gas chromatography–flame ionisation detector equipped with highly polar ionic liquid capillary column. Food Chem 160:39–45. https://doi.org/10.1016/j.foodchem.2014.03.069

84. Lock AL, Bauman DE (2004) Modifying milk fat composition of dairy cows to enhance fatty acids beneficial to human health. Lipids 39:1197–1206. https://doi.org/10.1007/s11745-004-1348-6

85. Aldai N, Dugan MER, Rolland DC, Kramer JKG (2009) Survey of the fatty acid composition of Canadian beef: backfat and longissimus lumborum muscle. Can J Anim Sci 89:315–329. https://doi.org/10.4141/CJAS08126

86. Cruz-Hernandez C, Kramer JKG, Kennelly JJ, Glimm DR, Sorensen BM, Okine EK, Goonewardene LA, Weselake RJ (2007) Evaluating the conjugated linoleic acid and *trans* 18:1 isomers in milk fat of dairy cows fed increasing amounts of sunflower oil and a constant level of fish oil. J Dairy Sci 90:3786–3801. https://doi.org/10.3168/jds.2006-698

87. Bauman DE, Griinari JM (2003) Nutritional regulation of milk fat synthesis. Annu Rev Nutr 23:203–227. https://doi.org/10.1146/annurev.nutr.23.011702.073408

88. Gómez-Cortés P, Rodríguez-Pino V, Juárez M, de la Fuente MA (2017) Optimization of milk odd and branched-chain fatty acids analysis by gas chromatography using an extremely polar stationary phase. Food Chem 231:11–18. https://doi.org/10.1016/j.foodchem.2017.03.052

89. Fievez V, Colman E, Castro-Montoya JM, Stefanov I, Vlaeminck B (2012) Milk odd- and branched-chain fatty acids as biomarkers of rumen function—an update. Anim Feed Sci Technol 172:51–65. https://doi.org/10.1016/j.anifeedsci.2011.12.008

90. Tyburczy Srigley C, Rader JI (2014) Content and composition of fatty acids in marine oil omega-3 supplements. J Agric Food Chem 62:7268–7278. https://doi.org/10.1021/jf5016973

91. Ando Y, Sasaki T (2011) GC separation of *cis*-eicosenoic acid positional isomers on an ionic liquid SLB-IL100 stationary phase. J Am Oil Chem Soc 88:743–748. https://doi.org/10.1007/s11746-010-1733-4

92. Hammann S, Vetter W (2015) Gas chromatographic separation of fatty acid esters of cholesterol and phytosterols on an ionic liquid capillary column. J Chromatogr B 1007:67–71. https://doi.org/10.1016/j.jchromb.2015.11.007

93. Moreau RA, Whitaker BD, Hicks KB (2002) Phytosterols, phytostanols, and their conjugates in foods: structural diversity, quantitative analysis, and health-promoting uses. Prog Lipid Res 41:457–500. https://doi.org/10.1016/S0163-7827(02)00006-1

94. Phillips KM, Ruggio DM, Toivo JI, Swank MA, Simpkins AH (2002) Free and esterified sterol composition of edible oils and fats. J Food Compos Anal 15:123–142. https://doi.org/10.1006/jfca.2001.1044

95. Evershed RP, Male VL, Goad LJ (1987) Strategy for the analysis of steryl esters from plant and animal tissues. J Chromatogr A 400:187–205. https://doi.org/10.1016/S0021-9673(01)81612-5

96. Fan H, Smuts J, Bai L, Walsh P, Armstrong DW, Schug KA (2016) Gas chromatography–vacuum ultraviolet spectroscopy for analysis of fatty acid methyl esters. Food Chem 194:265–271. https://doi.org/10.1016/j.foodchem.2015.08.004

97. Nan H, Anderson JL (2018) Ionic liquid stationary phases for multidimensional gas chromatography. Trends Anal Chem 105:367–379. https://doi.org/10.1016/j.trac.2018.03.020

98. Nolvachai Y, Kulsing C, Marriott PJ (2015) Thermally sensitive behavior explanation for unusual orthogonality observed in comprehensive two-dimensional gas chromatography comprising a single ionic liquid stationary phase. Anal Chem 87:538–544. https://doi.org/10.1021/ac5030039

99. Gu Q, David F, Lynen F, Vanormelingen P, Vyverman W, Rumpel K, Xu G, Sandra P (2011) Evaluation of ionic liquid stationary phases for one dimensional gas chromatography–mass spectrometry and comprehensive two dimensional gas chromatographic analyses of fatty acids in marine biota. J Chromatogr A 1218:3056–3063. https://doi.org/10.1016/j.chroma.2011.03.011

100. Nosheen A, Mitrevski B, Bano A, Marriott PJ (2013) Fast comprehensive two-dimensional gas chromatography method for fatty acid methyl ester separation and quantification using dual ionic liquid columns. J Chromatogr A 1312:118–123. https://doi.org/10.1016/j.chroma.2013.08.099

101. Zeng AX, Chin S-T, Marriott PJ (2013) Integrated multidimensional and comprehensive 2D GC analysis of fatty acid methyl esters. J Sep Sci 36:878–885. https://doi.org/10.1002/jssc.201200923

102. Pojjanapornpun S, Nolvachai Y, Aryusuk K, Kulsing C, Krisnangkura K, Marriott PJ (2018) Ionic liquid phases with comprehensive two-dimensional gas chromatography of fatty acid methyl esters. Anal Bioanal Chem 410:4669–4677. https://doi.org/10.1007/s00216-018-0944-7

103. Villegas C, Zhao Y, Curtis JM (2010) Two methods for the separation of monounsaturated octadecenoic acid isomers. J Chromatogr A 1217:775–784. https://doi.org/10.1016/j.chroma.2009.12.011

104. Qi M, Armstrong DW (2007) Dicationic ionic liquid stationary phase for GC-MS analysis of volatile compounds in herbal plants. Anal Bioanal Chem 388:889–899. https://doi.org/10.1007/s00216-007-1290-3

105. Cagliero C, Bicchi C, Cordero C, Liberto E, Sgorbini B, Rubiolo P (2012) Room temperature ionic liquids: New GC stationary phases with a novel selectivity for flavor and fragrance analyses. J Chromatogr A 1268:130–138. https://doi.org/10.1016/j.chroma.2012.10.016

106. Blumberg LM (2011) Metrics of separation performance in chromatography. Part 1. Definitions and application to static analyses. J Chromatogr A 1218:5375–5385. https://doi.org/10.1016/j.chroma.2011.06.017

107. Serôdio P, Nogueira JMF (2006) Considerations on ultra-trace analysis of phthalates in drinking water. Water Res 40:2572–2582. https://doi.org/10.1016/j.watres.2006.05.002

108. Polo M, Llompart M, Garcia-Jares C, Cela R (2005) Multivariate optimization of a solid-phase microextraction method for the analysis of phthalate esters in environmental waters. J Chromatogr A 1072:63–72. https://doi.org/10.1016/j.chroma.2004.12.040

109. Nassar N, Abeywardana P, Barker A, Bower C (2010) Parental occupational exposure to potential endocrine disrupting chemicals and risk of hypospadias in infants. Occup Environ Med 67:585–589. https://doi.org/10.1136/oem.2009.048272

110. Trasande L, Sathyanarayana S, Spanier AJ, Trachtman H, Attina TM, Urbina EM (2013) Urinary phthalates are associated with higher blood pressure in childhood. J Pediatr 163:747–753.e1. https://doi.org/10.1016/j.jpeds.2013.03.072

111. Werner EF, Braun JM, Yolton K, Khoury JC, Lanphear BP (2015) The association between maternal urinary phthalate concentrations and blood pressure in pregnancy: the HOME study. Environ Health 14:75. https://doi.org/10.1186/s12940-015-0062-3

112. Jaakkola JJK, Knight TL (2008) The role of exposure to phthalates from polyvinyl chloride products in the development of asthma and allergies: a systematic review and meta-analysis. Environ Health Perspect 116:845–853. https://doi.org/10.1289/ehp.10846

113. Net S, Delmont A, Sempéré R, Paluselli A, Ouddane B (2015) Reliable quantification of phthalates in environmental matrices (air, water, sludge, sediment and soil): a review. Sci Total Environ 515–516:162–180. https://doi.org/10.1016/j.scitotenv.2015.02.013

114. Li D, Stevens R, English C, GC-MS analysis of phthalates: comparison of GC stationary phase performance. https://www.restek.com/Technical-Resources/Technical-Library/General-Interest/gen_GNAN2380B-UNV, last time accessed: 19 Dec 2018

115. Sanchez-Prado L, Lamas JP, Garcia-Jares C, Llompart M (2012) Expanding the applications of the ionic liquids as GC stationary phases: plasticizers and synthetic musks fragrances. Chromatographia 75:1039–1047. https://doi.org/10.1007/s10337-012-2280-y

116. de Boer J, Blok D, Ballesteros-Gómez A (2014) Assessment of ionic liquid stationary phases for the determination of polychlorinated biphenyls, organochlorine pesticides and polybrominated diphenyl ethers. J Chromatogr A 1348:158–163. https://doi.org/10.1016/j.chroma.2014.05.001

117. Kimbrough RD, Jensen AA (1989) Halogenated biphenyls, terphenyls, naphthalenes, dibenzodioxins and related products, Elsevier. Amsterdam. https://doi.org/10.1016/C2009-0-12642-9

118. Ballschmiter K, Bacher R, Mennel A, Fischer R, Riehle U, Swerev M (1992) The determination of chlorinated biphenyls, chlorinated dibenzodioxins, and chlorinated dibenzofurans by GC-MS. J High Resolut Chromatogr 15:260–270. https://doi.org/10.1002/jhrc.1240150411

119. Ros M, Escobar-Arnanz J, Sanz ML, Ramos L (2018) Evaluation of ionic liquid gas chromatography stationary phases for the separation of polychlorinated biphenyls. J Chromatogr A 1559:156–163. https://doi.org/10.1016/j.chroma.2017.12.029

120. Wilson WB, Sander LC, Oña-Ruales JO, Mössner SG, Sidisky LM, Lee ML, Wise SA (2017) Retention behavior of isomeric polycyclic aromatic sulfur heterocycles in gas chromatography on stationary phases of different selectivity. J Chromatogr A 1485:120–130. https://doi.org/10.1016/j.chroma.2017.01.024

121. Poster DL, Schantz MM, Sander LC, Wise SA (2006) Analysis of polycyclic aromatic hydrocarbons (PAHs) in environmental samples: a critical review of gas chromatographic (GC) methods. Anal Bioanal Chem 386:859–881. https://doi.org/10.1007/s00216-006-0771-0

122. Zeigler C, Schantz M, Wise S, Robbat Jr A (2012) Mass spectra and retention indexes for polycyclic aromatic sulfur heterocycles and some alkylated analogs. Polycyc Aromatic Compds 32:154–176. https://doi.org/10.1080/10406638.2011.651679

123. Antle P, Zeigler C, Robbat Jr A (2014) Retention behavior of alkylated polycyclic aromatic sulfur heterocycles on immobilized ionic liquid stationary phases. J Chromatogr A 1361:255–264. https://doi.org/10.1016/j.chroma.2014.08.010

124. Krupčík J, Gorovenko R, Špánik I, Bočková I, Sandra P, Armstrong DW (2013) On the use of ionic liquid capillary columns for analysis of aromatic hydrocarbons in low-boiling petrochemical products by one-dimensional and comprehensive two-dimensional gas chromatography. J Chromatogr A 1301:225–236. https://doi.org/10.1016/j.chroma.2013.05.075

125. Nizio KD, Harynuk JJ (2012) Analysis of alkyl phosphates in petroleum samples by comprehensive two-dimensional gas chromatography with nitrogen phosphorus detection and post-column Deans switching. J Chromatogr A 1252:171–176. https://doi.org/10.1016/j.chroma.2012.06.070

126. Weber BM, Harynuk JJ (2013) Gas chromatographic retention of alkyl phosphates on ionic liquid stationary phases. J Chromatogr A 1271:170–175. https://doi.org/10.1016/j.chroma. 2012.10.052
127. Dutriez T, Borras J, Courtiade M, Thiébaut D, Dulot H, Bertoncini F, Hennion M-C (2011) Challenge in the speciation of nitrogen-containing compounds in heavy petroleum fractions by high temperature comprehensive two-dimensional gas chromatography. J Chromatogr A 1218:3190–3199. https://doi.org/10.1016/j.chroma.2010.10.056
128. Mahé L, Dutriez T, Courtiade M, Thiébaut D, Dulot H, Bertoncini F (2011) Global approach for the selection of high temperature comprehensive two-dimensional gas chromatography experimental conditions and quantitative analysis in regards to sulfur-containing compounds in heavy petroleum cuts. J Chromatogr A 1218:534–544. https://doi.org/10.1016/j.chroma. 2010.11.065
129. Cappelli Fontanive F, Souza-Silva ÉA, Macedo da Silva J, Bastos Caramão E, Alcaraz Zini C (2016) Characterization of sulfur and nitrogen compounds in Brazilian petroleum derivatives using ionic liquid capillary columns in comprehensive two-dimensional gas chromatography with time-of-flight mass spectrometric detection. J Chromatogr A 1461:131–143. https://doi. org/10.1016/j.chroma.2016.07.025
130. Padivitage NLT, Smuts JP, Armstrong DW (2014). Water determination. In: Riley CM, Rosanske TW, Rabel Riley SR (eds.) Specification of drug substances and products: development and validation of analytical methods. Elsevier, Oxford, pp 223–241. https://doi.org/ 10.1016/B978-0-08-098350-9.00011-4
131. Armstrong DW (2017) Measuring water: the expanding role of gas chromatography. LC GC 35:503–506. http://www.chromatographyonline.com/measuring-water-expanding-role-gas-chromatography
132. Weatherly CA, Woods RM, Armstrong DW (2014) Rapid analysis of ethanol and water in commercial products using ionic liquid capillary gas chromatography with thermal conductivity detection and/or barrier discharge ionization detection. J Agric Food Chem 62:1832–1838. https://doi.org/10.1021/jf4050167
133. Gallina A, Stocco N, Mutinelli F (2010) Karl Fischer titration to determine moisture in honey: a new simplified approach. Food Control 21:942–944. https://doi.org/10.1016/j.foodcont.2009. 11.008
134. Isengard H-D, Präger H (2003) Water determination in products with high sugar content by infrared drying. Food Chem 82:161–167. https://doi.org/10.1016/S0308-8146(02)00538-1
135. Inagaki S, Morii N, Numata M (2015) Development of a reliable method to determine water content by headspace gas chromatography/mass spectrometry with the standard addition technique. Anal Methods 7:4816–4820. https://doi.org/10.1039/C5AY00832H
136. Hanna WS, Johnson AB (1950) Water content of hydrocarbons. Modified Karl Fischer method. Anal Chem 22:555–558. https://doi.org/10.1021/ac60040a012
137. Ivanova PG, Aneva ZV (2006) Assessment and assurance of quality in water measurement by coulometric Karl Fischer titration of petroleum products. Accred Qual Assur 10:543–549. https://doi.org/10.1007/s00769-005-0056-x
138. Margolis SA, Hagwood C (2003) The determination of water in crude oil and transformer oil reference materials. Anal Bioanal Chem 376:260–269. https://doi.org/10.1007/s00216-003-1865-6
139. Cedergren A, Nordmark U (2000) Determination of water in NIST reference material for mineral oils. Anal Chem 72:3392–3395. https://doi.org/10.1021/ac9913006
140. Larsson W, Jalbert J, Gilbert R, Cedergren A (2003) Efficiency of methods for Karl Fischer determination of water in oils based on oven evaporation and azeotropic distillation. Anal Chem 75:1227–1232. https://doi.org/10.1021/ac026229+
141. Frink LA, Weatherly CA, Armstrong DW (2014) Water determination in active pharmaceutical ingredients using ionic liquid headspace gas chromatography and two different detection protocols. J Pharm Biomed Anal 94:111–117. https://doi.org/10.1016/j.jpba.2014.01.034

142. Frink LA, Armstrong DW (2016) The utilisation of two detectors for the determination of water in honey using headspace gas chromatography. Food Chem 205:23–27. https://doi.org/10.1016/j.foodchem.2016.02.118

143. Frink LA, Armstrong DW (2017) Using headspace gas chromatography for the measurement of water in sugar and sugar-free sweeteners and products. LC GC 30:6–10.   http://www.chromatographyonline.com/using-headspace-gas-chromatography-measurement-water-sugar-and-sugar-free-sweeteners-and-products-1

144. Xu B-Q, Rao C-Q, Cui S-F, Wang J, Wang J-L, Liu L-P (2018) Determination of trace water contents of organic solvents by gas chromatography-mass spectrometry-selected ion monitoring. J Chromatogr A 1570:109–115. https://doi.org/10.1016/j.chroma.2018.07.068

145. Zhang Y, Wang C, Armstrong DW, Woods RW, Jayawardhana DA (2011) Rapid, efficient quantification of water in solvents and solvents in water using an ionic liquid-based GC column. LC GC EUROPE 24:516–529

146. Knight HS, Weiss FT (1962) Determination of traces of water in hydrocarbons. A calcium carbide-gas liquid chromatography method. Anal Chem 34:749–751. https://doi.org/10.1021/ac60187a010

147. Quiram ER (1963) Applications of wide-diameter open tubular columns in gas chromatography. Anal Chem 35:593–595. https://doi.org/10.1021/ac60197a019

148. Kolb B, Auer M (1990) Analysis of water in liquid and solid samples by headspace gas chromatography. Part II: Insoluble solid samples by the "suspension approach". Fresenius J Anal Chem 336:297–302. https://doi.org/10.1007/BF00331387

149. O'Keefe WK, Ng FTT, Rempel GL (2008) Validation of a gas chromatography/thermal conductivity detection method for the determination of the water content of oxygenated solvents. J Chromatogr A 1182:113–118. https://doi.org/10.1016/j.chroma.2007.12.044

150. Streim HG, Boyce EA, Smith JR (1961) Determination of water in 1,1-dimethylhydrazine, diethylenetriamine, and mixtures. Anal Chem 33:85–89.   https://doi.org/10.1021/ac60169a025

151. Jayabalan R, Malbaša RV, Lončar ES, Vitas JS, Sathishkumar M (2014) A review on kombucha tea—microbiology, composition, fermentation, beneficial effects, toxicity, and tea fungus. Compr Rev Food Sci Food Safety 13:538–550. https://doi.org/10.1111/1541-4337.12073

Chapter 7
Ionic Liquid–Liquid Chromatography: A Novel Separation Method

Leslie Brown, Martyn J. Earle, Manuela A. Gilea, Natalia V. Plechkova and Kenneth R. Seddon

Abstract There is a wide selection of ionic liquid/(water/organic solvents) biphasic mixtures. These mixtures could be utilized for liquid–liquid extractions in countercurrent chromatography for the separation of organic, inorganic, and bio-based materials. A customized countercurrent chromatography has been designed and constructed specifically to adapt to the more viscous character of ionic liquid-based solvent systems to be used in a broad variety of separations (including transition metal salts, arenes, alkenes, alkanes, bio-oils, and sugars).

Keywords Ionic liquid–liquid chromatography · IL-Prep™ · Liquid–liquid extraction · Purification · Separations

7.1 Introduction

Liquid chromatography (LC) [1], particularly liquid–liquid chromatography (LLC) and countercurrent chromatography (CCC) [2], with the latter considered to be a continuous, automated form of liquid–liquid separation, has continued to make progress over the last 20 years. The industrial aspects have been discussed by Sutherland et al. [3] and Berthod and Serge [4]. These LC techniques rely on the ability of solutes to distribute themselves between two mutually immiscible liquid phases: a mobile phase (MP) and a stationary phase (SP) [5]. Current LC techniques have several drawbacks in terms of solubility, low scale of separation, the reactivity of

---

This chapter is dedicated to the memory of our brilliant and inspirational co-author Professor Kenneth R. Seddon.

L. Brown
AECS-QuikPrep Ltd., 55 Gower Street, London WC1 6HQ, UK

M. J. Earle (✉) · M. A. Gilea · N. V. Plechkova (✉) · K. R. Seddon
School of Chemistry, The QUILL Research Centre, The Queen's University of Belfast, Belfast BT9 5AG, Northern Ireland, UK
e-mail: m.earle@qub.ac.uk

N. V. Plechkova
e-mail: n.plechkova@qub.ac.uk

© Springer Nature Switzerland AG 2020
M. B. Shiflett (ed.), *Commercial Applications of Ionic Liquids*, Green Chemistry and Sustainable Technology, https://doi.org/10.1007/978-3-030-35245-5_7

solutes, and high costs of chromatography columns, partially due to their limited lifetimes. In LLC, the range of solvent systems that can be considered includes a narrow range of molecular solvents, such as water, methanol, ethanol, butanol, ethanenitrile, ethyl ethanoate, dichloromethane, trichloromethane, poly(ethylene) oxide (PEO), and hexane, mixed in varying proportions to result in the formation of a biphasic solvent system [6]. Additionally, if compounds are not soluble or react with the stationary or mobile phase, they are unsuitable for LLC. From a safety standpoint, the solvent systems can have high vapor pressures from the use of volatile organic compounds (VOCs) [7]. This limitation is avoided with ionic liquids, which have negligible vapor pressures [8].

In the last ten years, ionic liquids have become widely used in solvent extraction [9, 10]. The potential of ionic liquids in both countercurrent chromatography (CCC) [11] and centrifugal partition chromatography (CPC) [11, 12] has been evaluated by Berthod and was found to have a number of problems [13]. The instrumentation available at that time was not able to cope with high backpressures due to the high viscosities of many ionic liquids. After this study, ionic liquids have been found to be useful in separations in CCC and CPC by a number of authors [14a, b]. A custom-designed countercurrent chromatography instrument was manufactured that allowed ionic liquids to be used as the stationary phase or mobile phase [15]. This new type of instrument was utilized for industrially relevant separations, including the separation and isolation of transition metal salts, monosaccharides from disaccharides, alkanes from aromatics, and the refining of essential oils [14a, 16].

7.2   Chromatographic Separations

Chromatography techniques operate through the distribution of analytes between different phases (gas, liquid, solid). Figure 7.1 shows the various forms of liquid chromatography.

For liquid–liquid chromatography, two immiscible liquids are used, and the separations are based on the distribution ratio of solutes between the two liquid solvents [17].

7.3   Countercurrent Chromatography

There are various implementations of CCC that are in use [18], and these have been applied in separations on a large scale [19]. HPCCC (high-performance countercurrent chromatography) [19] is one of the forms of the instrumentation available currently. In CCC, the stationary and mobile phases used must be immiscible liquids. The main principle on which a CCC instrument works involves one fluid mixing and separating from a second fluid, often in a column or coiled pipe. This column comprises an open-ended tube coiled around a bobbin. In HPCCC, the bobbin is rotated

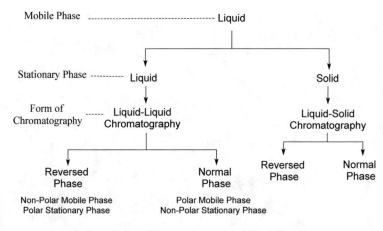

Fig. 7.1 Distribution of analytes in different forms of liquid chromatography

in a double-axis gyratory motion (tracing out a cardioid), where the fluids undergo varying acceleration forces which act on the column during each rotation (Fig. 7.2). This causes the two immiscible stationary and mobile phases to go through mixing and settling step per rotation of the machine. The coils are connected to the rest of the CCC apparatus, without using rotating connectors and special seals, since no overall rotation of the coil connections occurs. This is achieved by attaining planetary motion of the coils, as in Fig. 7.3 [20]. The head and tail of the pipe coils flex but do not rotate. The coils of the machine are attached to the pumps and detector using flexible flying leads. The pioneer of modern CCC with the coil planet centrifuge [21] is

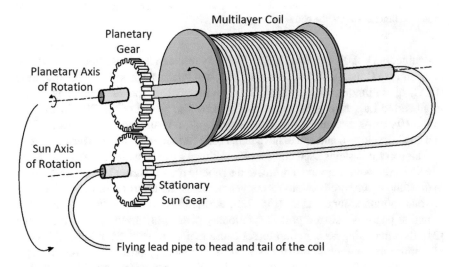

Fig. 7.2 Schematic of the J-type synchronous planetary motion centrifuge used in countercurrent chromatography [20]

Fig. 7.3 The external view of a Quattro chassis CCC instrument (or IL-Prep^{TM}) instrument. For HPCCC/HPLLC (high-performance liquid–liquid chromatography), solid supports are not used, and therefore the coil cannot clog up and result in inefficiency or damage because of impurities. If a new stationary phase is essential, more of the stationary phase is pumped in the instrument to regain the functionality of the coil

undoubtedly Dr. Yoichiro Ito [18, 22]. His early research led to establishment of two companies: PC Inc. (CCC) of Potomac, MD, and Pharmatech Research Corporation (CCC) of Baltimore, MD, in the USA. The innovative Japanese Company, Sanki Engineering Ltd., Kyoto, brought out a stacked-disc design of the CPC instrument in the 1980s. As a result of this pioneering work, a resurgence of interest in CCC and CPC developed, and since, have undergone further evolution and development.

The solvent systems used consist of a stationary phase, which is retained within a helical coil wound around a drum, and a mobile phase. Through a combination of hydrostatic and hydrodynamic forces generated as the drum rotates in a planetary or solar motion around a shaft (Fig. 7.2), the stationary phase is held in place and a mobile phase is pumped past the stationary phase [23]. In the J-type centrifuge [24], the alternating acceleration forces cause regions of ballistic mixing and zones of settling as the mobile phase passes by the stationary phase. This occurs typically at 400–4000 rpm, depending on the design and rotation speed of the rotor. For

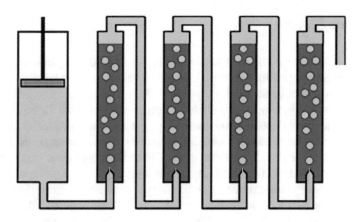

Fig. 7.4 Droplet type HPCPC or CPC where the mobile phase (green) is pumped through the stationary phase (blue), which is held in place by centripetal forces

larger separations, droplet-type high-performance centrifugal partition chromatography (HPCPC) systems are preferential and are based on the principle shown in Fig. 7.4. The instruments based on this principle generally are used for larger scale separations than with hydrodynamic HPLLC/HPCCC machines, and can provide better resolution of solutes for slow-settling biphasic eluents, for instance, for isolation of bioactive compounds from anthocyanins [25].

For centrifugal partition chromatography, the separation occurs in a series of chambers, where the stationary phase is retained by rotation around a single sun axis. As the mobile phase is pumped through the stationary phase, solutes that have a preference for the stationary phase are retained, and compounds with preferred solubility in the mobile phase pass through. Also, in liquid–liquid chromatography, far less solvent is used when compared to flash-chromatography or HPLC.

At the end of most chemical processes, a separation stage is required to produce a pure product. The performance and number of the separation stages can strongly affect the production costs for the industry. This problem is ubiquitous, across every sector of the chemical industry, from smaller-scale pharmaceutical industry separations to the large scales of the oil and gas industries. The ionic liquid-containing solvent systems represent a significant technological advance in separation science which: (1) is applicable to many liquids and solid separations, from the analytical scale of a few mg to the industrial multi-kilo scale, and (2) perform separations that cannot currently be achieved by any other technique [15, 26]. Given the millions of potential ionic liquid phases, there are no conceptual constraints to the use of this technique for any selected system.

7.4 Ionic Liquids

Ionic liquids have some distinct advantages when compared to organic solvents [9c, 27]. Although they are salts, they have low melting points (conventionally < 100 °C), or in some cases even below room temperature. They have interesting characteristics in terms of solubility; they behave either like polar solvents (with the ability to dissolve a huge range of organic compounds, polymers, inorganic molecules, and salts) [28] or like nonpolar solvents (dissolving molecules, such as hexane and benzene) [10b]. Because ionic liquids are salts, most of them have negligible vapor pressure [29] resulting in their non-flammability, hence ionic liquids were coined as "green solvents". However, caution must be applied here, as ionic liquids are not intrinsically green, and some may be either toxic or non-biodegradable, or both [30].

Currently, ionic liquids are used in different areas of chemistry and chemical engineering and in a number of industrial processes mainly due to their "designer" properties. Ionic liquids can be tuned by the choice of the anion and cation allowing them to be optimized for a particular application. An example of design is to vary the structure of an ionic liquid such that it phase-separates from the compound of interest, making product isolation straightforward [31]. Another approach is to make their building blocks (cation or the anion, or both) either acidic or basic (either Lewis or Brønsted) [32]. In this way, the ionic liquid–liquid interactions and distribution ratios can be modified and tuned to a particular end use [33].

A serious problem of ionic liquids is their high viscosities, especially for their application in chromatography [34]. However, this disadvantage can be overcome by specially designing and constructing CCC machines to eliminate pressure bottlenecks. Examples of the machinery suitable for ionic liquid–liquid chromatography (ILLC) from AECS-QuikPrep Ltd. (Fig. 7.3) can be found elsewhere [15, 26]. The aim in ILLC research is to develop a generic methodology for separation of practically all soluble organic and inorganic target compounds. The techniques used are based on the principles of conventional HPLLC, CCC, CPC, HPCCC, or HPCPC [35]. Given that industry uses several hundred organic solvents, and it is estimated that there are more than one million simple ionic liquids attainable, the permutations of possible mobile and stationary phases are immense [36]. In contrast, conventional LLC only uses about ten different molecular solvents, which are mixed in various proportions to produce biphasic systems.

7.5 Ionic Liquid–Liquid Chromatography (ILLC)

For ILLC, with a J-type centrifuge, the instrument that is applied must be able to withstand high working pressures (minimum 50 bar) for extended periods of time. For this reason, the coils are made from either stainless steel or, in the case of highly corrosive compounds (concentrated acids/bases), titanium, or polytetrafluoroethy-lene (PTFE). Additionally, the centrifuge must be capable of running at up to 50 °C, which can dramatically improve the ILLC separation performance [15, 26].

The inherent tunability ("designer solvents") of ionic liquids allows control over their physicochemical properties: viscosity, hydrophobicity, hydrophilicity, density, and corrosivity. Moreover, and of vital importance, the distribution ratios of a solute between an ionic liquid phase and a second phase (either an organic solvent or another ionic liquid) can be controlled. Indeed, ionic liquids can dissolve materials which are conventionally considered to be insoluble, such as kerogens [37]. As will be demonstrated here, this versatility will enable a generic methodology for a wide range of potential separations to be developed, especially the ones that were previously thought to be difficult to implement, for instance, bio-polymers, proteins, saccharides, polysaccharides, metal salts, and petrochemicals. Moreover, the ionic liquids are easily recyclable and can be used for multiple separations, which justifies the initial higher costs of using ionic liquid media.

To some extent, the early publications stating that ionic liquids could not be easily utilized with CCC and CPC [13] may have held back the development of ionic liquid solvent systems in CCC and CPC. AECS-QuikPrep Ltd./Quattro CCC (UK), in col-laboration initially with the QUILL Research Centre within the Queen's University of Belfast (UK), were able to successfully employ biphasic ionic liquid-based sol-vent systems in CCC [14a]. Here, ionic liquid-based solvent systems were utilized in the separation of a mixture of $CoCl_2$, $NiCl_2$, and $CuCl_2$, the separation of two monosaccharides form a disaccharide, and the extraction of cumene from hexane [14a]. By the use of a hexane/ionic liquid solvent system, the separation of hydro-carbons from oxygenated hydrocarbons in vetiver oil was achieved [16]. In addition, ionic liquid solvent systems have been studied in the countercurrent chromatographic separation of nonpolar lipids [38]. Cao et al. have used dilute solutions of ionic liq-uids in CCC in the determination of *Alternaria* mycotoxins and the determination of chlorophenols in red wine [14b, 39]. Also, four patents have been published on the use of ionic liquids in CCC solvent systems in general [15], in bio-organic [26b], in organic [26a], and in inorganic separations [26c]. Two examples of chiral sepa-rations by CCC are in the use of solutions of chiral salts [40] (similar to an ionic liquid), derived from cinchona alkaloid, in the separation of *N*-(3,5-dinitrobenzoyl)-(±)-leucine (Fig. 7.5). Similarly, cyclodextrin and a chiral copper complex have been used in the enantioseparation of naringenin (Fig. 7.5) by CCC [41].

7.6   ILLC and ILLE Instrumentation

The Quattro IL-Prep™ ILLC instrument that we are using at QUILL (Fig. 7.6) incorporates four coils, as shown in Table 7.1 and Fig. 7.7, and has been tested for several separations, in order to evaluate the effectiveness of ILLC for various mixtures. Here, we concentrate on four examples of ionic liquid biphasic systems based on two or three components.

Fig. 7.5 The structure of *N*-(3,5-dinitrobenzoyl)-(±)-leucine (left and center) [40] and naringenin (right) [41]

Fig. 7.6 The outside view of the four coils and PTFE flying leads installed in the IL-Prep instrument

Table 7.1 Details of the J-type centrifuge coils installed in the IL-Prep ILLC instrument

Coil number	Bore (mm)	Length (m)	Number of turns	Capacity (cm^3)	L/D ratio
1	1.0	13.35	26	12	13,350
2	2.1	36.02	76	133	16,000
3	1.0	39.89	78	34	39,890
4	3.7	23.83	52	236	6440

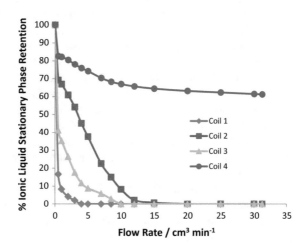

Fig. 7.7 The phase retention curves within the coils for a 4:1:1 mixture of water/ethyl ethanoate/and [P$_{6\,6\,6\,14}$][Cl], respectively, at 35 °C and 865 rpm

7.7 ILLC Separations

A range of separations using ionic liquid-containing solvent systems has been carried out to illustrate the versatility of ILLC methodology. Three example separations include separation of inorganic transition metal salts, separation of monosaccharides from disaccharides, and separation of low-polarity bio-organic molecules.

7.7.1 Transition Metal Separations

Copper(II), nickel(II), and cobalt(II) salts can be separated from each other using a number of techniques. Examples include a nine-stage process [42] involving solvent extraction and electro-refining or a five-stage continuous process involving the ionic liquid trihexyltetradecylphosphonium chloride [P$_{6\,6\,6\,14}$][Cl] [43]. Chromatographic separation of nickel(II) and cobalt(II) chloride was achieved by means of anion-exchange chromatography [44], TLC [45], HPLC of stilbene complexes [46], the use of amino acid chelates [47], micellar electrokinetic chromatography [48], the use of a guar-based chelating ion-exchange resin chromatography [49], and the use of a supported dibenzyl sulfoxide solution on silica [50].

Copper(II), nickel(II), and cobalt(II) chloride salts in aqueous solution can be separated from each other using ILLC. The solvent system used for the stationary phase was based on $[P_{6\,6\,6\,14}][Cl]$, which is a viscous, hydrophobic ionic liquid, with added ethyl ethanoate to reduce the viscosity of the ionic phase. Other co-solvents, such as dichloromethane, propanone, or butanol, were also found to be effective. Typically, the scale of the separation was 0.5–6.0 g (at *ca.* 0.6–7.0 M concentration) on coil 2 and 2–10 g on coil 4 (see Table 7.1 for details).

In order to achieve an effective separation of the aqueous salts, it is necessary to prepare the coils with a biphasic solvent system (degassed) consisting of $[P_{6\,6\,6\,14}][Cl]$ (250 cm^3), ethyl ethanoate (125–250 cm^3), and pure water (1000 cm^3). The ionic phase contains the majority of the ethyl ethanoate and forms the upper liquid layer. The aqueous phase contains a small amount of ethyl ethanoate, but no ionic liquid (<0.5 mol%) was detected by ^{1}H NMR spectroscopy [51]. For most chromatographic separations, it is necessary to obtain a phase retention curve for the two phases used in the separation. This shows how much stationary phase remains in the coils as the mobile-phase flow rate is increased, for a given temperature and instrument rotation rate. The performance of the four coils was tested using water, $[P_{6\,6\,6\,14}][Cl]$, and ethyl ethanoate (4:1:1 ratio) solvent system at 35 °C, and the phase retention for the four coils shown in Table 7.1 was determined (Fig. 7.7).

In order to obtain the phase retention curve, the coils were initially pumped full of the stationary ionic liquid phase. The coil's rotation rate was set to 865 rpm, and the flow rate versus stationary phase retention curves were measured at a range of flow rates ranging from 0.5 to 31.5 cm^3 min^{-1}. The stationary phase retention curve for the four coils is shown in Fig. 7.7. This experiment was carried out in the apparatus shown in Fig. 7.8. The amount of stationary phase remaining in the coil was calculated by measuring changes in the position of the mobile stationary phase boundary in a graduated measuring cylinder (see Fig. 7.7), where its variation can

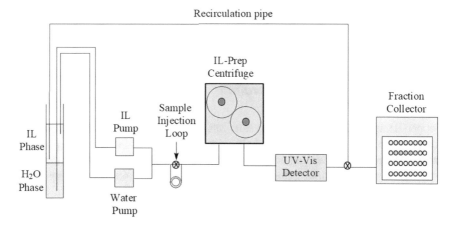

Fig. 7.8 The configuration of the IL-Prep instrument for metal separations and phase retention studies

be used to calculate phase volumes in the coil. A dead volume of 4.0 cm^3 must be taken into account, and the ratio of the phase volumes is accurate to $\pm0.5\%$.

In Fig. 7.7, it can be seen that there are considerable differences between the behavior of the two phases in the four coils. The rate of loss of ionic liquid stationary phase is dependent on both the coil diameter and length. For example, coils 1 and 3 have the same internal diameter but have completely different phase retention curves (Fig. 7.7). This observation indicates that the stationary phase is not evenly distributed in the coil during operation, and is concentrated at the head of the coil.

Ionic liquids are significantly more viscous than solvents conventionally used in CCC. This results in greater pressure drop due to the connecting pipework and the coils. Additionally, this affects the design of the coils, which were made from stainless steel rather than the more conventional PTFE tubing [52]. As a result, the IL-Prep machine is able to operate at pressures of up to 70 bar and allows the use of high flow rates. The pressures encountered during the measurement of the phase retention curves in Fig. 7.7 are shown in Fig. 7.9. In the smaller coil 2, surface, film, and interfacial tension effects become more prominent. For good separations, the instrument should be operated in the plateau region, which in Fig. 7.7 corresponds to 0.5–2.5 cm^3 min^{-1} for coil 4, but for coil 2 this stable region is much smaller. These conditions need to be optimized for every combination of mobile and stationary phases, and it is important to collect the ionic liquid lost from the coil so that it can be recycled.

ILLC allows the direct single-step chromatographic separation of these metal(II) salts on a preparative scale with high sample loadings, under neutral conditions, and without complex chelating agents. When $[M(H_2O)_6]Cl_2$ salts (M = Co, Ni, Cu) [53] are mixed with the biphasic water/$[P_{6\,6\,6\,14}][Cl]$/ethyl ethanoate (4:1:1 v/v/v) solvent system (Fig. 7.10) [54], the metals distribute themselves as shown in Table 7.2 [14a].

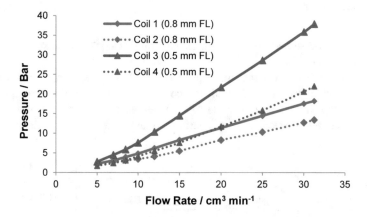

Fig. 7.9 The mobile phase operating pressures encountered during the experiment to determine the phase retention curves in Fig. 7.7, for the water/$[P_{6\,6\,6\,14}][Cl]$/ethyl ethanoate (4:1:1) solvent system with water as the mobile phase. Error $= \pm1.4$ bar (20 psi)

Fig. 7.10 A photograph of nickel(II) chloride (left), cobalt(II) chloride (center), and copper(II) chloride (right) dissolved in a water/[P$_{6\,6\,6\,14}$][Cl]/ethyl ethanoate (4:1:1 v/v/v) biphasic solvent system. The upper phase is the ionic liquid [P$_{6\,6\,6\,14}$][Cl]/ethyl ethanoate mixture and the lower phase is water saturated with ethyl ethanoate (7.6 wt%)

Table 7.2 The distribution ratios of the metal(II) chloride complexes in the biphasic water/[P$_{6\,6\,6\,14}$][Cl]/ethyl ethanoate (4:1:1 v/v/v) solvent system

Metal	Temp. (°C)	Peak maximum (min)	DR from ILLC separation	DR from UV-Vis spectrometer
Ni	30	23.5	0.31	0.2
Co	30	29.5	0.62	0.6
Cu	30	56.8	2.22	2.2
Ni	40	20.8	0.16	0.2
Co	40	29.8	0.66	0.6
Cu	40	55.3	2.07	2.1

DR = Distribution Ratio

The metals are presumed to exist as the tetrahedral [MCl$_4$]$^{2-}$ complexes in the ionic phase (Fig. 7.11) [55].

The separation was carried out at 30 and 40 °C and is shown in Fig. 7.12. A partial separation was obtained with the cobalt(II) chloride peak overlapping the start of the copper peak and the end of the nickel peak. At 40 °C, the performance of the separation improved, and the separation time was reduced from 68 to 61 min.

$$\left[\begin{array}{c} OH_2 \\ OH_2 \\ H_2O-M-OH_2 \\ H_2O \quad OH_2 \end{array}\right]^{2+} + 4\ Cl^- \rightleftharpoons \left[\begin{array}{c} Cl \\ Cl\text{''''}M \\ Cl \quad Cl \end{array}\right]^{2-} + 6\ H_2O$$

Fig. 7.11 The behavior of metal(II) chloride complexes (Co, Ni, or Cu) in the biphasic water/[P$_{6\,6\,6\,14}$][Cl]/ethyl ethanoate (4:1:1 v/v/v) solvent system

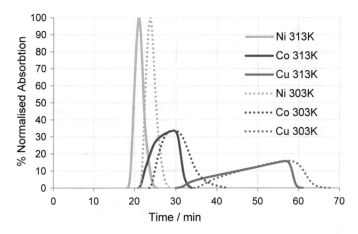

Fig. 7.12 The concentration of aqueous $CoCl_2$ (red line), $NiCl_2$ (green line), and $CuCl_2$ (blue line) at 30 °C (dotted lines) and 40 °C (solid lines) eluted on coil 2, showing the normalized absorption of the metal halide salts where aqueous $NiCl_2 = 100$ (equal areas under each curve). Error $= \pm 2\%$ in UV absorption readings

The peak shape of the eluting copper(II) chloride is triangular and is showing anti-Langmuirian behavior [56]. This is thought to be due to the nickel(II) and cobalt(II) complexes affecting the phase equilibria of the copper(II) chloride complexes, and the time taken for the multistage change from the hydrated metal complex in the water phase to the tetrachlorometalate(II) complex in the ionic phase.

7.7.2 Separation of Saccharides

The separation of saccharides has been investigated using various chromatographic methodologies [57] but is not used industrially because the separation costs are too high [58]. There are several HPLC methods for saccharide separation that can be used on the analytical scale [59]. Preparative-scale HPLC separations of saccharides are possible [60], such as with a PL-Hi-Plex Ca 300 $\times$ 25 mm HPLC column with a water eluent, but these methods suffer from the problem of small sample sizes [61]. The use of CCC for separations of saccharides using aqueous biphasic solvent systems, such as aqueous salt/ethanol systems, or an ethanenitrile/1.0 M $NaCl_{(aq)}$ (5:4) solvent system has been carried out [62], but it was only tested on 2 mg scale.

The ionic liquid-based solvent system $[C_4C_1im][Cl]/(3.0$ M aqueous $K_2[HPO_4]$, 1:1 v/v) was used in the separation of sucrose (distribution ratio (DR) = 0.90), from either fructose (DR = 0.22) or glucose (DR = 0.26) (see Fig. 7.13 for stationary phase retention curve). All three saccharides are optically active, so polarimetry was used to measure the saccharide concentration in the fractions eluted from the coil and is shown in Fig. 7.14.

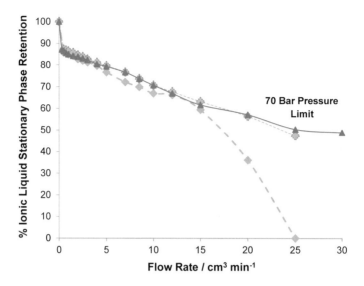

Fig. 7.13 The variation of % stationary phase (ionic liquid containing) retention plotted against the mobile phase (aqueous phosphate) flow rate for the [C_4mim][Cl]/(K_2[HPO_4]–H_2O; 3.0 M; 1:1) two-phase solvent system at 30 °C (dashed green line), 40 °C (dotted light blue line), and 45 °C (solid blue line), on coil 2

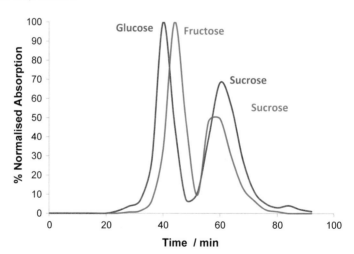

Fig. 7.14 The separation of glucose (red solid line) and fructose (blue solid line) from sucrose by ILLC using the [C_4C_1im][Cl]/(3.0 M aqueous K_2[HPO_4], 1:1 v/v biphasic solvent system at 45 °C, on coil 2). The stationary phase retention curve for the solvent system is shown in Fig. 7.13 solid blue line

The complete separation of glucose or fructose from sucrose was achieved on a 1.00 g scale (500 mg of monosaccharide + 500 mg of sucrose). A similar saccharide separation by Shinomiya and Ito was carried out on a 2.5 mg scale on a 34 cm³ coil [62]. Using the IL-Prep instrument with an ionic liquid system, on the 133 cm³ coil (Table 7.1) [14b], 500 mg quantities of saccharides could be separated. This gives a much-increased degree of process intensification (or space-time yield) of *50 times* and shows what could be achieved with ionic liquid-containing solvent systems. Commercially, the separation of fructose or glucose from sucrose is carried out with crystallization methodologies, since chromatographic techniques are not economically viable [63]. However, ILLC technology can be used successfully on high-value saccharides and polysaccharides [26b].

7.7.3 Separation of Vetiver Oil

Vetiver oil is used in the perfume industry [64] and is extracted from the roots of *Chrysopogon zizanioides* [65] by azeotropic (or steam) distillation [66]. Annually, worldwide vetiver oil production is approximately 250 tons [67]. Of the 300+ chemical components of vetiver oil, compounds of interest to the fragrance industry include polycyclic alkenes, such as α- and δ-cadinene, and polycyclic alcohols, such as khusimol [68]. The separation of vetiver oil into each individual compound is not feasible; however, a separation of vetiver oil into classes of sesquiterpenes can be carried out by ILLC (Fig. 7.15).

The ILLC separation of vetiver oil into alkene sesquiterpenes (Fig. 7.16) and oxygenated sesquiterpenes was carried out using the $[C_{12}C_1im][NTf_2]$/hexane solvent system [16, 33]. The ionic liquid is insoluble in the hexane mobile phase, but hexane is soluble in the $[C_{12}C_1im][NTf_2]$. The solutes are eluted from the coil dissolved in hexane, which are easily recovered by evaporation or distillation of the hexane from the sesquiterpenes that elute from the coil.

GCMS analysis of the fractions in the fraction collector revealed that they were separated into two main classes [69]. The fractions in the T6–T25 range contained sesquiterpene cyclic alkenes with the formula $C_{15}H_{24}$ or $C_{15}H_{26}$ [69b]. Five examples of the isolated alkenes include γ-muurolene, α-cadinene, humulene, β-vetivenene,

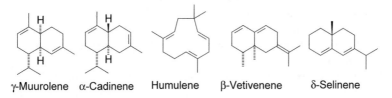

γ-Muurolene   α-Cadinene   Humulene   β-Vetivenene   δ-Selinene

Fig. 7.15 Five literature examples of alkenes present in vetiver oil and five GCMS software-identified compounds found in tubes 6–29 in the ILLC separation of vetiver oil [68]

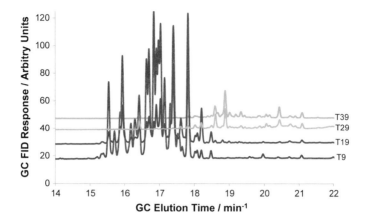

Fig. 7.16 The GC response for four fractions (T9, T19, T29, and T39, where T = tube number) for the ILLC separation of vetiver oil with a hexane/[$C_{12}C_1$im][NTf_2] two-phase solvent system with hexane as the mobile phase. GCMS analyses of T6 to T25 contain C_{15} hydrocarbon sesquiterpenes (polycyclic alkenes) and tubes over number 26 contain predominately oxygenated C_{15} sesquiterpenes [16]

and δ-selinene which are shown in Fig. 7.15. From fraction T26 onwards, the compounds isolated were mostly oxygenated hydrocarbons, such as ethers, epoxides, and alcohols, eluting in order of increasing distribution ratio. The chemical formulas of these oxygenated products were $C_{15}H_{24}O$ and $C_{15}H_{26}O$ based on GCMS analysis.

7.7.4 Purification of Lentinan

The first commercial application of the ionic liquid-CCC/CPC process has come about due to the ability of ionic liquids to solubilize bio-polymers without denaturing them. This has allowed the direct purification of a very complex mixture of saccharides and polysaccharides contained within hot water extracts of shiitake mushrooms. The product (lentinan) was isolated in its biologically active form, which is a triple-helix 1,3-1,6-β-glucan polysaccharide. It should be noted that lentinan is easily denatured in hot water or in DMSO solution [70]. The ionic liquid-CCC/CPC process uses a 2.5 M $K_2[HPO_4]$/[C_4C_1im][Cl] solvent system with [C_4C_1im][Cl] as the mobile phase and allows lentinan to be isolated in high yield while retaining its full bioactivity in treating cancer [26b]. The isolation of the products from ionic liquids (particularly with an ionic liquid-containing mobile phase) can be very straightforward. In the case of the lentinan, the addition of ethanol to the solution of lentinan in the [C_4C_1im][Cl]/H_2O mobile phase resulted in the precipitation of colorless lentinan. The [C_4C_1im][Cl], used in lentinan separation, was recycled by filtration of an ethanol solution of [C_4C_1im][Cl] through charcoal and silica followed by recovery of the ethanol.

7.8 Solvent Engineering and Ionic Liquid–Liquid Chromatography

The chemical structure of ionic liquids can be designed to give a particular set of physicochemical properties, which can be optimized for a particular end use [9c]. Factors, such as viscosity, hydrophobicity, hydrophilicity, density, acidity or basicity, surface or interfacial tensions, and corrosivity, can be adjusted, which permits considerable control over phase behavior and therefore separation performance. The distribution ratios of compounds between an ionic liquid-containing phase and a second immiscible phase (which may be either water, an organic solvent, or other ionic liquid) can be manipulated by altering the chemical makeup of the ionic liquid. For example, if a solute, dissolved in a two-phase hexane-ionic liquid solvent system has a solubility that is too low in the ionic liquid phase, then the structure of the ionic liquid cation or anion can be changed to increase the solubility in the ionic phase. The use of longer alkyl chains attached to the charged parts of the anion or cation can be used to improve solubility in the ionic phase. This great flexibility makes ILLC a general-purpose separation technology, useful for the separations of a very wide range of compounds including compounds that were previously thought to be too difficult to separate. We call this approach solvent engineering, a complement to the chemical engineering required in the design of ILLC instrumentation.

7.9 Conclusions

The use of molecular solvents as biphasic eluents has been shown to be excellent for a huge variety of applications in laboratory-scale preparations of up to a kilo or multi-kilo targets. However, the complexity of molecular-solvent biphasic eluents with three, four, or on occasion five solvents becomes less favored in larger-scale processes. For these larger-scale process applications of 10s of kilos to many tons per annum, the unique properties of ionic liquids can provide many advantages, which we have discussed. This can make the additional effort of developing ionic liquid eluents well worthwhile. The use of ionic liquids in countercurrent chromatography instruments was initially found to be problematic due to high backpressures and consequently low mobile-phase flow rates. With careful design of the fluid flow paths and by eliminating pressure bottlenecks, this problem has now been solved. Currently, there are very few papers in the literature describing the use of solvent systems containing a high percentage of ionic liquids in CCC or LLC. Control over the structure and design of ionic liquids and ionic liquid-containing solvent systems (solvent engineering) has enabled us to select or alter solute distribution ratios of solutes that are to be separated.

The testing of these new ionic liquid solvent systems was performed in a custom-designed ILLC instrument capable of operating with higher viscosity solvent systems and at considerably higher backpressures (70 bar) than are usually found in conventional CCC and LLC instruments. The ILLC equipment uses coils that are made from stainless steel (rather than PTFE/poly(vinyl alcohol) (PVA)), and the main pressure bottlenecks in the solvent flow pathways have been removed [15]. ILLC has been found to be a very versatile separation technology and allows a very wide range of separations to be performed. Examples include the direct separation of metal(II) salts, saccharides, polysaccharides, triglycerides, petrochemicals, and terpenes [14b, 16, 38].

By the modification of the structure and therefore properties of ionic liquids, they can be designed or selected for use in any given separation. The viscosity, density, density difference, interfacial tension, and relative solubility of solutes can all be adjusted through alterations to the structure of the ionic liquid anion or cation. Hence, ILLC allows the separation of practically any soluble mixture, provided that a suitable two-phase solvent system can be designed or found. ILLC can, therefore, be described as using "designer solvents for designer separations". Whilst this area of research is still relatively new, ILLC has the potential to allow separations to be carried out which are currently considered to be either too difficult or too expensive to perform on a large scale. Compounds that are thought to be too insoluble or too immiscible with biphasic molecular solvent systems can now be separated by ILLC. An example of this is with the purification of lentinan [26b], in which the scale of the separation was boosted from the tens of milligrams [71] to the gram scale, using a similar-sized apparatus. With larger capacity ILLC instruments, including CPC instruments, much larger scale separations can be achieved.

There can be no doubt that having been a Cinderella project for many decades, CCC/CPC is now maturing into not only a valuable laboratory preparative technique but also into a pilot and large-scale process technique [25, 72]. Instrumentation has been developed capable of laboratory, pilot, and large-scale process applications with both standard biphasic solvents and ionic liquid biphasic solvents [15, 26b]. Similarly, ionic liquid research and technology has moved from an electrochemical curiosity in the 1980s to a new scientific discipline in its own right, and it is being deployed in industry on process scales [9b, 73].

The full potential of combining both ionic liquid and countercurrent chromatography technologies is beginning to be realized. This union massively increases the potential number of solvent systems available in CCC separations, and concomitantly leads to a greater understanding of ionic liquid phase behavior that can be applied to separations that were once not thought possible on a large scale. Liquid–liquid CCC/CPC research has been carried out by many non-ionic liquid specialists, and they have expressed concern to AECS-QuikPrep Ltd. that certain ionic liquids can degrade various chemical structures and compounds. As in the case with lentinan, there are a vast number of options for the choice of ionic liquids and usually many can be found that do not react with or denature compounds being separated. Ionic liquids once purchased off-the-shelf or custom-synthesized can be recycled and cleaned up at minimal cost [74].

Peptides, proteins, monoclonal antibodies (mAb's), enzymes, bioactive compounds, precious metals, actinides, and lanthanides are all ideal candidates for ionic liquid-CCC/CPC separation processes, as all are high value but can be potentially difficult to purify cost-effectively from their starting crude matrices. Ionic liquid-CCC/CPC/CCE/CPE science and research are likely going to be driven by commercial applications rather than in the academic arenas. Research groups are already in place in the USA and Europe to develop the processes necessary for very large-scale process applications.

References

1. Snyder LR, Kirkland JJ (1979) Introduction to modern liquid chromatography, 2nd edn. Wiley, New York
2. Conway WD, Petroski RJ (eds) (1995) Modern countercurrent chromatography. In: ACS Symposium Series 593. Washington, DC
3. Sutherland IA, Brown L, Forbes S, Games G, Hawes D, Hostettmann K, McKerrell EH, Marston A, Wheatley D, Wood P (1998) Countercurrent chromatography (CCC) and its versatile application as an industrial purification & production process. J Liq Chromatogr Relat Technol 21:279. https://doi.org/10.1080/10826079808000491
4. Berthod A, Serge A (2005) Industrial applications of CCC. In: Encyclopedia of chromatography, 2nd edn. (ed: Cazes JE). CRC Press, Boca Raton, p 833
5. Oka F, Oka H, Ito Y (1991) Systematic search for suitable two-phase solvent systems for high-speed counter-current chromatography. J Chromatogr 538:99. https://doi.org/10.1016/s0021-9673(01)91626-7
6. Friesen JB, Pauli GF (2007) Rational development of solvent system families in counter-current chromatography. J Chromatogr A 1151:51. https://doi.org/10.1016/j.chroma.2007.01.126
7. Rafson HJ (ed) (1998) Odor and VOC control handbook. McGraw-Hill, New York
8. Brennecke JF, Maginn EJ (2001) Ionic liquids: innovative fluids for chemical processing. AIChE J 47:2384. https://doi.org/10.1002/aic.690471102
9. a: Wasserscheid P, Welton T (eds) (2008) Ionic liquids in synthesis, vol 1, 2nd edn. Wiley-VCH, Weinheim, Germany; b: Plechkova NV, Seddon KR (2008) Applications of ionic liquids in the chemical industry. Chem Soc Rev 37:123 http://doi.org/10.1039/b006677j; c: Freemantle M (2010) An introduction to ionic liquids. Royal Society of Chemistry, Cambridge, UK
10. a: Anderson JL, Armstrong DW, Wei G-T (2006) Ionic liquids in analytical chemistry. Anal Chem 78:2892 http://doi.org/10.1021/ac069394o; b: Arce A, Earle MJ, Rodríguez H, Seddon KR, Soto A (2009) Bis{(trifluoromethyl)sulfonyl}amide ionic liquids as solvents for the extraction of aromatic hydrocarbons from their mixtures with alkanes: effect of the nature of the cation. Green Chem 11:365. http://doi.org/10.1039/b814189d
11. Sutherland IA (2007) Recent progress on the industrial scale-up of counter-current chromatography. J Chromatogr A 1151:6. https://doi.org/10.1016/j.chroma.2007.01.143
12. Marchal L, Legrand J, Foucault A (2003) Centrifugal partition chromatography: a survey of its history, and our recent advances in the field. Chem Rec 3:133. https://doi.org/10.1002/tcr. 10057
13. Berthod A, Carda-Broch S (2004) Use of the ionic liquid 1-butyl-3-methylimidazolium hexafluorophosphate in countercurrent chromatography. Anal Bioanal Chem 380:168. https://doi.org/10.1007/s00216-004-2717-8
14. a: Brown L, Earle MJ, Gîlea MA, Plechkova NV, Seddon KR (2017) Ionic liquid–liquid separations using countercurrent chromatography: a new general-purpose separation methodology. Aust J Chem 70:923 https://doi.org/10.1071/CH17004; b: Fan C, Cao L, Liu M, Wang W (2016), Determination of *Alternaria* mycotoxins in wine and juice using ionic liquid modified countercurrent chromatography as a pretreatment method followed by high-performance liquid chromatography. J Chromatogr A 1436:133 http://doi.org/10.1016/j.chroma.2016.01.069
15. Earle MJ, Seddon KR, Self R, Brown L (2013) Ionic liquid separations. WO2013121218A1

16. Brown L, Earle MJ, Gîlea MA, Plechkova NV, Seddon KR (2017) Ionic liquid–liquid chromatography: a new general purpose separation methodology. Top Curr Chem 375:74. https://doi.org/10.1007/s41061-017-0159-y
17. Ito Y, Conway W (1986) High-speed countercurrent chromatography. Crit Rev Anal Chem 17:65. https://doi.org/10.1080/10408348608542792
18. Ito Y, Bowman RL (1970) Countercurrent chromatography: liquid-liquid partition chromatography without solid support. Science 167:281. https://doi.org/10.1126/science.167.3916.281
19. Rubio N, Ignatova S, Minguillón C, Sutherland IA (2009) Multiple dual-mode countercurrent chromatography applied to chiral separations using a (S)-naproxen derivative as chiral selector. J Chromatogr A 1216:8505. https://doi.org/10.1016/j.chroma.2009.10.006
20. Ito Y (1996) High-speed countercurrent chromatography. In: Chemical analysis, vol 132 (eds: Ito Y, Conway WD). Wiley, New York, p 3
21. Ito Y, Weinstein M, Aoki I, Harada R, Kimura E, Nunogaki K (1966) The coil planet centrifuge. Nature 212:985. https://doi.org/10.1038/212985a0
22. Ito Y, Bowman RL (1970) Countercurrent chromatography: liquid-liquid partition chromatography without solid support. J Chromatogr Sci 8:315. https://doi.org/10.1093/chromsci/8.6.315
23. Ignatova S, Hawes D, van den Heuvel R, Hewitson P, Sutherland IA (2010) A new non-synchronous preparative counter-current centrifuge—the next generation of dynamic extraction/chromatography devices with independent mixing and settling control, which offer a step change in efficiency. J Chromatogr A 1217:34. https://doi.org/10.1016/j.chroma.2009.10.055
24. Sutherland IA, Heywood-Waddington D, Ito Y (1987) Counter-current chromatography: applications to the separation of biopolymers, organelles and cells using either aqueous—organic or aqueous—aqueous phase systems. J Chromatogr 384:197. https://doi.org/10.1016/s0021-9673(01)94671-0
25. Margraff R, Intes O, Renault JH, Garret P (2005) Partitron 25, a multi-purpose industrial centrifugal partition chromatograph: rotor design and preliminary results on efficiency and stationary phase retention. J Liq Chromatogr Relat Technol 28:1893 https://doi.org/10.1081/jlc-200063539
26. a: Earle M, Seddon K (2013) Ionic liquid separations WO2013121219A1; b: Earle MJ, Gilea MA (2013) Lentinan extraction process from mushrooms using ionic liquid WO2013140185A1; c: Earle MJ, Seddon KR (2013) Ionic liquid separations WO2013121220A1
27. Earle MJ, Seddon KR (2000) Ionic liquids. Green solvents for the future. Pure Appl Chem 72:1391. https://doi.org/10.1351/pac200072071391
28. Reichardt C, Welton T (2011) Solvents and solvent effects in organic chemistry, 4th edn. Wiley-VCH, Weinheim
29. Earle MJ, Esperança JMSS, Gilea MA, Canongia Lopes JN, Rebelo LPN, Magee JW, Seddon KR, Widegren JA (2006) The distillation and volatility of ionic liquids. Nature 439:831. https://doi.org/10.1038/nature04451
30. Petkovic M, Seddon KR, Rebelo LPN, Silva Pereira C (2011) Ionic liquids: a pathway to environmental acceptability. Chem Soc Rev 40:1383. http://doi.org/10.1039/c004968a
31. Carmichael AJ, Earle MJ, Holbrey JD, McCormac PB, Seddon KR (1999) The Heck reaction in ionic liquids: a multiphasic catalyst system. Org Lett 1:997. https://doi.org/10.1021/ol9907771
32. Keim W, Korth W, Wasserscheid P (2000) Ionic liquids, WO/2000/016902
33. Arce A, Earle MJ, Rodríguez H, Seddon KR (2007) Separation of benzene and hexane by solvent extraction with 1-alkyl-3-methylimidazolium bis{(trifluoromethyl)sulfonyl}amide ionic liquids: effect of the alkyl-substituent length. J Phys Chem B 111:4732. https://doi.org/10.1021/jp066377u
34. Berthod A, Carda-Broch S (2003) A new class of solvents for CCC: the room temperature ionic liquids. J Liq Chromatogr Relat Technol 26:1493. https://doi.org/10.1081/jlc-120021262
35. Berthod A, Maryutina T, Spivakov B, Shpigun O, Sutherland IA (2009) Countercurrent chromatography in analytical chemistry (IUPAC technical report). Pure Appl Chem 81:355. https://doi.org/10.1351/pac-rep-08-06-05

36. Seddon KR (1999) Proceedings of the International George Papatheodorou Symposium (eds: Boghosian S, Dracopoulos V, Kontoyannis CG, Voyiatzis GA). Institute of Chemical Engineering and High Temperature Chemical Processes, Patras, Greece, p 131

37. Patell Y, Seddon KR, Dutta L, Fleet A (2003) The dissolution of kerogen in ionic liquids. In: Green industrial applications of ionic liquids, NATO Science Series II, vol 92 (eds: Rogers RD, Seddon KR, Volkov S). Springer, Dordrecht, p 499

38. Müller M, Englert M, Earle MJ, Vetter W (2017) Development of solvent systems with room temperature ionic liquids for the countercurrent chromatographic separation of very nonpolar lipid compounds. J Chromatogr A 1488:68. https://doi.org/10.1016/j.chroma.2017.01.074

39. Fan C, Li N, Cao X (2015) Determination of chlorophenols in red wine using ionic liquid countercurrent chromatography as a new pretreatment method followed by high-performance liquid chromatography. J Sep Sci 38:2109. https://doi.org/10.1002/jssc.201500172

40. Franco P, Blanc J, Oberleitner WR, Maier NM, Lindner W, Minguillón C (2002) Enantiomer separation by countercurrent chromatography using cinchona alkaloid derivatives as chiral selectors. Anal Chem 74:4175. https://doi.org/10.1021/ac020209q

41. Wang S, Han C, Wang S, Bai L, Li S, Luo J, Kong L (2016) Development of a high speed countercurrent chromatography system with Cu(II)-chiral ionic liquid complexes and hydroxypropyl-β-cyclodextrin as dual chiral selectors for enantioseparation of naringenin. J Chromatogr A 1471:155. https://doi.org/10.1016/j.chroma.2016.10.036

42. Zhu Z, Zhang W, Pranolo Y, Cheng CY (2012) Separation and recovery of copper, nickel, cobalt and zinc in chloride solutions by synergistic solvent extraction. Hydrometallurgy 127:1. https://doi.org/10.1016/j.hydromet.2012.07.001

43. Wellens S, Goovaerts R, Möller C, Luyten J, Thijs B, Binnemans K (2013) A continuous ionic liquid extraction process for the separation of cobalt from nickel. Green Chem 15:3160. https://doi.org/10.1039/c3gc41519h

44. Kauffman GB, Adams ML (1989) The separation of cobalt from nickel by anion exchange chromatography. J Chem Educ 66:166. https://doi.org/10.1021/ed066p166

45. a: Bhave NS, Dhudey SR, Kharat RB (1978) Separation of copper(II), nickel(II), palladium(II), and cobalt(II) chelates with 4-S-benzyl-1-p-Cl-phenyl-5-phenyl-2,4-isodithiobiuret (BPPTB) from their binary mixture by adsorption thin-layer chromatography. Sep Sci Technol 13:193. http://doi.org/10.1080/01496397808057101; b: Mohammad A, Iraqi E, Sirwal YH (2003) New TLC system for simultaneous separation of iron, cobalt, and nickel ions from acidic and ammoniacal solutions. Sep Sci Technol 38:2255. http://doi.org/10.1081/ss-120021623

46. Khuhawar MY, Soomro AI (1993) Gas and liquid chromatographic studies of copper(II), nickel(II), palladium(II) and oxovanadium(IV) chelates of some fluorinated ketoamine Schiff bases. J Chromatogr A 639:371. https://doi.org/10.1016/0021-9673(93)80279-H

47. Toyota E, Itoh K, Sekizaki H, Tanizawa K (1996) Chromatographic separation of diastereomeric schiff base copper(II), nickel(II), and zinc(II) chelates from α-amino acid racemates. Bioorg Chem 24:150. https://doi.org/10.1006/bioo.1996.0013

48. Mirza MA, Khuhawar MY, Arain R, Choudhary MA, Kandhro AJ, Jahangir TM (2013) Micellar Electrokinetic Chromatographic Separation/Determination of Uranium, Iron, Copper and Nickel From Environmental Ore Samples Using Bis(salicylaldehyde)meso-stilbenediimine as Chelating Reagent. Asian J Chem 25:3719. http://dx.doi.org/10.14233/ajchem.2013.13728

49. Loonker S, Sethia JK (2009) Use of newly synthesized guar based chelating ion exchange resin in chromatographic separation of copper from nickel ions. Bulg Chem Commun 41:19

50. Vláčil F, Khanh HD, (1980) Extraction-chromatographic separation of iron from cobalt, nickel, and copper using dibenzyl sulphoxide solution as the stationary phase. Fresenius Z Anal Chem 302:36. https://doi.org/10.1007/bf00469760

51. Anderson K, Rodríguez H, Seddon KR (2009) Phase behaviour of tri-hexyl(tetradecyl)phosphonium chloride, nonane and water. Green Chem 11:780. https://doi.org/10.1039/b821925g

52. a: Wood PL, Hawes D, Janaway L, Sutherland IA (2003) Stationary phase retention in CCC: modelling the J-type centrifuge as a constant pressure drop pump. J Liq Chromatogr Relat

Technol 26:1373. http://doi.org/10.1081/jlc-120021256; b: He C-H, Zhao C-X (2007) Retention of the stationary phase for high-speed countercurrent chromatography. AIChE J 53:1460. http://doi.org/10.1002/aic.11185

53. Barrera NM, McCarty JL, Dragojlovic V (2002) Effects of Concentration on Hexaaquacobalt(II)/Tetrachlorocobalt(II) Equilibrium. A Discovery-Oriented Experiment for Chemistry Students. Chem Educ 7:142. https://doi.org/10.1007/s00897020559a

54. Arce A, Earle MJ, Katdare SP, Rodríguez H, Seddon KR (2006) Mutually immiscible ionic liquids. Chem Commun 2548. http://doi.org/10.1039/b604595b

55. a: Winter A, Thiel K, Zabel A, Klamroth T, Pöppl A, Kelling A, Schilde U, Taubert A, Strauch P (2014) Tetrahalidocuprates(II)—structure and EPR spectroscopy. Part 2: tetrachloridocuprates(II). New J Chem 38:1019. http://doi.org/10.1039/c3nj01039b. b: Ruhlandt-Senge K, Müller U (1990) Kristallstrukturen der Tetrachloroniccolate (PPh$_4$)$_2$[NiCl$_4$] und [Na-15-Krone-5]$_2$[NiCl$_4$]/Crystal structures of the tetrachloroniccolates (PPh$_4$)$_2$[NiCl$_4$] and [Na-15-Crown-5]$_2$[NiCl$_4$]. Z Naturforsch (B) 45:995. https://doi.org/10.1515/znb-1990-0714; c: Piecha-Bisiorek A, Bieńko A, Jakubas R, Boča R, Weselski M, Kinzhybalo V, Pietraszko A, Wojciechowska M, Medycki W, Kruk D (2016) Physical and structural characterization of imidazolium-based organic–inorganic hybrid: (C$_3$N$_2$H$_5$)$_2$[CoCl$_4$]. J Phys Chem A 120:2014. http://doi.org/10.1021/acs.jpca.5b11924

56. Williamson Y, Davis JM (2005) Modeling of anti-Langmuirian peaks in micellar electrokinetic chromatography: benzene and naphthalene. Electrophoresis 26:4026. https://doi.org/10.1002/elps.200500245

57. Verzele M, Simoens G, Van Damme F (1987) A critical review of some liquid chromatography systems for the separation of sugars. Chromatographia 23:292. https://doi.org/10.1007/BF02311783

58. Nitsch E (1974) Method of producing fructose and glucose from sucrose. US patent 3,812,010

59. a: Shaw PE, Wilson III CW (1983) Separation of fructose, glucose and sucrose in fruit by high performance liquid chromatography using UV detection at 190 nm. J Sci Food Agric 34:109. http://doi.org/10.1002/JSFA.2740340116; b: Filip M, Vlassa M, Coman V, Halmagyi A (2016) Simultaneous determination of glucose, fructose, sucrose and sorbitol in the leaf and fruit peel of different apple cultivars by the HPLC–RI optimized method. Food Chem 199:653. http://doi.org/10.1016/j.foodchem.2015.12.060; c: Schmid T, Baumann B, Himmelsbach M, Klampfl CW, Buchberger W (2016) Analysis of saccharides in beverages by HPLC with direct UV detection. Anal Bioanal Chem 408:1871. http://doi.org/10.1007/s00216-015-9290-1

60. González J, Remaud G, Jamin E, Naulet N, Martin GG (1999) Specific natural isotope profile studied by isotope ratio mass spectrometry (SNIP−IRMS): ^{13}C/^{12}C ratios of fructose, glucose, and sucrose for improved detection of sugar addition to pineapple juices and concentrates. J Agric Food Chem 47:2316. https://doi.org/10.1021/JF981093V

61. Véronèse T, Bouchu A, Perlot P (1999) Rapid method for trehalulose production and its purification by preparative high-performance liquid chromatography. Biotechnol Tech 13:43. https://doi.org/10.1023/a:1008857613103

62. Shinomiya K, Ito Y (2006) Countercurrent chromatographic separation of biotic compounds with extremely hydrophilic organic-aqueous two-phase solvent systems and organic-aqueous three-phase solvent systems. J Liq Chromatogr Relat Technol 29:733. https://doi.org/10.1080/10826070500509298

63. Doremus RH (1985) Crystallization of sucrose from aqueous solution. J Colloid Interface Sci 104:114. https://doi.org/10.1016/0021-9797(85)90015-3

64. Kim H-J, Chen F, Wang X, Chung HY, Jin ZY (2005) Evaluation of antioxidant activity of vetiver (*Vetiveria zizanioides* L.) oil and identification of its antioxidant constituents. J Agric Food Chem 53:7691. https://doi.org/10.1021/JF050833E

65. a: Adams RP, Zhong M, Turuspekov Y, Dafforn MR, Veldkamp JF (1998) DNA fingerprinting reveals clonal nature of *Vetiveria zizanioides* (L.) Nash, Gramineae and sources of potential new germplasm. Mol Ecol 7:813. http://doi.org/10.1046/J.1365-294X.1998.00394.X; b: Guenther E (1950) The essential oils, vol 4 (ed: Guenther E). Van Nostrand, New York, p 156

66. Guenther E (1948) The essential oils: history—origin in plants—production—analysis, vol 1 (ed: Guenther E). Van Nostrand, New York, p 153
67. Massardo DR, Senatore F, Alifano P, Del Giudice L, Pontieri P (2006) Vetiver oil production correlates with early root growth. Biochem Syst Ecol 34:376. https://doi.org/10.1016/J.BSE. 2005.10.016
68. Fahlbusch K-G, Hammerschmidt F-J, Panten J, Pickenhagen W, Schatkowski D, Bauer K, Garbe D, Surburg H (2005) Flavors and fragrances. In: Ullmann's Encyclopedia of Industrial Chemistry, vol 1–3 (ed: Elvers B). Wiley-VCH, Weinheim, Germany, p 1. https://doi.org/10. 1002/14356007.a11_141
69. a: Chahal KK, Bhardwaj U, Kaushal S, Sandhu AK (2015) Chemical composition and biological properties of *Chrysopogan zizanoides* (L.) Roberty syn. *Vetiveria zizanoides* (L.) Nash—a review, Indian J Nat Prod Resour 6:251; b: Mallavarapu GR, Syamasundar KV, Ramesh S, Rajeswara Rao BR (2012) Constituents of south Indian vetiver oils. Nat Prod Commun 7:223. https://doi.org/10.1177/1934578X1200700228
70. a: Maeda YY, Watanabe ST, Chihara C, Rokutanda M (1988) Denaturation and renaturation of a *β*-1,6;1,3-glucan, lentinan, associated with expression of T-cell-mediated responses, Cancer Res 48:671; b: Xu X, Zhang X, Zhang L, Wu C (2004) Collapse and association of denatured lentinan in water/dimethylsulfoxide solutions. Biomacromolecules 5:1893. http://doi.org/10. 1021/bm049785h; c: Xu X, Wang X, Cai F, Zhang L (2010) Renaturation of triple helical polysaccharide lentinan in water-diluted dimethylsulfoxide solution. Carbohydr Res 345:419. http://doi.org/10.1016/j.carres.2009.10.013
71. Jiang ZG, Du QZ, Sheng LY (2009) Separation and purification of lentinan by preparative high speed counter current chromatography. Chinese J Anal Chem 37:412
72. a: Ignatova S, Wood P, Hawes D, Janaway L, Keay D, Sutherland I (2007) Feasibility of scaling from pilot to process scale. J Chromatogr A 1151:20. https://doi.org/10.1016/j.chroma.2007. 02.084; b: Sutherland IA (2010) Encyclopedia of chromatography, vol. 3, 3rd edn. (ed: Cazes J). Taylor & Francis, Boca Raton, p 2116; c: Sutherland IA, Booth AJ, Brown L, Kemp B, Kidwell H, Games D, Graham AS, Guillon GG, Hawes D, Hayes M, Janaway L, Lye GJ, Massey P, Preston C, Shering P, Shoulder T, Strawson C, Wood P (2001) Industrial scale-up of countercurrent chromatography. J Liq Chromatogr Relat Technol 24:1533. https://doi. org/10.1081/jlc-100104362; d: Sutherland IA, Brown L, Graham AS, Guillon GG, Hawes D, Janaway L, Whiteside R, Wood P (2001) Industrial scale-up of countercurrent chromatography. J Chromatogr Sci 39:21. https://doi.org/10.1093/chromsci/39.1.21; e: Sutherland I, Hawes D, Ignatova S, Janaway L, Wood P (2005) Review of progress toward the industrial scale-up of CCC. J Liq Chromatogr Relat Technol 28:1877. https://doi.org/10.1081/jlc-200063521; f: Sutherland I, Ignatova S, Hewitson P, Janaway L, Wood P, Edwards N, Harris G, Guzlek H, Keay D, Freebairn K, Johns D, Douillet N, Thickitt C, Vilminot E, Mathews B (2011) Scalable technology for the extraction of pharmaceutics (STEP): the transition from academic knowhow to industrial reality. J Chromatogr A 1218:6114. https://doi.org/10.1016/j.chroma.2011.01.016
73. Abai M, Atkins MP, Hassan A, Holbrey JD, Kuah Y, Nockemann P, Oliferenko AA, Plechkova NV, Rafeen S, Rahman AA, Ramli R, Shariff SM, Seddon KR, Srinivasan G, Zou Y (2015) An ionic liquid process for mercury removal from natural gas. Dalton Trans 44:8617. https:// doi.org/10.1039/C4DT03273J
74. a: Earle MJ, Gordon CM, Plechkova NV, Seddon KR, Welton T (2007) Decolorization of ionic liquids for spectroscopy. Anal Chem 79:758. http://doi.org/10.1021/ac061481t; b: Earle MJ, Gordon CM, Plechkova NV, Seddon KR, Welton T (2007) Welton, Decolorization of ionic liquids for spectroscopy. Anal Chem 79:4247. https://doi.org/10.1021/ac070746g

Chapter 8
Commercial Production of Ionic Liquids

Thomas J. S. Schubert

Abstract By 2019, a large number of different ionic liquids (ILs) had been synthesized. There is a large uncertainty about the correct number, but there must be at least a couple thousand different ones described in the scientific literature and a similar number in labs of chemical companies. To synthesize ionic liquids, numerous organic and inorganic compounds can be used and combined. A major goal of chemists who are designing novel ionic liquids is to predict ionic compounds that may be liquid at room temperature (or are at least liquid below a melting point of 100 °C, which is the common definition for ILs). A lot of experience is necessary to avoid the synthesis of too many substances that do not have sufficient properties. It is also possible to use efficient physico-chemical tools to design ionic liquids, such as the "molecular volume approach" released by Krossing et al. in 2007 [1]. Using this method, it was possible to predict fundamental data, such as melting points, viscosities, conductivities, and densities, which matched very well with experimental data. As a consequence, over the past two decades ionic liquids became indeed "designer solvents" but also "designer electrolytes" and "designer (functional) materials." Powerful tools combined with experience made it possible after more than 20 years of intensive research in many fields for ionic liquids to become industrially important products. As a consequence, the importance of methods for the commercial production became more and more relevant. This overview for understandable reasons cannot be complete because the producers of ionic liquids are surely using numerous lab secrets and non-published information for their processes. Thus, in this chapter, the most prominently known production methods are summarized.

Keywords Commercial production · Environmental impact · Life cycle analysis and costing · Microreaction technology · Regulatory issues

T. J. S. Schubert (✉)
IOLITEC Ionic Liquids Technologies GmbH, Salzstrasse 184, 74076 Heilbronn, Germany
e-mail: schubert@iolitec.de

© Springer Nature Switzerland AG 2020   191
M. B. Shiflett (ed.), *Commercial Applications of Ionic Liquids*, Green Chemistry
and Sustainable Technology, https://doi.org/10.1007/978-3-030-35245-5_8

8.1   General Aspects of Ionic Liquids Production Methods

8.1.1   Aspects of Purity

For many R&D-related purposes, researchers will typically—for quite understand-able reasons—use the highest available quality. In physical chemistry and electro-chemistry, it is often essential for many reasons to operate with the highest available purity. Nevertheless, early investigations of published data about melting points or viscosities of common ionic liquids led to some viscosity deviations of more than 30% depending on the source of the material. In addition, in our own labs, we had a couple of surprises when ionic liquids for some customers were synthesized. In one case, triethylsulfonium bis(trifluoromethylsulfonyl)imide ($[S_{222}][TFSI]$ or $[S_{222}][NTf_2]$) was described in the literature as a room-temperature ionic liquid, and it also appeared as a liquid after bottling it. However, during transport to the customer, it became a solid, and it never became liquid again at room temperature. The reason for the higher melting point was a higher purity because impurities reduce the melting point. As one can imagine, it led to a customer complaint, since they expected to receive a liquid material.

In another case, a very common ionic liquid $[C_1C_6im][Cl]$ was described in the literature to be a liquid at room temperature. When it was synthesized in a 25 kg batch size following an advanced synthesis protocol (in terms of purification of the corresponding starting materials), it had a nearly colorless appearance. During the work-up, after removing the solvent, it surprisingly crystallized overnight causing a lot of trouble to remove it from the reaction vessel.

If someone needs commercial quantities of an ionic liquid, it is essential that the customer has a clear understanding with the producer what quality is required or what quality is sufficient for the application. This is an important question because any additional purification step in general increases the production costs.

8.1.2   Aspects of Price

The price often determines the scope of applications. This is of importance, especially if an ionic liquid-based technology is in competition with established technologies, where typically a high cost pressure exists. A lower sensitivity to the cost of the IL can be observed in fields where an ionic liquid is part of a system or process creating a completely novel field of application ("disruptive technologies").

In order to save valuable time in applied science and even more in corporate R&D, it could be useful to consider the following questions before starting experiments for implementing a proof of concept of an ionic liquid-based technology:

I. Added value: Is there really an added value?

Look at the product through the eyes of a user of your technology.

II. "Life-Cycle-Costing": Are the costs for the complete lifetime of a technology lower?

This question is of interest if processes are involved. Can the operating costs for a process combined with energy savings and/or lower disposal costs, and so forth, justify higher costs for the ionic liquid?

III. Regulatory issues: Can your technology replace a risky or hazardous process?

The search for novel compounds can become important because of regulatory issues. Well-known examples include replacements for perfluorooctylsulfonic acid and its derivatives or replacing metals, such as cadmium or chromium. In some cases, there are exemptions if a more environmentally friendly and/or less toxic solution does not exist.

IV. Would you be willing to pay the price for this novel technology?

Put yourself in the user's shoes and ask if it is worth it to pay the price for your technology. It can be beneficial to convince people in your own personal environment first. If you can answer at least one of these questions with "yes," then you should proceed!

8.1.3 The Aspect of Ionic Liquids as "Green Solvents": Toxicity and Environmental Impact

Originally, ionic liquids were often called "green solvents," namely because of their low volatility. Since in principle, a broad variety of organic and inorganic materials can be combined, this label is of course problematic. In fact, some ionic liquids can be green, but some of them can be toxic or nonbiodegradable or even persistent and bio-accumulative. As a consequence, each ionic liquid has to undergo the region-specific test protocols as soon as it becomes a commercial product (e.g., REACH is the European Regulation on Registration, Evaluation, Authorization, and Restriction of Chemicals aimed at protecting human health and the environment).

Unfortunately, the assumption that the combination of a cation, which is known to be nontoxic in one compound, with an anion, which is known to be nontoxic in another compound, necessarily yields a nontoxic compound is not valid. Thus, each novel combination of cation and anion must be tested separately.

8.2   Methods for the Commercial Production of Ionic Liquids

From the thousands of ionic liquids reported in the literature over the past two decades, only a few percent (less than 50) have become commercially relevant. In addition, because of their complexity, it is not possible to include all methods within this chapter.

8.2.1   Purification of Starting Materials

As mentioned in Sect. 8.1.1, to achieve sufficient quality for ionic liquids, it is necessary to work with the highest quality starting materials. Once a reaction is completed, it is typically difficult and cost-intensive to remove all nonvolatile impurities from the ionic liquid because of its ultra-low vapor pressure. As a consequence, it is essential to work with purified starting materials with a purity greater than 98%.

8.2.1.1   Purifications of Amines and *N*-Heterocycles

By far, most ionic liquids are based on nitrogen-containing cations, such as ammonium, pyrrolidinium, pyridinium, and imidazolium being the most prominent. Most of these materials show a tendency to react with CO_2 in the air to form carbonates. Thus, to achieve the best results, those impurities can be removed by distillation over potassium hydroxide. Since amines and in particular *N*-heterocycles often have high boiling points, the distillation should typically be operated under reduced pressure. To avoid any further contamination with CO_2, the use of an inert atmosphere is beneficial.

8.2.1.2   Purifications of Alkyl Halides

Many alkyl halides typically contain stabilizers, such as hydroquinone, that can be removed easily by stirring with 5–10 vol.% concentrated sulfuric acid. After settling, a dark liquid can be separated.

8.2.1.3   Purification of Inorganic Salts

Most of the inorganic salts (sodium, potassium, and ammonium cations and hexafluorophosphate, tetrafluoroborate, and bis(trifluoromethylsulfonyl)imide anions) can be used without any further purification. If higher purities are needed, recrystallization or zone melting are the methods of choice.

8.2.1.4   Purification of Solvents

For many synthesis protocols, the use of solvents is beneficial. Typical solvents are dichloromethane, ethylacetate, acetonitrile, or methanol. To achieve the best ionic liquid quality, it is required to operate with starting materials having the highest commercially available purity or to purify them by following common purification protocols.

8.2.2   Industrial Relevant Types of Reactions

It is difficult to draw a line where industrially relevant production starts and lab scale ends; this depends on each application. The smallest batches we produce for commercial purposes are in the range of 5 kg. So, for purposes of this book, lab scale means below 5 kg, and commercial scale is 5 kg and above.

In the following, the most important examples of reactions are described.

8.2.2.1   Alkylation Reactions

Intermediate Ionic Liquids

To produce ionic liquid intermediates, numerous alkylating agents can be used. Many ionic intermediates may also fall under the common definition of an ionic liquid. Those ionic intermediates, such as $[C_1C_4im][Cl]$, can be converted in a second step, typically via an anion-exchange reaction, into other ionic liquids.

The most important commercially relevant intermediates are chlorides, bromides, methylsulfates, hydroxides, and methylcarbonates.

Ionic Liquids Generated by Alkylation

In some cases, an industrially relevant ionic liquid can be directly synthesized by alkylation. This is the case if the leaving group of the alkylating agent also becomes the "final" anion. Good examples are $[C_1C_2im][OTf]$ and $[C_1C_2im][C_2SO_4]$.

Quaternization of Amines and Derivatives

The quaternization of amines and N-heterocycles is by far the most important and most frequently used synthesis in the field of ionic liquids. For this synthesis in particular, alkyl halides are used, but of course other alkylating agents, such as dialkylsulfates or alkyl triflates, just to name a few, can be used as well. The most important issue for achieving good results is to purify all starting materials before the quaternization reaction is performed, as described in Sect. 8.2.1.

At lab scale (<5 kg), reactions are typically described in the literature as being carried out in glass flasks equipped with a reflux condenser. To achieve the best results, it is beneficial to use the Schlenk technique to exclude oxygen and moisture as much as possible because *N*-heterocycles have the tendency to form colored side products in the presence of oxygen.

Less often, quaternization reactions are performed in autoclaves, though the advantages of this type of technique are generally underestimated. The reaction rates and purity are typically much higher because there is a reduced tendency for side reactions to occur.

At commercial scale (>5 kg), three different reaction setups are feasible.

The first setup uses larger reflux reactors (up to 100 L reaction volume), operated at atmospheric pressure for reactions having a sufficient reaction rate. Consider the alkylation of 1-methylimidazole with 1-bromobutane. It takes approximately 10 h at 40 °C for completion of the reaction, while it takes three weeks at 55 °C for the higher homologue 1-chlorobutane. Because of this long reaction time, the latter case is better performed under pressure using an autoclave, leading to higher reaction rates.

The second setup uses larger autoclaves. While lab autoclaves can typically be operated up to pressures of 200 bar, at a larger scale of 1 metric ton, 10 bar is the upper pressure limit. The scale-up of reactions performed in autoclaves is not trivial. Since some alkylation reactions often have a strong tendency to show an autocatalytic acceleration of the reaction rate, important reaction parameters must be determined, such as the temperature dependence of the reaction rate, the reaction enthalpy, and the activation energy. By monitoring these important parameters on a smaller scale, it is possible to scale up to a higher reaction volume.

The third setup, which so far is not well-established in organic synthesis and generally underestimated, is the use of continuous-flow microreaction technology. For example, an amine (including nitrogen-containing heterocycles) and an alkylating agent are continuously taken from separate reservoirs, and each is pumped at elevated pressures (approximately 5–10 bar) into a microreactor. To achieve these pressures, high-performance liquid chromatography (HPLC) pumps can provide the best results. Both starting materials are mixed within a mixer having a reaction-specific, tailor-made geometry. After the mixing process, the reaction mixture has to remain at a specific reaction temperature until the reaction is complete. In some cases, product and starting materials may form a 2-phase system, leading to deceleration of the reaction rate. If such a 2-phase system occurs, the use of a so-called split-and-recombine mixer often leads to good results. Take for instance, the synthesis of 1-butyl-3-methylimidazolium bromide, the use of this technique leads to superior results if operated at elevated temperature and pressure (Fig. 8.1).

(a) **(b)**

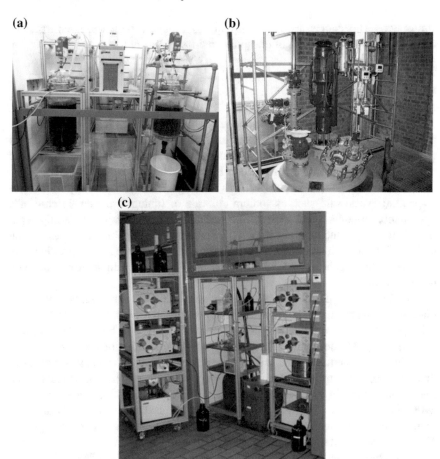

(c)

Fig. 8.1 (a) Glass reactors, batch size, 20 kg (© IOLITEC, 2019). (b) Stainless steel reactor, batch size 1 metric ton, operating pressure up to 10 bar (© IOLITEC, 2019). (c) Continuous-flow microreaction technology, 50 kg per day (© IOLITEC, 2019)

Alkylation of Phosphine Derivatives

At the present time, phosphonium-based ionic liquids are underrepresented in the scientific literature compared with imidazolium-based ionic liquids. Phosphonium-based salts (many of them are commonly defined as ionic liquids) are used in commercial processes, such as phase-transfer catalysts or as process chemicals in the production of semiconductors. Consequently, these materials are already produced on a commercial scale, but the corresponding knowledge is often not publicly available. For example, substances of the type PR_3R'-Cl (R is butyl or hexyl; R' is hexyl, octyl, decyl, or dodecyl) can be synthesized in autoclaves from pyrophoric trialkylphosphine and the corresponding alkyl halide in acetonitrile or toluene under an inert atmosphere. This dangerous operation requires a lot of experience. Even traces of oxygen lead to the formation of phosphine oxides as side products.

8.2.2.2 Anion Metathesis

The anion metathesis, which should not be confused with ring-closing metathesis, is a type of reaction where the anion of an intermediate (as defined in Sect. 8.2.2.1) is replaced by a different anion. It is important to differentiate whether a hydrophobic or a hydrophilic ionic liquid is formed as a product.

The better the anion metathesis works, the higher the lattice energy of the corresponding waste salt, such as sodium chloride or lithium bromide. In case of a hydrophobic product, these highly water-soluble side products can be easily separated from the product-containing phase by extraction or separation. Depending on each product, a small amount of cross-contamination by the reaction-specific waste salts occurs that can be reduced to values below 10 ppm by washing the product with deionized water.

The situation is different for hydrophilic ionic liquids. In this case, both intermediate and product ionic liquids as well as the waste salt are soluble in water, so that it is not possible to separate the product from the waste salt by phase separation. Consequently, it is necessary to choose an organic solvent that dissolves the target ionic liquid sufficiently but has only a poor tendency to dissolve the waste salt. For this purpose, polar aprotic solvents with low boiling points give good results, but the best solvent must be identified for each case. Nevertheless, the halide content of hydrophilic ionic liquids produced by using anion metathesis is in nearly all cases higher than that for hydrophobic ionic liquids because of the comparably high solubility of waste salts in the product phase. Therefore, to achieve low halide content for hydrophilic ionic liquids, further purification is necessary (see Sect. 8.3).

An alternative method for performing an anion metathesis is the use of ion-exchange resins. This method has not been applied to the commercial production of ionic liquids, but has, if automated, a significant potential because it is of interest for synthesizing hydrophilic ionic liquids. Furthermore, it is also a very good approach for preparing hydroxide-based ionic liquids, which can be used as intermediates and transferred into a broad variety of ionic liquids by a simple acid–base reaction.

8.2.2.3 Acid–Base Reactions

Acid–base reactions are another commonly used type of synthesis for commercially relevant ionic liquids.

Ionic Liquids Formed by Brønsted Acids and Bases

Historically, Brønsted acid–base reactions were the first type of reactions leading to ionic liquids, such as Walden's "red oil" ethylammonium nitrate. In this case, ionic liquids can be generated by a simple proton transfer from a Brønsted acid to a Brønsted base. To form stable ionic liquids, the difference in the pK_a values of the acid and base should be above 10, otherwise the temperature-dependent equilibrium between the starting materials and product will shift more and more to the side of the starting materials. By this type of reaction, commercially relevant ammonium-based ionic liquids, such as ethylammonium nitrate and formate, are prepared on a larger scale [2].

Ionic Liquids Formed by Lewis Acids and Bases

In addition to Brønsted-type reactions, Lewis acid–base reactions are also of technical relevance for the synthesis of tetrachloroaluminate-based ionic liquids, which are used in catalysis [3] as well as for electrodeposition of metals [4]. In terms of safety, these reactions are extremely challenging. The addition of a Lewis acid, such as $AlCl_3$, must be made in small portions under adequate stirring to avoid hot spots or overheating. Furthermore, it is important to exclude even traces of water, since $AlCl_3$ reacts violently and produces toxic and corrosive HCl. This can be avoided by performing the complete process under an inert atmosphere.

Creating Anions via Reactive Intermediates

Some ionic liquids can be produced by reactions of reactive ionic intermediates, which form neutral, nonionic side products during the reaction. One of the major advantages of this route is that it is halogen-free. A disadvantage is the fact that the solvent (often water or methanol) has to be removed by using time-consuming vacuum technology.

Quaternary phosphonium, ammonium, numerous types of N-heterocycles, and R,R',R''-imidazolium hydroxides (R,R',R'' is H or -alkyl) can be converted into numerous types of ionic liquids with alternative anions just by neutralization with acetates, triflates, or hydrogen sulfates, just to name a few. Some less important approaches involve hydroxide-based ionic liquids being synthesized from halide-based ionic liquids via ion-exchange resins (Sect. 8.2.2.2). In this context, it is worth noting that an aqueous solution of 1,3-dialkylimidazolium hydroxide, which is of particular commercial interest, is only stable up to concentrations of approximately 5–10 wt%. At higher concentrations, the C2 position is deprotonated to form the corresponding carbenes, which typically undergo further reactions, such as dimerization (Scheme 8.1).

An elegant route is to synthesize (methyl) carbonate-based ionic liquids (CBILS©), which are typically provided as 30 wt% solutions in methanol. They

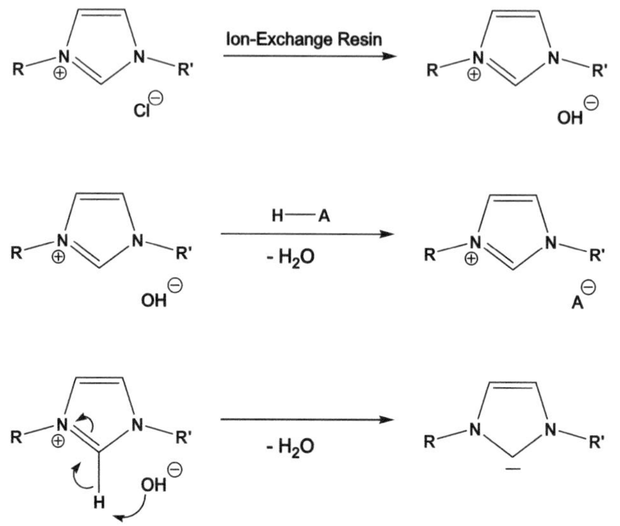

Scheme 8.1 Synthesizing ionic liquids using hydroxides as intermediates

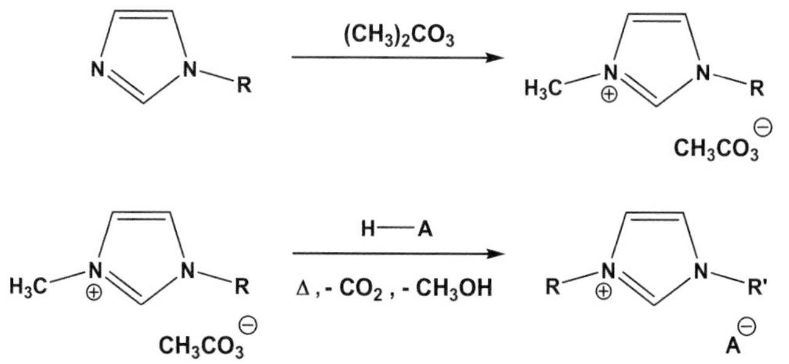

Scheme 8.2 CBILS© technology, for more information see also: www.proionic.com

are useful intermediates for introducing an anion through the reaction of an acid with the methylcarbonate anion. A new ionic liquid is formed, while methanol and CO_2 are formed as side products (Scheme 8.2).

8.2.2.4   Other Types of Reactions

Synthesis of Heterocycles

Ionic liquids based on nitrogen-containing heterocycles, such as imidazolium, pyridinium, and pyrrolidinium, dominate the scientific literature in terms of the number of publications. The 1-methyl derivatives of these heterocycles are commercially available on larger scale, while longer chain lengths are only available for the mid or lab scale. Consequently, if there is an increasing demand, especially for chain lengths

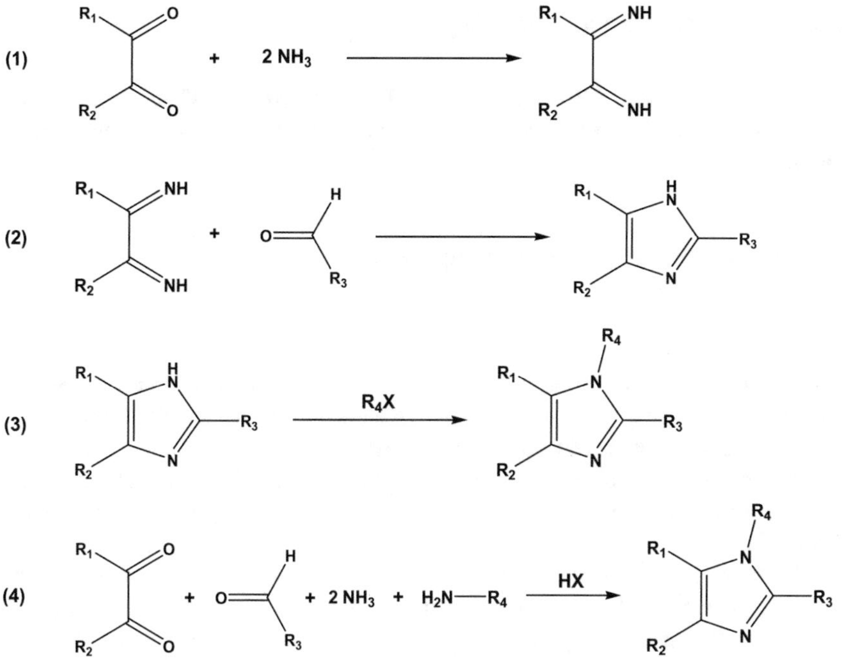

Scheme 8.3 Strategies to synthesize heterocycles: (1) Debus reaction [5], (2) Radziszewski reaction [6], (3) deprotonation/alkylation of the heterocycle, (4) synthesis of 1-methylimidazoles by Arduengo et al. [7]

larger than methyl, they have to be synthesized on a larger scale. In principle, this can be achieved by two different approaches. The first one is to perform a cyclization reaction. Unfortunately, the yields of these reactions are sometimes comparably low. In addition to that, the starting materials are sometimes expensive or are only available in small quantities. Therefore, in terms of costs, it may be easier and cheaper to alkylate the corresponding deprotonated heterocycle (Scheme 8.3).

Task-Specific ILs (TSILs)

The beautiful concept of task-specific ionic liquids (TSILs) was introduced by Davis in 2001 [8]. This effort led to numerous novel ionic liquids, but though they sometimes have unique properties, it has not been reported that they have been produced on a relevant commercial scale. In principle, a plethora of organic reactions can be applied to create ionic liquids having functionalized groups in their side chains, but because of the complexity of this issue, it is not possible to consider them in detail.

8.2.3   Purification of Ionic Liquids

As mentioned in Sect. 8.2.1, once an ionic liquid is synthesized, its purification is often challenging. Nevertheless, there are methods available, but they may require a significant effort.

8.2.3.1   Removing Typical Ionic Impurities

Extraction

For hydrophobic ionic liquids, it is relatively easy to remove ionic impurities, such as inorganic salts, using the anion-exchange process and washing with purified water. The halide content of the water determines the quality of each ionic liquid. Depending on the specific water uptake of each hydrophobic ionic liquid at a given temperature, water must be removed at reduced pressure after the washing procedure (see also Sect. 8.2.3.2).

Removing ionic impurities from hydrophilic ionic liquids is of enormous interest, but it is generally difficult to achieve since washing with water is of course not an option. In our company (IOLITEC), many different approaches are currently under investigation at the lab scale, but until they are evaluated in terms of potential submission of intellectual rights, it is not possible to report on them.

On a larger scale, extraction processes that can be continuously operated are of interest. In our own labs, we investigated the potential of counterflow extraction for hydrophobic ionic liquids with the result that it becomes feasible on larger scale when compared to batch extraction as shown in Fig. 8.2, which is in principle just a larger version of a separatory funnel.

Another option, which can be operated continuously, is mixer–settler systems, which are commonly used in mineral processing.

Recrystallization

If a product is not a room-temperature ionic liquid or an ionic liquid intermediate, recrystallization and zone melting are useful purifications methods. For any kind of recrystallization, a suitable solvent must be identified for each ionic liquid. In terms of costs, this method should be avoided if possible, since it suffers from a significant yield loss and extra cost associated with solvents and drying.

Zone Melting

Zone melting is a process that is typically applied in purifications of semiconductors to produce ultrapure silicon. It is an alternative to recrystallization and can thus

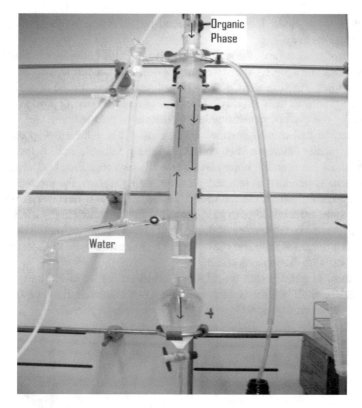

Fig. 8.2 Pilot production using a counterflow extraction column (© IOLITEC, 2019)

be applied for the purification of similar ionic liquids. Though so far it has not been applied on an industrial scale, it is an interesting and cost-effective purification method.

8.2.3.2 Removing Water and Organic Solvents

Solvents with low boiling points, such as water and organic solvents, can be removed by distillation at reduced pressure. Since ionic liquids are good solvents for numerous types of solids including adsorption media, such as SiO_2 or zeolites, adsorption drying techniques are not favored.

On the lab scale, rotary evaporators are commonly used for pre-drying and operate at pressures of 0.1–0.5 bar. To meet higher specifications by reducing the amount of solvents, a higher vacuum of $(1-5) \times 10^{-3}$ bar has to be applied and can be achieved by using oil pumps.

In principle, these steps can also be applied to the mid-scale (1–50 kg), but on a scale of 50 kg and above, larger rotary evaporators, which so far can only be operated in batch mode, have to be replaced by alternative methods that can be operated continuously.

In terms of process time, the removal of solvents is the bottleneck in commercial production of ionic liquids. Otherwise, every synthesis procedure and any purification step can be performed faster. Consequently, developments for the optimization of production processes are currently concentrating on faster drying processes, which are typically achieved by generating larger surfaces of the medium.

One method that can be operated continuously is a thin-film evaporator. This device is typically built up of two columns forming a double wall. A heat-transfer fluid circulates between the inner and outer walls for controlling temperature. The inner column is operated at reduced pressure. At its surface, the product phase is poured continuously from the top to the bottom, and it is spread, for example, by rotating stainless steel paddles to generate a thin film as shown in Figs. 8.3 and 8.4.

By this operation, the surface of the product phase, having an interphase with the vacuum, is enlarged significantly resulting in an enhanced mass transport from the product phase into the gaseous phase.

An important point is that for effective industrial drying of ionic liquids there are no "off-the-shelf" solutions. New techniques other than thin-film evaporators are currently under investigation.

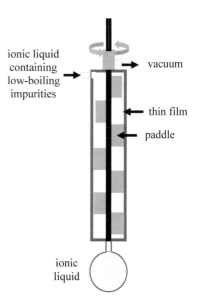

Fig. 8.3 Schematic of falling-film evaporator with rotating paddles

Fig. 8.4 Laboratory falling-film evaporator (© IOLITEC, 2019)

8.2.3.3   Removing Unreacted Amines and Heterocycles

Generally, removing polar starting materials having high boiling points must be avoided. While solvents and many alkylating agents can be removed by distillation under reduced pressure, unreacted amines and heterocycles show a strong tendency to remain within the ionic liquid phase. Consequently, it is easier to perform each synthesis process in a way that a reaction of these compounds is nearly 100% complete.

8.3   Outlook

Though ionic liquids have already made their way into commercial processes and applications, they will have the chance to enter even more fields if the prices drop significantly. In many cases, ionic liquid producers see themselves faced with a "chicken-and-egg problem," in other words, if there is more demand the prices drop, but on the other side there is more demand only if the prices drop.

8.3.1 Enhanced Production Technologies

One key to the success of ionic liquid technologies is the development of production processes that are optimized for the specific challenges of their production. A unique class of materials also needs unique production and especially purification technologies. As mentioned in Sect. 8.2.3.2, if fast and cost-effective processes are available, the prices of ionic liquids will drop significantly. For the lower ton scale, significant advancements for alkylation reactions were made by replacing batch processes with continuous-flow technology. In addition, if further progress can be made for extraction and drying processes, it will definitely reduce time and therefore cost for implementing and commercializing ionic liquids for numerous other technologies.

8.3.2 Starting Materials

In terms of commercialization of ionic liquids, a detailed analysis of the availability of materials is important. This means that each starting material should be fragmented into its initial raw materials available from natural feedstocks. This analysis leads to a clear picture of the raw materials on which each ionic liquid is dependent, the sustainability for the production of each ionic liquid, and where future bottlenecks may appear.

It is quite obvious that a higher demand for ionic liquids will also affect prices of relevant starting materials. If we take the price for lithium bis(trifluoromethylsulfonyl)imide as an example, in 2003, the price was approximately three times higher than today—just because of the fact that the market demand increased. If a producer moves its production from the scale of 1 ton to the scale of 100 tons, the percentage of personnel costs remains nearly the same, but at 100 times higher batch size. In addition, the prices for these specific starting materials also decrease. In conclusion, the economy of scale surely works for ionic liquids.

8.3.3 "Green Solvents" and Life Cycle Analysis (LCA)

Ionic liquids were often called "green solvents," which is a dangerous descriptor, since in terms of general toxicity not all of them are necessarily "green." In this context, it is important to point out that a fair balance must also consider the potential of each compound to reduce CO_2 emissions or to save raw materials by replacing or enhancing existing technologies. Once this information is assembled, the important CO_2 footprint can be determined. In view of the first major industrial applications, life cycle analyses were investigated for a couple of manufacturing processes. Because of the fact that such knowledge typically is the property of companies, it has been reported only in few cases [9].

If the non-negligible success of ionic liquids is to be extended to other fields, manufacturers must support the scientific community not only with technical data but also with information from life cycle analysis (LCA) on their CO_2 footprint and their overall environmental impact. This will give researchers motivation and also arguments to justify and to promote their work in the future.

8.4   Conclusions

In the world of chemistry, ionic liquids are today not a footnote anymore. Nevertheless, if we look back to the predictions made in the years 2000–2005 concerning their future role, many of them have not been fulfilled. On the other hand, several smaller applications not foreseen earlier are now being developed.

Therefore, typical production capacity today is producing on a scale from kilograms to a few metric tons—but not more, and the technologies described in this chapter are optimized for handling such amounts. It is also obvious that an overview of commercial production methods cannot be complete at this point because most manufacturers will not give a detailed insight into their procedures.

From a personal point of view, we are currently witnessing a number of applications under development. The question is—will ionic liquids become a major technology with significant increases in production volumes or will ionic liquids only be used at small scale in niche applications?

Sometimes the answer will be influenced by price, sometimes the decision will be influenced by people who are in a position to explain to the decision makers the added value, and in a few cases regulatory issues will lead to the use of ionic liquids in new markets. Let's wait and see!

References

1. Slattery JM, Daguenet C, Dyson PJ, Schubert TJS, Krossing I (2007) How to predict the properties of ionic liquids: a volume-based approach. Angew Chem 119:5480. https://doi.org/10.1002/ange.200700941
2. Information by IOLITEC Ionic Liquids Technologies GmbH, Heilbronn, Germany
3. Wasserscheid P, Schulz P (2008) Transition metal catalysis in ionic liquids, Chap 5.3. In: Wasserscheid P, Welton T (eds) Ionic liquids in synthesis, 2nd edn. Wiley-VCH, Weinheim, p 369ff. https://doi.org/10.1002/9783527621194
4. Endres F, Abbott A, MacFarlane D (eds) (2017) Electrodeposition from ionic liquids, 2nd edn. Wiley-VCH, Weinheim. https://doi.org/10.1002/9783527682706
5. Debus H (1858) Ueber die Einwirkung des Ammoniaks auf Glyoxal. Justus Liebigs Ann Chem 107(2):199. https://doi.org/10.1002/jlac.18581070209
6. Radziszewski B (1882) Ueber Glyoxalin und seine Homologe. Ber Dtsch Chem Ges 15(29):2706. https://doi.org/10.1002/cber.188201502245
7. Arduengo III AJ, Gentry Jr. FP, Taverkere PK, Simmons III HE (2001) Process for manufacture of imidazoles. US patent 6,177,575, issued 23 Jan 2001

8. Davis Jr. JH (2004) Task-specific ionic liquids. Chem Lett 33(9):1072–1077. https://doi.org/10.1246/cl.2004.1072
9. Project SCAIL-UP. Electrodeposition of aluminum on a pilot-plant scale, funded by the European Union. https://cordis.europa.eu/project/rcn/109186/factsheet/en

Part IV
Future Ionic Liquid Applications

Chapter 9
Natural Fiber Welding

Luke M. Haverhals, David P. Durkin and Paul C. Trulove

Abstract Ionic liquid-based (IL-based) manufacturing has the potential to revolutionize the materials industry and disrupt overdependence on petroleum-based plastics. Nature provides amazing materials at large scale; however, scalable techniques to mold and shape biomaterials have not existed at scale. Natural Fiber Welding, Inc. (NFW) has developed commercially viable processes and patent-protected materials that are scalable to meet modern challenges while reducing pollution and emissions. This chapter discusses a number of demonstrations that are being scaled for global markets as well as reviews several examples of new functionalities that can be achieved. Practical applications that create composites from waste textiles and new indigo dye processes are discussed. Examples of "exotic" materials that perform catalytic waste-water treatment and wearable energy storage are also reviewed. In all cases, NFW is able to make natural materials, such as cotton and silk, perform in new and unexpected ways. Prospects for scaling commercial applications are also discussed. With economically viable methods to reclaim, recycle, and reuse IL-based solvents, the future looks extremely bright. In the near future, industry-relevant complex natural composites will be produced at cost points that compete with incumbent synthetic plastics. This new way of manufacturing has significant potential to reduce emissions, eliminate pollution, and bring new circularity into, for example, the textile industry.

Keywords Biodegradable · Biomass · Cotton · Natural composites · Sustainability

9.1 Introduction

Ionic liquid-based solvents offer revolutionary opportunities to design and manufacture high-performance composites. Natural Fiber Welding, Inc. (NFW), located in Peoria, IL, USA, is developing an extremely flexible and powerful fabrication

L. M. Haverhals (✉)
Natural Fiber Welding, Inc., Peoria, IL, USA
e-mail: luke.haverhals@naturalfiberwelding.com

D. P. Durkin · P. C. Trulove
Department of Chemistry, United States Naval Academy, Annapolis, MD, USA

© Springer Nature Switzerland AG 2020 211
M. B. Shiflett (ed.), *Commercial Applications of Ionic Liquids*, Green Chemistry
and Sustainable Technology, https://doi.org/10.1007/978-3-030-35245-5_9

platform [1, 2] that has its roots in the discoveries of IL efficacy towards biopolymers—notably by Swatloski and Rogers [3], as well as Trulove, DeLong, and Mantz [4]. NFW is peerless as it develops efficient manufacturing processes for high-performance composite materials using abundant natural materials while preserving complex structures that are unique to natural materials [5–7]. Using tunable IL-based chemistries, NFW is pioneering fast, scalable fabrication processes that solve systemic problems within large industries. For example, NFW has patented processes that achieve zero-waste indigo dyeing while simultaneously recycling "waste" cotton fiber into denim fabrics [8]. In this chapter, we will discuss several examples of how fabrication through welding fiber fills important manufacturing gaps and is poised to provide renewable, biodegradable natural materials that outperform petroleum-based plastic incumbents across many types of applications.

It is well known and documented that petroleum-based synthetics are simultaneously wonderful and terrible [9]. This is particularly obvious within the textile industry where, during the past 60+ years, humanity has become increasingly reliant on petroleum-based synthetic plastics [10]. Presently, the textile industry uses around 100 billion pounds of polyester annually, which represents about two-thirds of the tonnage of all textiles. Innovation around polyester has been rapid because this polymer can be extruded into a variety of form factors ("formats") that are desirable for both manufacturers and end-users. Unfortunately, polyester is not biodegradable. Polyester fibers, both virgin and recycled, break loose from fabrics while being worn and during laundering. Recent studies have shown that as many as 100 million pounds of polyester microfiber are released into watersheds each year by more than one billion washing machines currently in operation around the globe [10, 11]. The rate synthetic plastic microfibers are released from textiles, tires, packaging, etc., continues to increase, and data suggest there may be more tonnage of non-biodegradable plastics in the oceans than fish by 2050 [10]. Compounding these problems is the fact that synthetics, such as polyester, absorb and concentrate toxins, such as microcystins [12]. Biologists are now documenting that many species of aquatic life that form the base of the food chain are consuming and concentrating toxin-laden plastics sometimes with detrimental effects [13–16]. Plastic microfiber pollution has been discovered in seafood [17], sea salt [18], and tap water [19] samples from around the world. Whereas microplastics in facial cleaners have been banned due to their known deleterious effects, microplastics from synthetic textiles is a more entrenched problem due to the scale of the textile industry [10].

Of course, there are explicable reasons why synthetic plastics have displaced market share from natural materials, such as cotton. The combination of performance and availability are the chief reasons synthetics have taken market share from cotton. For example, whereas cotton fabrics are produced from relatively short staple fibers, synthetics can be extruded to produce continuous filaments. Filament "format" morphologies are advantageous both from the standpoint of production efficiency (e.g., compatible with warp knitting) and performance (e.g., strength and durability even when fabrics are sheer). As demand for textiles has increased in the past few decades, synthetics produced from relatively inexpensive feedstocks have filled the gap. Despite the increasing usage of synthetics, global demand for, and ultimate production of cotton has actually grown (but at a much slower growth rate than synthetics). Despite limitations of short staple fibers, cotton continues to be valuable because it is generally considered to be very comfortable and does not generally pick up bad odors, a significant customer complaint about polyester textiles.

Presently, about 59 billion pounds of cotton fiber are grown annually on 33 million hectares of farmland [20]. For context, around 700 million hectares of grain are grown each year globally [21]. Cotton is a relatively high-value crop, and cotton prices are directly proportional to the length of fibers. Cotton agriculture is poised to be substantially more valuable as nutritious edible gossypol-free cotton seed varieties become available [22, 23]. Today, significant fractions of virgin cotton fibers are too short to be effectively spun into yarns that are knit or woven into fabrics. These short fibers are removed at great aggregate cost and utilized for applications, such as rags and Q-tips. While cotton fabrics can be mechanically broken down back into reusable fiber, both post-industrial and post-consumer recycled cotton fibers exhibit significantly reduced length, thus greatly lowering their value and utility. Small amounts of short mechanically recycled fibers can be blended into (re)spun yarns, but this often requires either long staple virgin fibers and/or synthetic fiber tows in order to make the yarns suitable (e.g., strong enough) for efficient fabric construction. Intimately blended yarns composed of polyester mixed with cotton fibers are particularly difficult to recycle [10].

9.2 The Gap—Lack of Sustainable + Scalable Textile Manufacturing

It is well understood that the textile industry needs a revolution both of materials and fabrication techniques to continue to support billions of consumers—let alone to grow as new wealth enables larger populations to participate in the global economy. "Cradle to Cradle" and "Circular" are terms that presently receive much-needed consideration [24–26]. Unfortunately, to date, there have been few scalable technological solutions that can meaningfully engage calls for action at global scales. It has been suggested that biotechnology and so-called "biofabrication" techniques might offer new circularity; however, these platforms are and will continue to be extremely limited from the perspective of delivering relevant performance with meaningful unit economics that can be scaled [27]. Simply put, sustainable materials manufacturing must be scalable (e.g., unit economics that produces materials at low single-digit dollars per pound or less) to meaningfully address global plastic pollution from the world's largest industries (e.g., textiles). Technologies that cannot deliver scalable economics are simply not credible "answers" to address global sustainability issues.

Manufactured cellulose fibers are the most notable existing scalable technologies that promote greater circularity. Both the viscose and Lyocell processes produce regenerated cellulose fibers at cost-competitive price points and have been scaled to around 10 billion pounds of combined annual production [28]. However, the viscose process produces significant waste, and both processes require relatively pure cellulose pulp inputs. These types of processes fully denature and dissolve the starting cellulosic materials. In addition to causing confusion over the source of cellulose [29] and traceability (e.g., "sustainable" bamboo versus "sustainable" beechwood or eucalyptus), full dissolution often results in deleterious materials properties [30] that have limited the breadth of adoption. Lastly, the ability to functionalize manufactured cellulose and to create complex (multi-material) composites is limited.

Life processes produce remarkably diverse composite materials that cannot be easily produced by any other means at a relevant scale. For example, similar to the annular growth rings of trees, cotton fibers exhibit daily growth rings. These ring structures reveal exquisitely controlled orientation of cellulose microstructures that enables cotton to exhibit wet strength that is important for durability during laundering of fabrics. Full dissolution of these structures destroys native order and structure. Upon regeneration, entropy dictates that complex microstructures are not remade. Loss of structure often yields suboptimal consequences and includes lack of wet strength for rayon (viscose) [30]. Cotton is the primary plant-based fiber produced for textiles today, but it is just one of many fibers that can be produced in overwhelming abundance by sustainable agriculture. Flax (linen), jute, kapok (*Ceiba pentandra*), bamboo, ramie, kenaf, industrial hemp, etc. are all examples of plant-based fibers that can grow in different climates and exhibit diverse microstructures and useful macromorphologies. Likewise, various differentiated types of animal-based materials, such as leather, wool, silk, and chitin (e.g., purified from shrimp and crab shells), have important utility within higher price sectors of the textile industry. In particular, the abundance of cellulosic fibers is practically inexhaustible with estimates approaching 100 gigatons of global annual production of cellulose biopolymers alone [31, 32]. The 100 billion pounds of staple and filament polyester fiber produced and utilized annually represents just ~0.045 gigatons, and thus, a tiny ~0.00045 fraction compared to the estimated annual cellulose production by life processes [33]. Of course, plants harness sunlight to produce these wondrously diverse materials while sequestering carbon dioxide. Natural systems are well balanced and have evolved to thrive with 100s of gigatons of cellulose and lignocellulose fibers in various stages of growth and decomposition (in all forms of biodegradable macro and "microfiber" formats) within global ecosystems. This global system essentially constitutes the original cradle-to-cradle materials manufacturing technology on earth. So long as natural fibers, such as cotton, are not treated with toxic chemistries (e.g., fluorinated water repellants), it is clear that sustainably grown, biodegradable, biopolymer-based fibers are sufficiently abundant and circular to drive the global materials economy for textiles and beyond, as shown in Fig. 9.1.

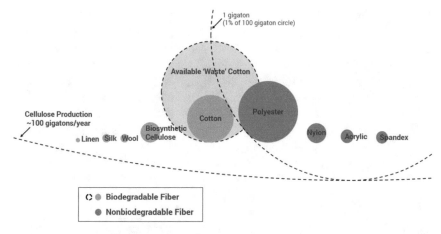

Fig. 9.1 The size of circles (for linen, silk, wool, biosynthetic cellulose, cotton, polyester, nylon, acrylic, and spandex, respectively) represents the tonnage of materials used by the global textile industry per year. Polyester is the most used at 100 billion pounds annually. The dotted line "gigaton" arcs represent the 100% and 1% of annual cellulose production by life on earth per year. The filled dotted line circle, "available waste cotton", represents the approximate aggregate amount of cotton waste that is available for recycling

9.3   Filling the Gap—Sustainable + Scalable Fabrication Technologies

To be "sustainable", new technologies must also be scalable to address global issues. Outside of legislation, market forces demand that biodegradable products outperform nonbiodegradable synthetic incumbents in order to displace them. That is to say, biodegradable materials must exhibit superior value relative to synthetics for broadest adoption. NFW's fabrication technologies can be viewed as a new hub that synergistically unifies the diversity and availability of natural materials with existing scaled industrial infrastructure and techniques. NFW is demonstrating scalable cost-effective ways to produce high-performance materials by leveraging abundant natural inputs. This is significant because there are few technological alternatives to, for example, meaningfully address global issues, such as plastic microfiber pollution.

NFW is developing an extremely tunable, automated fabrication platform that fills important manufacturing gaps that exist within the textile, paper, and composites industries. NFW uses proprietary closed-loop processes that leverage abundant sustainably sourced natural materials in ways that can cut manufacturing costs relative to conventional approaches. Instead of full dissolution and full denaturation of natural substrates, natural polymers are swelled and are mobilized only at fiber surfaces. This greatly reduces chemistry costs while preserving native structures and extending key intermolecular associations (e.g., hydrogen bonding) between neighboring fibers, as shown in Fig. 9.2. This approach effectively enables short fibers to act like long fibers and has immediate utility to recycling and even upcycling existing

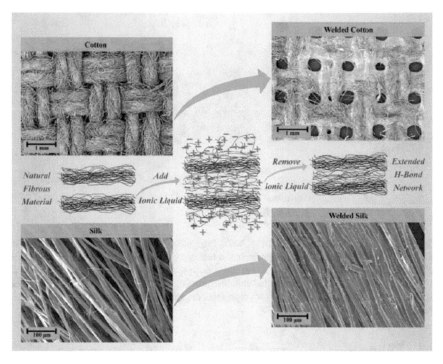

Fig. 9.2 Images show natural fibers, cotton (top) and silk (bottom), respectively, before (at left) and after (at right) transformation. The cartoons in the center detail the basic concept of extending hydrogen-bonding networks by controllably adding and then removing (and recycling) IL-based chemistries to yield robust composites. Figure is the journal cover art from Haverhals et al. [1]

natural materials (e.g., upcycling short cotton fiber to create new high-performance textiles).

NFW enables "low"-utility (lower cost) agricultural fibers to find new service in applications that typically require higher cost fiber. In addition, NFW produces biodegradable composites that can *be* and *do* more because different natural materials can be combined to create unique hierarchies of structure that are not possible to produce by any other means. By preserving key structures of biopolymers while introducing new types of macroscopic "formats" (morphologies), substrates can be tuned to perform "super-natural" functions that go far beyond what natural and biosynthetic materials can achieve on their own. For example, growing and/or entrapping nanomaterials within welded fiber enables fabrics to span wide sets of properties from intrinsically safe antibacterial properties and resistance to flame spread, to conductivity, and to catalytic activity for water treatment applications.

9.4   Functional "Welded" Composites

Previous research publications have detailed some of the basic operating principles for "vanilla" welding processes. (Here we use "vanilla" to describe processes that impart beneficial morphological modifications to natural fiber substrates.) For the past few years, NFW has been further inventing, refining, and scaling platforms that will have global impacts across large markets. NFW is particularly focused on the ways in which fabrication processes can be extended and do more for the textile industry. For example, NFW has developed and patented "indigo flavors" of welding processes that not only increase the utility of short cotton fibers by accessing new morphologies, but indigo dyeing can be simultaneously accomplished with a zero-waste closed-loop operation [8]. The remainder of this chapter will be devoted to detailing several examples that show platform breadth and diversity of applications.

9.4.1   Furniture from Waste Textiles

Billions of pounds of textiles are discarded annually. Much of this waste goes to landfills, and, unfortunately, too often textile waste is improperly disposed of and compounds the problems of the release of nonbiodegradable plastic microfibers into the environment. NFW has demonstrated products that convert waste fabrics (e.g., waste denim) into functional composites that can have a second life as building materials for furniture, as shown in Fig. 9.3.

Fig. 9.3 A natural fiber-welded composite tabletop created from waste denim. The tabletop is smooth (physically) but displays visual depth from the "welded" cut scrap and seams of the original denim garments. Image courtesy of turnstone®, a Steelcase brand

The composite tabletop shown in Fig. 9.3 is made only of scrap denim without any glues or resins. Instead, cotton fibers have been fused by the extension of hydrogen-bonding networks to span neighboring cotton fibers. The basic process begins when a controlled amount of IL-based solvent is applied to the substrate. The materials are heated, pressed, and then IL-based solvents, which enable tunable physical changes to the substrate, are recovered and recycled for reuse. After drying, the final composite is found to have properties similar to wood. The fabrication process that produced this denim composite is immune to the fact that some of the denim scraps can contain spandex and polyester. In fact, synthetic fibers are safely encapsulated within the composite. As with wood, welded composites can be painted or coated with varnishes that create water-repellant surfaces. The tabletop shown has had a natural wax applied for water resistance.

9.4.2   Waterless Indigo Dye Processes

Traditional ring dye processes are designed to be accomplished in water. As shown in Fig. 9.4, indigo is reduced to the water-soluble leucoindigo anion, which is yellow in color. Cotton is subsequently dipped into aqueous solutions containing leucoindigo. By controlling ionic strength, pH, time of dip(s), type of washing and rinsing, et cetera, unique versions of the so-called "ring dye" effect are obtained.

Indigo dyeing is accomplished at a massive scale and is the backbone of the nearly $60 billion per year denim industry. It is well documented that indigo dye processes tend to consume large quantities of water. Although newly developed foam dye processes consume less water, water-based indigo dye processes often use large amounts of reducing agents to produce the water-soluble anion (leucoindigo). These reducing agents along with surfactants and other chemicals used during rinsing steps create a toxic effluent that is often released into the environment.

NFW's proprietary platform is able to produce ring dyed effects using a zero-waste (closed-loop) approach, as shown in Fig. 9.5 [8]. Instead of designing processes around the chemical properties of water, IL-based solvents are tuned to take indigo dye in its molecular form (not derivatized). This eliminates the need for harsh reducing agents and eliminates both the costs of chemistry and the costs of cleaning chemistry out of wastewater streams. Moreover, ring dye effects are controlled by

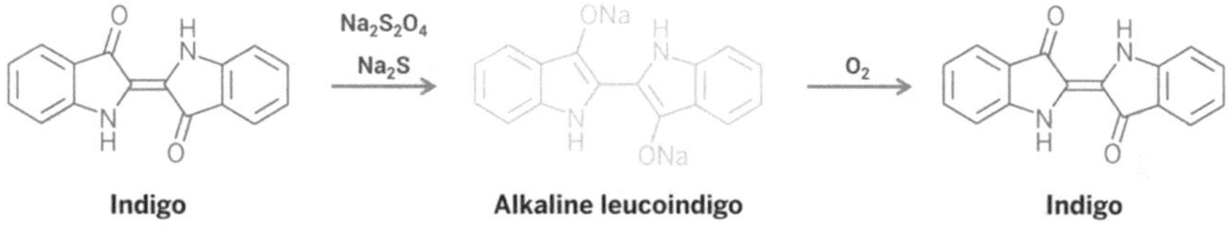

Fig. 9.4 Scheme of the indigo dye process for denim production

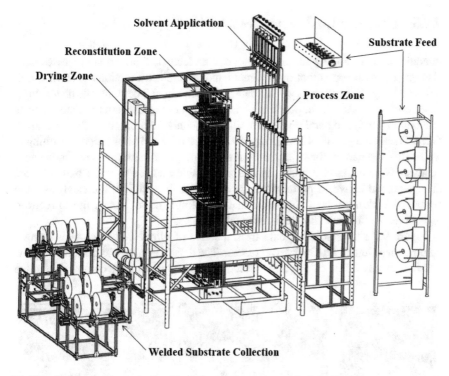

Fig. 9.5 Schematic of a prototype closed-loop indigo dye range built by NFW

adjustments of IL-based solvents and improve process control. All of these advantages are accomplished while polymers are controllably swelled by control of other relevant process conditions. This has the effect of simultaneously creating morphological effects within yarns that improve their evenness, strength, and abrasion resistance.

Figure 9.5 details a prototype machine that transforms conventional yarns into welded composites. Cotton yarn substrates (far right) feed into an apparatus that applies a controlled amount of IL-based solvents. The yarn substrates continue to feed into the process zone where temperature, atmosphere, and other conditions are precisely controlled. Yarn continues to matriculate into the reconstitution zone where IL-based solvents are recovered and then recycled. After the washing (reconstitution) step, the wet yarn moves through a drier and is collected and wound onto packages (at far left). The "welded" yarn is now ready to be converted into fabrics either by knitting or weaving. Welded yarns are typically stronger and finer than their conventional counterparts. In particular, increases in strength are beneficial for subsequent fabric conversion steps as well as the performance of fabrics. NFW is in the process of building much larger scale versions of the apparatus shown in Fig. 9.5 and plans to begin selling selected Welded Cotton[TM] yarn and fabric products in 2020.

9.4.3 Catalytic Wastewater Treatment

In addition to entrapping sub-nanometer dye molecules, fabrication processes have also been demonstrated that incorporate nanometer-sized functional materials [34, 35]. The image in Fig. 9.6 as well as data plotted in Fig. 9.7 demonstrate how nature-based composites can offer peerless advantages over plastics. Natural fibers, such as linen (in Fig. 9.6) and bamboo, were modified to contain Pd-based catalytic nanoparticles. These catalysts nucleated and grew within the natural fiber support resulting in well-dispersed nanoparticles throughout the biopolymer matrix. Moreover, because catalysts were encased in matrices that still enabled access to water and solution, they remained active to perform nitrate reduction. In addition, the catalysts were contained within a robust composite that was simple at the end of life to recover, regenerate, and recycle the precious metal catalysts.

Figure 9.7 shows selected data from a series of nitrate reduction tests using a natural fiber-welded Pd-Cu catalyst reactor. The data in Fig. 9.7a and Table 9.1 show that the activity of the catalysts in welded fiber composites were comparable to the

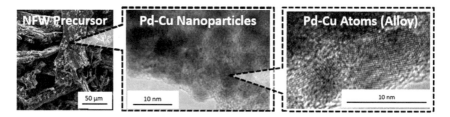

Fig. 9.6 Scanning electron microscopy (left) and transmission electron microscopy (center, right) of a linen "precursor" substrate modified with entrapped Pd-Cu nanoparticle catalysts. Data adapted from Durkin et al. [35]

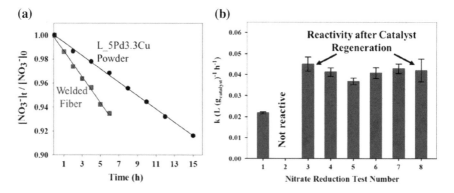

Fig. 9.7 At left (**a**) is a (not normalized) plot of nitrate reduction kinetics for a welded fiber composite and a powder catalyst (slurry in solution). At right (**b**) is a plot of calculated rate constants for the welded fiber composite during 114 h of nitrate reduction tests. Data adapted from Durkin et al. [35]

Table 9.1 Data that normalizes the leftmost plot in Fig. 9.7

Material type	Catalyst loading (g L^{-1})	Reaction rate (L g^{-1} h^{-1})
Catalyst powder	0.5	2.2 ($\pm$0.2) $\times$ 10^{-2}
Welded fiber	0.265	2.2 ($\pm$0.05) $\times$ 10^{-2}

A catalyst-containing welded fiber substrate was found to have a similar rate constant as catalyst powder alone (not entrapped).

activity of catalysts dispersed in powder slurries. Of course, nanoparticle slurries are extremely difficult to recover. Figure 9.7b shows the performance of welded fiber composites across a series of tests. Test 1 is an initial nitrate reduction test (with hydrogen sparging). After test 1, the composite was exposed to air and upon test 2, no reactivity was observed following sparging the catalyst with hydrogen at room temperature. However, the catalyst within the composite was restored after heat treatment of 105 °C under nitrogen and hydrogen, each for 2 h, respectively. Tests 3–7 were conducted after the catalyst regeneration step. The catalyst reduced nitrate for 14 h per day for 5 consecutive days. Between tests 3–7, the catalyst was rinsed with water and stored at 60 °C in air. After test 7, the regenerative heat treatment of 105 °C under nitrogen and hydrogen, each for 2 h, respectively, was performed again. Test 8 shows the catalyst performance following this second regeneration. In all tests, pseudo-first-order rate constants were normalized to catalyst loading and calculated as the mass of catalyst in the total volume of the reaction solution. The data demonstrate the unique advantages of using tunable fiber welding processes to entrap catalysts within the natural fiber (linen) composite.

In a second study on Pd-based nanoparticles, we produced a more reactive, robust, and sustainable catalyst for water treatment created through welding of lignocellulose-supported palladium-indium (Pd-In) nanoparticles onto linen yarns [34]. Again, the Pd-In catalysts were synthesized to preserve the lignocellulose and yielded small (5–10 nm), near-spherical crystalline nanoparticles of Pd-In alloy, and a uniform Pd-In metal composition throughout the fibers. Nitrate reduction tests identified the existence of an optimum Pd-In catalyst composition for maximum reactivity; the most reactive Pd-In catalyst was 10 times more reactive than the best performing Pd-Cu system, as shown in Figs. 9.6 and 9.7. Nitrate reduction tests and X-ray photoelectron spectroscopy depth profiling of aged Pd-In catalysts showed that they remained stable and lost no reactivity during extended storage in air at room temperature. Next, the optimized Pd-In catalyst was fiber-welded onto linen yarns using a novel, scalable fabrication process that controlled catalyst loading and delivered a Pd-In catalyst coating onto the yarn surface. These fiber-welded Pd-In catalyst yarns were integrated into a novel water treatment reactor and evaluated for four months and more than 180 h of nitrate reduction tests in ultrapure water, as shown in Fig. 9.8a. During this evaluation, the fiber-welded catalysts maintained their reactivity with negligible metal leaching due to the robust integration of the catalyst into the support. When tested in raw or (partially) treated drinking water and wastewater, the fiber-welded catalysts were robust and stable, and their performance was

Fig. 9.8 Nitrate reduction performance of fiber-welded Pd-In catalyst yarn (**a**) in ultrapure water, as well as water matrices from the Frederick P. Griffith Jr. Water Treatment Plant (GWP) (using source water from the Occoquan reservoir, VA) and the Broad Run Water Reclamation Facility (BRWRF), and (**b**) before and after 24 h activation in ultrapure water with H_2/CO_2 bubbling. Following activation, testing occurred over 5 consecutive days in the ultrapure water matrix. Data adapted from Durkin et al. [34]

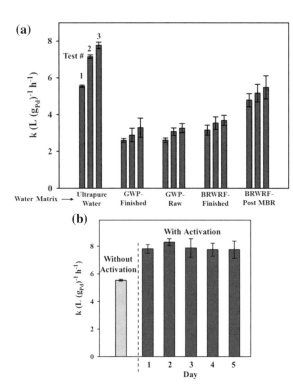

not significantly impacted by constituents in the complex waters (e.g., alkalinity and organic matter), as shown in Fig. 9.8b. This research demonstrated an innovative, scalable approach for designing and implementing robust, sustainable lignocellulose-supported catalysts with enhanced reactivity capable of water purification in complex water chemistries.

9.4.4 Energy Storage in Wearable Textiles

The electronic textiles (e-textiles, sometimes also called "smart textiles") market is expected to gain relevance in industry sectors ranging from healthcare and fitness to automotive, home goods, and military/defense [36]. It has been estimated that the e-textile industry will grow to greater than $5 billion per year by 2022 [37]. For these projections to become reality, a new "toolbox" of manufacturing techniques will be necessary to create and integrate appropriate functionalities (e.g., sensors, communications, et cetera) into textiles. NFW is building versatile tools necessary to drive new innovation within the e-textile industry.

Energy storage is an e-textile application that is accomplished with welded fiber composites that contain micron size-regime functional materials [38, 39]. In Fig. 9.9,

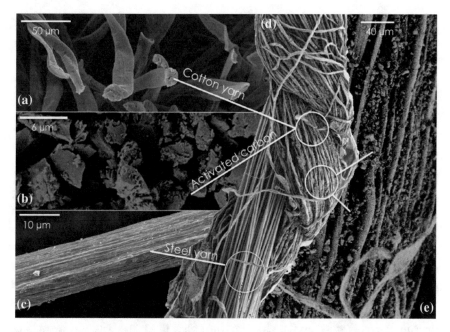

Fig. 9.9 Scanning electron microscopy of composite supercapacitor yarns. Cotton fibers (**a**) are controllably fused with high surface-area carbon (**b**) that becomes entrapped. A stainless-steel current collector (**c**) is plied with the welded fiber construct to create the composite yarn (**d** and **e**) that is capable of storing charge. Figure is data from Jost et al. [38]

high-surface area capacitive carbon materials are shown entrapped within cotton-based composite yarns. These composite yarns were plied with stainless-steel current collectors and subsequently coated with a flexible polymer electrolyte. The result was a flexible yarn-based supercapacitor that was able to be knit into fabrics, as shown in Fig. 9.10. In this seminal 2015 study, the capacitance of the yarns produced topped out at 37 mF cm^{-1} and was one of the highest values for carbon-based yarns ever reported. NFW has recently substantially improved on these results and with processes capable of mass production of energy storage yarns. As the e-textile and other textile submarkets develop, NFW is demonstrating value with general-purpose processes that deliver both "practical" and "exotic" performance that can be custom-tuned.

9.5 Conclusions

The development of commercially available ILs is a key factor that is enabling new tunable processes that produce robust, functional composites using natural materials. When necessary, IL-based solvents can be tuned to enable processes that are tolerant of, and even work synergistically with, synthetic materials. Of course, complete

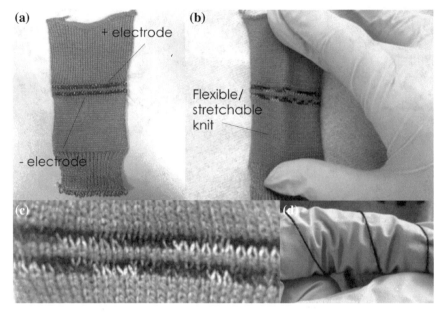

Fig. 9.10 Images of supercapacitor yarns knit into a fabric (**a**, **c**). The fabric was shown to be sufficiently stretchable (**b**) due to the flexibility of the composite yarn (**d**). Figure is data from Jost et al. [38]

biodegradability is lost for any composite that contains nonbiodegradable plastics, so applications must be thoughtfully considered. At the same time, fiber welding fabrication techniques can be utilized to incorporate functional materials at many different size regimes, ranging from molecular species to nano- and micron-sized particles. A wide range of functionalities can be imparted to natural fiber substrates. Natural materials are generally more complex than can be replicated synthetically. In particular, plants (utilizing photosynthesis) create a diversity of complex materials in abundance with unit economics that are extremely favorable. IL-based chemistries can be specifically tuned to preserve key natural hierarchical structures. NFW is poised to be a disruptive force in large markets by unlocking the potential of plentiful, high-performance natural materials to displace nonbiodegradable synthetic plastics.

References

1. Haverhals LM, Reichert WM, De Long HC, Trulove PC (2010) Natural fiber welding. Macromol Mater Eng 295(5):425–430. https://doi.org/10.1002/mame.201090008
2. Haverhals LM, Reichert WM, De Long HC, Trulove PC (2012) Natural fiber welding. U.S. patent no. 8202379. Awarded 19 June 2012
3. Swatloski RP, Spear SK, Holbrey JD, Rogers RD (2002) Dissolution of cellulose with ionic liquids. J Am Chem Soc 124:4974–4975. https://doi.org/10.1021/ja025790m

4. Phillips DM, Drummy LF, Conrady DG, Fox DM, Naik RR, Stone MO, Trulove PC, De Long HC, Mantz RA (2004) Dissolution and regeneration of *Bombyx mori* silk fibroin using ionic liquids. J Am Chem Soc 126:14350–14351. https://doi.org/10.1021/ja046079f
5. Haverhals LM, Sulpizio HM, Fayos ZA, Trulove MA, Reichert WM, Foley MP, De Long HC, Trulove PC (2012) Process variables that control natural fiber welding: time, temperature, and amount of ionic liquid. Cellulose 19:13–22. https://doi.org/10.1007/s10570-011-9605-0
6. Haverhals LM, Nevin LM, Foley MP, Brown EK, De Long HC, Trulove PC (2012) Fluorescence monitoring of ionic liquid-facilitated biopolymer mobilization and reorganization. Chem Commun 48:6417–6419. https://doi.org/10.1039/C2CC31507F
7. Haverhals LM, Foley MP, Brown EK, Fox DM, De Long HC, Trulove PC (2012) Natural fiber welding: ionic liquid facilitated biopolymer mobilization and reorganization. In: Visser A, Bridges N, Rogers R (eds) Ionic liquids: science and applications, ACS Symposium Series 1117, American Chemical Society. Washington, DC, Chap. 6, pp 145–166. Alternatively: ACS Symp Ser 2012, 1117:145–166. https://doi.org/10.1021/bk-2012-1117.ch006
8. Haverhals LM, Amstutz AK, Choi J, Tang X, Molter M, Null SJ (2018) Methods, processes, and apparatuses for producing dyed and welded substrates. U.S. patent no. 10011931. Awarded 3 July 2018
9. Ellen MacArthur Foundation (2017) The new plastics economy: rethinking the future of plastics and catalyzing action, pp 1–66
10. Ellen MacArthur Foundation (2017) A new textiles economy: redesigning fashion's future, pp 1–150
11. Browne MA, Crump P, Niven SJ, Teuten E, Tonkin A, Galloway T, Thompson R (2011) Accumulation of microplastic on shorelines worldwide: sources and sinks. Environ Sci Technol 45(21):9175–9179. https://doi.org/10.1021/es201811s
12. Kohoutek J, Babica P, Bláha L, Maršálek B (2008) A novel approach for monitoring of cyanobacterial toxins: development and evaluation of the passive sampler for microcystins. Anal Bioanal Chem 390(4):1167–1172. https://doi.org/10.1007/s00216-007-1785-y
13. Cole M, Lindeque P, Fileman E, Halsband C, Goodhead R, Moger J, Galloway TS (2013) Microplastic ingestion by zooplankton. Environ Sci Technol 47(12):6646–6655. https://doi.org/10.1021/es400663f
14. McCormick A, Hoellein TJ, Mason SA, Schluep J, Kelly JJ (2014) Microplastic is an abundant and distinct microbial habitat in an urban river. Environ Sci Technol 48(20):11863–11871. https://doi.org/10.1021/es503610r
15. Rochman CM, Parnis JM, Browne MA, Serrato S, Reiner EJ, Robson M, Young T, Diamond ML, Teh SJ (2017) Direct and indirect effects of different types of microplastics on freshwater prey (*Corbicula fluminea*) and their predator (*Acipenser transmontanus*). PLoS ONE 12(11):e0187664. https://doi.org/10.1371/journal.pone.0187664
16. Jeong C-B, Won E-J, Kang H-M, Lee M-C, Hwang D-S, Hwang U-K, Zhou B, Souissi S, Lee S-J, Lee J-S (2016) Microplastic size-dependent toxicity, oxidative stress induction, and p-JNK and p-p38 activation in the monogonont rotifer (*Brachionus koreanus*). Environ Sci Technol 50(16):8849–8857. https://doi.org/10.1021/acs.est.6b01441
17. Smith M, Love DC, Rochman CM, Neff RA (2018) Microplastics in seafood and the implications for human health. Curr Environ Health Rep 5(3):375–386. https://doi.org/10.1007/s40572-018-0206-z
18. Yang D, Shi H, Li L, Li J, Jabeen K, Kolandhasamy P (2015) Microplastic pollution in table salts from China. Environ Sci Technol 49(22):13622–13627. https://doi.org/10.1021/acs.est.5b03163
19. https://orbmedia.org/stories/Invisibles_plastics/. Site visited 10 Jan 2019
20. US Department of Agriculture (2018) Cotton: world markets and trade, 11 Dec 2018 report. https://apps.fas.usda.gov/psdonline/circulars/cotton.pdf. Site visited 10 Jan 2019
21. US Department of Agriculture (2019) World agricultural production, 11 Dec 2018 report. https://apps.fas.usda.gov/psdonline/circulars/production.pdf. Site visited 10 Jan 2019
22. Wedegaertner T, Rathore K (2015) Elimination of gossypol in cottonseed will improve its utilization. Procedia Environ Sci 29:124–125. https://doi.org/10.1016/j.proenv.2015.07.212

23. https://www.npr.org/sections/thesalt/2018/10/17/658221327/not-just-for-cows-anymore-new-cottonseed-is-safe-for-people-to-eat. Site visited 10 Jan 2019
24. https://mbdc.com/. Site visited 10 Jan 2019
25. https://www.ellenmacarthurfoundation.org/. Site visited 10 Jan 2019
26. https://fashionforgood.com/. Site visited 10 Jan 2019
27. Edlund AM, Jones J, Lewis R, Quinn JC (2018) Economic feasibility and environmental impact of synthetic spider silk production from *Escherichia coli*. New Biotechnol 42:12–18. https://doi.org/10.1016/j.nbt.2017.12.006
28. https://www.textileworld.com/textile-world/fiber-world/2015/02/man-made-fibers-continue-to-grow/. Site visited 10 Jan 2019
29. https://www.ftc.gov/news-events/press-releases/2015/12/nordstrom-bed-bath-beyond-backcountrycom-jc-penney-pay-penalties. Site visited 10 Jan 2019
30. Das A, Ishtiaque SM, Singh S, Meena HC (2009) Tensile characteristics of yarns in wet condition. Indian J Fibre Text Res 34:338–344
31. Cox PM, Betts RA, Jones CD, Spall SA, Totterdell IJ (2000) Acceleration of global warming due to carbon-cycle feedbacks in a coupled climate model. Nature 408:184–187. Erratum ibid. (2000) 408:750. https://doi.org/10.1038/35041539, https://doi.org/10.1038/35047138
32. Griffith JD, Willcox S, Powers DW, Nelson R, Baxter BK (2008) Discovery of abundant cellulose microfibers encased in 250 Ma Permian halite: a macromolecular target in the search for life on other planets. Astrobiology 8(2):215–218. https://doi.org/10.1089/ast.2007.0196
33. https://www.plasticsinsight.com/resin-intelligence/resin-prices/polyester/. Site visited 10 Jan 2019
34. Durkin DP, Ye T, Choi J, Livi KJT, De Long HC, Trulove PC, Fairbrother DH, Haverhals LM, Shuai D (2018) Sustainable and scalable natural fiber welded palladium-indium catalysts for nitrate reduction. Appl Catal B 221:290–301. https://doi.org/10.1016/j.apcatb.2017.09.029
35. Durkin DP, Ye T, Larson E, Haverhals LM, Livi KJT, De Long HC, Trulove PC, Fairbrother DH, Shuai D (2016) Lignocellulose fiber- and welded fiber- supports for palladium based catalytic hydrogenation: a natural fiber welding application for water treatment. ACS Sustain Chem Eng 4(10):5511–5522. https://doi.org/10.1021/acssuschemeng.6b01250
36. Seymour S (2008) Fashionable technology, the intersection of design, fashion, science, and technology. Springer Wien, New York. https://doi.org/10.1007/978-3-211-74500-7
37. Sharma K (2019) Smart textile market by function (energy harvesting, sensing, thermoelectricity, luminescent, and others) and end user (healthcare, military and defense, entertainment, automotive, sport and fitness)—global opportunity analysis and industry forecast, 2014–2022. Allied market research, series: emerging and next generation technology. https://www.alliedmarketresearch.com/smart-textile-market. Site visited 10 Jan 2019
38. Jost K, Durkin DP, Haverhals LM, Brown EK, Langenstein M, De Long HC, Trulove PC, Gogotsi Y, Dion G (2015) Natural fiber welded electrode yarns for knittable textile supercapacitors. Adv Energy Mater 5:1401286. https://doi.org/10.1002/aenm.201401286
39. Durkin DP, Jost K, Brown EK, Haverhals LM, Dion G, Gogotsi Y, De Long HC, Trulove PC (2014) Knitted electrochemical capacitors via natural fiber welded electrode yarns. ECS Trans 61:17–19. https://doi.org/10.1149/06106.0017ecst

Chapter 10
Development of New Cellulosic Fibers and Composites Using Ionic Liquid Technology

Frank Hermanutz, Marc Philip Vocht and Michael R. Buchmeiser

Abstract One of the most important applications of ionic liquids (ILs) is their use as a solvent for natural polymers. In particular, solutions of cellulose, chitosan, and chitin in ILs are used for the production of fibers, coatings, composites, and new materials. The initiation of this new field of research was triggered by the publication of Swatloski et al., which reported in 2002 for the first time the solubility of cellulose in ILs. Numerous papers have been devoted to the search for new solvents for cellulose and for scientific and industrial applications. Depending on the application of the IL, it is necessary to compromise between the ecological (toxicity) and the economic parameters (cost of the IL). This chapter discusses the scope of this approach and the limits in the practical application of ILs for the dissolution of cellulose, a natural polymer. The rational choice of ILs for use in particular processes, the features of dissolution methods for natural polymers (including cellulose, chitin, and fibroin), and the preparation of blends from the solutions of polymers with ILs are discussed.

Keywords Biopolymers · Cellulosic blend fibers · All-cellulose composites · Recycling · Super-microfibers

10.1 Introduction

Cellulose is the most abundant biopolymer and consists of a linear chain of β-(1 → 4) linked D-glucose repeat units [1]. Cellulose has many attractive physical properties, such as thermal and chemical stability, biocompatibility, and biodegradability [2]. The intra- and intermolecular hydrogen-bonding interactions between the individual polysaccharide chains result in a semi-crystalline

F. Hermanutz (✉) · M. P. Vocht · M. R. Buchmeiser
German Institutes of Textile and Fiber Research (DITF), Körschtalstr. 26,
73770 Denkendorf, Germany
e-mail: Frank.Hermanutz@ditf.de

M. R. Buchmeiser
Institute of Polymer Chemistry (IPOC), University of Stuttgart, Pfaffenwaldring 55, 70569 Stuttgart, Germany

© Springer Nature Switzerland AG 2020
M. B. Shiflett (ed.), *Commercial Applications of Ionic Liquids*, Green Chemistry and Sustainable Technology, https://doi.org/10.1007/978-3-030-35245-5_10

polymer with both highly structured crystalline and amorphous regions. One of the most remarkable properties of cellulose is the insolubility in water as well as in most organic solvents [3]. The challenge in dissolving cellulose is to destroy the strong hydrogen bonds. Nevertheless, there are some solvent systems for cellulose described in the literature, such as *N*-methylmorpholine-*N*-oxide (NMMO), *N,N*-dimethylacetamide/lithium chloride (DMAc/LiCl), *N,N*-dimethylsulfoxide/tetrabutylammonium fluoride (DMSO/TBAF), and several ionic liquids (ILs) [4–8]. For industrial applications, NMMO is now the most common direct solvent system for cellulose and the corresponding NMMO/cellulose solutions are used for spinning cellulosic fibers. In the so-called Lyocell process, NMMO is used at concentrations between 10 and 14 wt%. The spinning dope has to be stabilized by additives, like isopropyl gallate, to prevent side reactions [9, 10]. By spinning through an air gap prior to coagulation in an aqueous bath, high stretch ratios can be realised, leading to high orientation of the cellulose chains. After drying, long and thin crystallites are formed, which align along the fiber axis in a highly oriented manner. Consequently, there is little lateral interaction between the individual macrofibrils, which, in the wet state, leads to a high tendency to fibrillation [9]. However, due to the economic and ecological drawbacks of both processes with regard to the dissolution and processing of cellulose, more efficient and environmentally friendly solvents are required [9].

Since Swatloski et al. discovered the ability of some ILs to dissolve cellulose, a new research field opened up and new possibilities to process cellulose are now imaginable [7]. This so-called IL-technology is an economically and environmentally friendly alternative process due to thermal and chemical stability, non-flammable nature, and miscibility with many other solvent systems of ILs. Besides the spinning of cellulosic fibers, this technology also offers new ways for the preparation of cellulose-based composites, coatings, and the chemical modification of cellulose [11–16].

10.2 Selection of Ionic Liquids

10.2.1 Ionic Liquids and Green Chemistry

ILs are salts in the liquid state and were first described by Walden [17]. Ionic liquids consist of a cation and an anion like "normal" salts. However, in contrast to these, ILs have melting points below 100 °C, and ILs with melting points below 25 °C, so-called room-temperature ionic liquids (RTILs), are known as well [18–22]. The low melting points of such ILs are a result of the selection of cation and anion [23–27].

A key feature of ILs is that their physical properties can be tailored by the selection of ions and substituents on the cations (R-group) [18]. Thus, their solubility in organic solvents and water can be controlled by the nature of the R-group [28, 29]. However, the miscibility of ILs with polar and non-polar solvents also depends on the

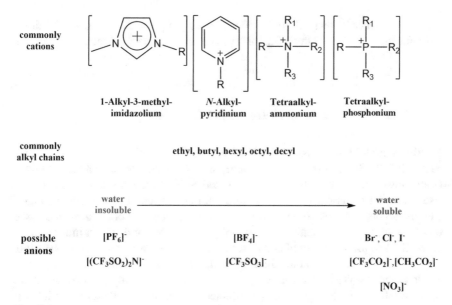

commonly cations

1-Alkyl-3-methyl-imidazolium   *N*-Alkyl-pyridinium   Tetraalkyl-ammonium   Tetraalkyl-phosphonium

commonly alkyl chains   **ethyl, butyl, hexyl, octyl, decyl**

water insoluble ⟶ water soluble

possible anions

$[PF_6]^-$ $[BF_4]^-$ Br^-, Cl^-, I^-

$[(CF_3SO_2)_2N]^-$ $[CF_3SO_3]^-$ $[CF_3CO_2]^-$,$[CH_3CO_2]^-$

$[NO_3]^-$

Fig. 10.1 Structural variations in ILs reproduced from Seddon et al., © IUPAC, 2000 [29]

anion [30]. For example, the miscibility of imidazolium-based ILs with water largely depends on the nature of the anion. 1-Butyl-3-methylimidazolium-based ($[C_4C_1im]^+$) ILs in combination with more hydrophobic anions, like $[PF_6]^-$, $[(CF_3SO_2)_2N]^-$, or $[C(CN)_3]^-$, are often immiscible with water at room temperature, while in combination with hydrophilic anions, such as $[Cl]^-$, $[CF_3SO_3]^-$, or $[BF_4]^-$, they are miscible with water. An overview of common ions of ILs is given in Fig. 10.1.

The structure of the cation in ILs can be varied substantially; the most common IL cations are imidazolium-, pyridinium-, tetraalkylammonium-, and tetraalkylphosphonium-based [29]. However, many other structural motifs also exist [29, 31–33]. In addition to their low melting points, ILs also have some other characteristic properties, such as a low vapor pressure, ionic conductivity, and thermal stability up to 450 °C [19, 26, 27, 29, 33–35]. The toxicity of many ILs is known; consequently, a tailored selection of ILs can be made for a given process [36–39]. Generally, for industrial use, any IL selected for cellulose dissolution and processing has to match specific economic and ecological criteria to create a sustainable process. Thus, the IL should be easily accessible, recyclable in large amounts (>99.5%), possess the lowest possible toxicity, have a low melting point, have literally no vapor pressure, have a low propensity to side reactions, and an excellent dissolution capability for different celluloses [40]. Luckily, due to the large structural diversity of ILs, the physical, chemical, and physiological properties can be adjusted. For example, by varying the anion, the dissolution behavior for cellulose can be influenced. The toxicity can be purposely reduced by varying the alkyl chain length in the cation or anion

[15, 41, 42]. Once IL-technology-based processes comply with the above conditions, a significant step toward Green Chemistry in cellulose-processing technology can be realized [43, 44].

10.2.2 Ionic Liquids as Solvents for Cellulose

Swatloski et al. used imidazolium-based ILs like $[C_4C_1im]^+$ with different anions, such as $[Cl]^-$, $[Br]^-$, and $[SCN]^-$ [7]. Depending on the temperature and IL, solutions containing up to 25 wt% cellulose were realized. Since then, there has been a dramatic rise in interest in dissolving cellulose in ILs for both scientific and industrial applications. So far, the most common ILs used for the processing of cellulose are based on imidazolium-derived cations [27, 45–59]. A summary of different imidazolium-based ILs and the solubility of cellulose within these ILs is given in Table 10.1 [60, 61]. A selection of non-imidazolium-based ILs is given in Table 10.2.

However, despite the vast number of ILs available, the only ILs of interest are those in which cellulose possesses a sufficient solubility. It is therefore of interest to understand how the solubility of cellulose depends on the structure of the IL and the mechanism for the dissolution of cellulose in ILs. Disappointingly, both the mechanism and the connection between IL structure and solubility of cellulose are still not fully understood in detail [58, 67]. Indeed, the high ionic strength of the ILs interrupts the inter- and intra-molecular hydrogen-bonding-based interactions in cellulose chains [50, 67–69]. Some studies show that the nature of the anion is responsible for the dissolution of cellulose. However, it has also been shown that the structure of the cation especially influences the solvation process [69–72]. Additionally, the degree of polymerization (DP) of the cellulose is a factor in the dissolution process [73].

10.3 Cellulosic Fiber Spinning Using IL-Technology

The most important and also oldest process used in the manufacturing of cellulosic fibers is the viscose process outlined in Fig. 10.2. In contrast to the Lyocell process, this process involves activation of pulp in aqueous sodium hydroxide solution followed by reaction with CS_2 to form a cellulose xanthate. After dissolution in alkaline solution, the cellulose xanthate is spun into a sulfuric acid bath. Along with other auxiliaries (e.g., carbon disulfide, sodium hydroxide, sulfuric acid, and zinc sulfate), a significant volume of fresh water is required and amounts to approximately one ton of water per kg of cellulosic fiber produced [74]. Also, the cellulose concentration in solution is limited to a range of 8–10 wt%, which is critical in terms of the environmental effects caused by the process [9, 75–77].

Table 10.1 Structure of imidazolium-based ionic liquids and the extent of cellulose solubility

IL	Chemical structure	Type of cellulose (DP)	Solubility (wt%)	Conditions	References
[C$_2$C$_1$im][Ace]		Avicel	16	90 °C	[53]
		Avicel (225)	15	110 °C	[62]
		Eucalyptus pulp (569)	13.5	90–130 °C	[48]
		Pulp linters	16	–	[63]
[C$_4$C$_1$im][Ace]		Avicel	12	100 °C	[50]
		Eucalyptus pulp (569)	13.2	90–130 °C	[48]
		MCC (229)	15.5	70 °C	[45]
[C$_8$C$_1$im][Ace]		Avicel (225)	<1	110 °C	[62]
[H(OC$_2$)$_2$-C$_1$im][Ace]		Avicel (225)	5	110 °C	[62]
[H(OC$_2$)$_3$-C$_1$im][Ace]		Avicel (225)	2	110 °C	[62]
[C$_1$(OC$_2$)$_2$-C$_2$im][Ace]		Avicel (225)	12	110 °C	[62]

(continued)

Table 10.1 (continued)

IL	Chemical structure	Type of cellulose (DP)	Solubility (wt%)	Conditions	References
[C$_1$(OC$_2$)$_3$-C$_2$im][Ace]		Avicel (225)	12	110 °C	[62]
[C$_1$(OC$_2$)$_4$-C$_2$im][Ace]		Avicel (225)	10	110 °C	[62]
[C$_1$(OC$_2$)$_7$-C$_2$im][Ace]		Avicel (225)	3	110 °C	[62]
[C$_1$(OC$_3$)$_7$-C$_2$im][Ace]		Avicel (225)	0.5	110 °C	[62]
[C$_2$C$_1$im][O$_2$P$_{1,0}$]		MCC (250)	10	55 °C	[57]

(continued)

Table 10.1 (continued)

IL	Chemical structure	Type of cellulose (DP)	Solubility (wt%)	Conditions	References
$[C_2C_1im][HO_2PO_1]$		MCC (250)	4	25 °C	[57]
		MCC (250)	10	45 °C	[57]
$[C_2C_1im][O_2PO_1\,O_1]$		MCC (250)	10	65 °C	[57]
$[C_2C_1im][O_2P\,O_2\,O_2]$		Avicel	>12	100 °C	[50]
$[C_1C_1im][O_2PO_1\,O_1]$		Cellulose	>5	90 °C	[53]
		Avicel	10	100 °C	[50]

(continued)

Table 10.1 (continued)

IL	Chemical structure	Type of cellulose (DP)	Solubility (wt%)	Conditions	References
[C$_2$C$_1$im][Cl]		Avicel	5	90 °C	[53]
		Avicel	10	100 °C	[54]
[C$_4$C$_1$im][Cl]		Avicel (225)	10	110 °C	[62]
		Avicel	20	100 °C	[54]
		Avicel (286)	18	83 °C	[47]
		Spruce sulfite pulp (593)	13	83 °C	[47]
		Cotton linters (1198)	10	83 °C	[47]
		Cotton linters	10	Sonication	[55]
[C$_5$C$_1$im][Cl]		Avicel	1–1.5	100 °C	[50, 54]
[C$_6$C$_1$im][Cl]		Avicel	6	100 °C	[50]
		Pulp cellulose (1000)	5	100 °C	[7]
[C$_7$C$_1$im][Cl]		Avicel	5	100 °C	[54]

(continued)

Table 10.1 (continued)

IL	Chemical structure	Type of cellulose (DP)	Solubility (wt%)	Conditions	References
[C$_8$C$_1$im][Cl]		Avicel	4	100 °C	[54]
[C$_9$C$_1$im][Cl]		Avicel	2–2.5	100 °C	[50, 54]
[C$_{10}$C$_1$im][Cl]		Avicel	0.5	100 °C	[54]
[(allyl)C$_1$im][Cl]		Pulp	14.5	80 °C	[46]
		Cotton linters	13	Sonication	[55]
[C$_2$C$_1$im][Br]		Avicel	2	100 °C	[50]

(continued)

Table 10.1 (continued)

IL	Chemical structure	Type of cellulose (*DP*)	Solubility (wt%)	Conditions	References
[C$_4$C$_1$im][Br]		Avicel	3	100 °C	[50]
		Pulp cellulose (1000)	7	Microwave	[7]
[C$_4$C$_1$im][I]		Avicel	2	100 °C	[50]
[C$_4$C$_1$im][SCN]		Pulp cellulose (1000)	7	Microwave	[7]
[C$_2$C$_1$im][C$_8$O$_2$]		Pulp cellulose	12	80–130 °C	[64]

DP degree of polymerization

Table 10.2 Structure of non-imidazolium-based ionic liquids and the extent of cellulose solubility

IL	Chemical structure	Type of cellulose (DP)	Solubility (wt%)	Conditions	References
[C₄C₁py][Cl]		Alicell super (599)	23	120 °C	[61]
		Avicel (286)	39	105 °C	[47]
		Spruce sulfite pulp (593)	37	105 °C	[47]
		Cotton linters (1198)	12	105 °C	[47]
[C₄C₁py][Br]		Alicell super (599)	3.5	110 °C	[61]
		Alicell super (599)	7	165 °C	[61]
[HN₁ ₁ ₂OH][Ace]		Avicel (225)	<0.5	110 °C	[62]
[HN₁ ₁ ₂O₁][Ace]		Avicel (225)	12	110 °C	[62]
[HN₁ ₂O₁ ₂O₁][Ace]		Avicel (225)	<0.5	110 °C	[62]

(continued)

Table 10.2 (continued)

IL	Chemical structure	Type of cellulose (DP)	Solubility (wt%)	Conditions	References
[H$_2$N$_2$o$_1$$_2o_1$][Ace]		Avicel (225)	<0.5	110 °C	[62]
[N$_2$$_2$$_2$$_2o_2o_1$][Ace]		Avicel (225)	10	110 °C	[62]
[N$_4$$_4$$_4$$_4$][HCO$_2$]		Avicel (225)	1.5	110 °C	[62]
[N$_{bz}$$_1$$_1$$_1$$_4$][Cl]		Avicel (286)	5	62 °C	[47]
		Spruce sulfite pulp (593)	2	62 °C	[47]
		Cotton linters (1198)	1	62 °C	[47]

(continued)

Table 10.2 (continued)

IL	Chemical structure	Type of cellulose (DP)	Solubility (wt%)	Conditions	References
[P$_{4444}$][HCO$_2$]		Avicel (225)	6	110 °C	[62]
[DBNH][Ace]		Eucalyptus pulp	7	100 °C	[65]
		Eucalyptus pulp (1026)	13	70 °C	[66]

DP Degree of polymerization

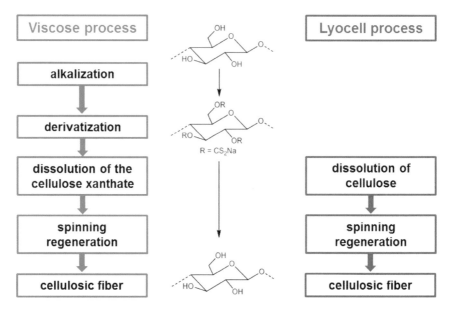

Fig. 10.2 Process principles in regenerated cellulose fiber spinning technologies

In contrast, the IL approach generates a new type of cellulosic fibers [7–9, 63]. As in the Lyocell process, cellulose is directly dissolved in an appropriate IL and spun into a water-based coagulation bath without any other auxiliaries. Notably, these IL-based spinning dopes can either be processed by wet spinning or by dry-jet wet spinning, allowing for a modification of the fiber properties according to the desired application. Parameters, such as crystallinity, the degree of polymerization, as well as the stabilizing influence of hydrogen bonds, strongly influence the mechanical properties of fibers [9, 78]. Due to the low vapor pressure, low flammability, low reactivity, and good recyclability of selected ILs, [63] the IL-technology offers environmental and process safety advantages over conventional processes, for example, in fiber production [48, 72, 79–86].

In addition, the cellulose concentration in an IL-based spinning dope can be increased up to 16.5 wt%, for example, by using $[C_4C_1im][Cl]$ [87]. An appropriate IL-system, $[C_2C_1im][Ace]$, was proposed for the wet spinning of cellulose including details on the rheology of the spinning dopes, characterization of the fiber, and recycling of ILs [63]. Hermanutz et al. reported that a $[C_2C_1im][Ace]$-based spinning dope (12 wt%) was stable up to 120 °C providing flexible process parameters for fiber property adjustments. Apart from several ILs of the imidazolium family, a spinning process based on the so-called Ioncell process, which uses 1,5-diazabicyclo[4.3.0]non-5-ene acetate ([DBNH][Ace]), was developed by Sixta and co-workers [66, 88–92]. Important parameters, like the influence of pulp source, degree of polymerization, mass distribution, as well as process parameters, like the draw ratio, air gap size, and temperature of the coagulation bath, were studied [90,

Table 10.3 Mechanical properties of cellulosic fibers spun from different ILs

Process	Solvent	σ[a] (cN/tex)	σ[a] (MPa)	E[a] (GPa)	ε_b[a] (%)	References
IL-based	[(allyl)C$_1$im][Cl]	–	74	–	17	[83]
IL-based	[C$_4$C$_1$im][Ace]	44–49	–	–	13–16	[48]
IL-based	[C$_4$C$_1$im][Cl]	53	–	–	13	[48]
IL-based	[C$_2$C$_1$im][Ace]	–	198	13	3	[84]
IL-based	[C$_2$C$_1$im][Ace]	46	–	–	11	[48]
IL-based	[C$_2$C$_1$im][Ace]	12–35	–	–	8–24	[85]
IL-based	[C$_2$C$_1$im][Cl]	53	–	–	13	[48]
IL-based	[C$_2$C$_1$im][O$_2$P$_{O2O2}$]	–	216–305	15–22	5	[86]
IL-based	[C$_2$C$_1$im][O$_2$P$_{O2O2}$]	–	305–660	22–41	5–7	[72]
IL-based	[C$_2$C$_1$im][O$_2$C$_8$]	17–31	250–460	15–29	6–13	[82, 94]
IL-based	[C$_2$C$_1$im][O$_2$C$_8$]	15–40	230–600	15–30	4–6	[10, 94]
Ioncell	[DBNH][Ace]	28–48	354–784	9.5–33	9–21	[66, 90–93]
Lyocell	NMMO	40–44	–	–	13–17	[9, 48, 79]
Viscose	Derivatization	20–25	–	7	18–23	[9, 66, 79]

[a]σ = tensile strength; E = Young's modulus; ε_b = elongation at break

92]. Furthermore, Ma et al. reported using recycled cellulose waste materials for Ioncell-based fiber production [93]. Compared to industrially produced cellulose fibers, the mechanical properties of the IL-based are similar to those of the Lyocell fibers. A summary of the mechanical properties of the IL-based cellulosic fibers is given in Table 10.3.

Apart from standard spinning, IL-technology opens new possibilities, like the in-situ chemical modification of cellulose in the course of the spinning process and the preparation of super-microfiber filaments [82, 87, 95, 96]. The processing of in-situ modified cellulose fibers is a good way to reduce waste and cut down on the overall number of process steps.

10.3.1   Spinning Super-Microfibers Using IL-Technology

Super-microfibers prepared in an IL-technology-based process can be used for new textile applications and products. Applications for filters and cleaning products are of special interest. By using cellulose IL-based spinning dopes, it is possible to spin filaments with finenesses down to 0.2 dtex[1]. Notably, such fineness for cellulosic fibers cannot be achieved by any conventional filament spinning processes [87]. Spinning fibers of such finenesses in a direct filament spinning process was made possible by

---

[1]dtex (deci-tex) is the linear density measured in grams per 10 km yarn.

newly developed spinnerets with hole diameters down to 20 μm. These new kinds of spinnerets were manufactured using laser drilling technology. Laser-drilled spinnerets with up to 1500 holes have been realized [97]. Different hole geometries were tested, and it was found that spinneret holes with smaller exit diameters than diameters at the drilling entrance are the best geometry for wet spinning of cellulosic super-microfibers, as shown in Fig. 10.3.

Super-microfibers are characterized by their extremely high surface area to fiber diameter ratio, as illustrated in Fig. 10.4. This is a huge advantage of super-microfibers <5 μm in diameter, corresponding to a fineness <0.5 dtex, compared to microfibers having fiber diameters between 5 to 9 μm. Based on the high fiber surface, these cellulosic super-microfibers have a silky touch and are characterized by high water absorption; however, there are even more technical opportunities. Cellulosic super-microfibers are an important precursor for carbon fibers if superfine carbon fibers (CFs) with diameters down to 2 μm are required. Furthermore, it is remarkable that IL-derived super-microfibers show virtually no fibrillation compared to their "standard" diameter analogues, Fig. 10.5, since the stress-induced orientation is adjustable [97].

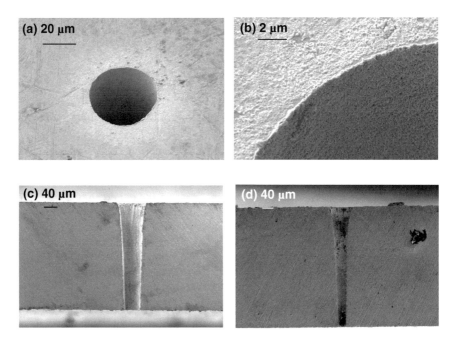

Fig. 10.3 Scanning electron micrograph (SEM) pictures of (**a**) the exit of a spinneret hole drilled into a Au–Pt alloy and (**b**) detailed picture of the bore rim. The average drilling diameter (exit) of the spinning nozzle is 25.4 μm. (**c**) SEM pictures of *positive conical* cross-sectional polishes from the drilling channels made out of stainless steel and (**d**) made of an Au–Pt alloy. Adapted with permission from dfv-Fachverlag [87]

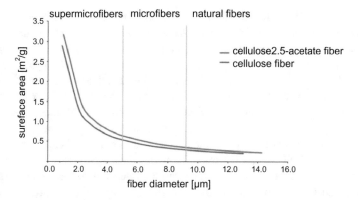

Fig. 10.4   Ratio of the surface area to fiber diameter of super-microfibers, microfibers, and natural fibers. Adapted with permission from dfv-Fachverlag [87]

Fig. 10.5   SEM pictures of cellulose super-microfibers. Adapted with permission from dfv-Fachverlag [87]

10.3.2   Ionic Liquid Technology for the Wet Spinning of Cellulose/Polymer Blends

IL-technology offers opportunities to produce new cellulosic fibers blended with other polymers, like aramides, poly(acrylonitrile) (PAN), aromatic poly(sulfonamide)s (PSA), chitin, and lignin [98–103]. In fact, the blending of cellulose with fiber polymers in a primary spinning process using ILs offers a unique opportunity to develop fiber/materials that combine the advantages of a renewable raw material and the possibility of tailor-made fiber properties for both textile and technical applications. For example, cellulose/PSA-blended fibers show improved flame-retardant properties, while cellulose/PAN-blended fibers can be used as a CF precursor [98, 102]. Finally, substitution of expensive PAN fibers by the cheaper cellulose as CF precursors could reduce process costs for CFs [98]. Among the numerous possibilities, the processing of cellulose/chitin and cellulose/chitosan blends from ILs especially stands out since chitin is the second-most abundant polymer in nature. Reviews about chitin and chitosan polymers, addressing their solubility, fiber formation, properties, and applications, have been provided by Silva et al., Pillai et al., and Rinaudo [104–106]. Different applications in the medical field, like chitin sutures, have been reported. Chitin itself is chemically stable, is biodegradable, and has the most appealing properties, such as the acceleration of wound healing and the suppression of tumor cell growth. It is remarkable that the IL-technology enables a direct fiber spinning process of chitin. This new research field is based on the wet spinning process because chitin decomposes before melting. A list of different ILs having high potential for dissolution and wet spinning of chitin was provided by Walther et al. Among the ILs tested, $[C_2C_1im][O_2C_3]$ and $[C_2C_1im][O_2C_4]$ were found the most promising for chitin. Chitin concentrations up to 14 wt% could be realised, which is high enough for technical conversion [107]. Special chemically purified chitin with purification steps to remove minerals and proteins was used in their study. Alternatively, Qin et al. used $[C_2C_1im][Ace]$ for the extraction and dissolution process using microwave irradiation heating for a much faster dissolution process. Purified, high molecular-weight chitin powder was obtained and directly spun into fibers [108, 109]. In addition to standard wet spinning, chitin was also electrospun into fibers [110]. In contrast, NaOH/urea was used as a solvent by Huang et al. for the preparation of chitin fibers and nonwovens [111]. This product showed better wound healing than traditional materials based on the inherent antibacterial properties and the ability to regulate inflammatory mediators of chitin. Chitosan produced by deacetylation is more soluble in ILs than chitin. However, it can also be processed from acetic acid. The use of IL for the processing of chitosan is not necessarily required as is the case for chitin and cellulose. Toskas et al. produced pure microfibers on an industrial scale using 8.5 wt% chitosan dissolved in acidic acid. These fibers were recently tested as medical products [112, 113]. Studies by Pillai et al. have shown that the dry–wet spinning of chitin in acetic acid results in highly deacetylated products [104]. This polymer-analogous reaction can be avoided by using ILs. For the preparation of spinning dopes, it is important that the solubility of chitin increases with increasing hydrogen-bond acceptance properties of the IL and also depends on the dissolution temperature [114]. Notably, it was found that pure chitin fibers are brittle due to the high crystallinity of chitin [108]. This issue can be improved by blending chitin with cellulose. Up to now, there are some cellulose/chitin blend materials, like membranes, films, and even multifilament fibers, described in the literature [100, 115, 116].

Fig. 10.6 SEM pictures of cellulose/chitin blend fiber spun from $[C_2C_1im][O_2C_3]$. Adapted with permission from Elsevier [100]

10.3.2.1   Chitin/Cellulose and Chitosan/Cellulose Fiber Blends

Cellulose/chitin blend fibers were manufactured by Mundsinger et al. using ILs. The focus was to apply recyclable and non-deacylating types of ILs to retain chitin properties [100]. Chitin and cellulose were dissolved in 1-ethyl-3-methylimidazolium propionate, $[C_2C_1im][O_2C_3]$. Stable spinning dopes were achieved without any phase separation tendency in the applied temperature range. Cellulose/chitin fiber blends were spun containing up to 25 wt% chitin, as shown in Fig. 10.6. Multifilament fibers were produced by a continuous wet spinning process. Walther et al. found that $[C_2C_1im][O_2C_3]$ dissolves cellulose/chitin mixtures better than dialkylimidazolium acetates or halides [107].

Silva et al. used 1-allyl-3-methylimidazolium bromide for the processing of chitin and chitosan [106]. Solutions of chitin and chitosan have also been used for the preparation of gels, coatings, and films [117]. For the development of new polymer materials, blends of biopolymers, cellulose, and chitosan are promising systems because the reactivity of chitosan is more versatile due to the presence of NH_2 groups [118]. Another way to improve the properties of the cellulose is to blend cellulose with two biopolymers. Interactions between cellulose and chitosan are shown to be attractive, as confirmed by Holmberg et al. [119]. The production of chitosan fibers using 1,3-dialkylimidazolium-based ILs solvents with different anions, like [Cl]⁻, $[ClCH_2COO]^-$, and [Ace]⁻, is possible [120]. It has also been reported that the blending of cellulose with chitosan is possible with the aid of a mixture of glycine hydrochloride and $[C_4C_1im][Cl]$ [121]. Such fibers show higher tensile strength and reduced elongation at break. Finally, Stefanescu et al. chose $[C_4C_1im][Ace]$ for producing chitosan/cellulose blend fibers [122].

10.4 Preparation of All-Cellulose Composites Using IL-Technology

Fiber-reinforced plastics (FRPs) are important materials for lightweight construction and are progressively replacing traditional materials. The properties of composites [9] depend on the matrix, the reinforcing fiber, and the matrix–fiber interface [123]. The matrix provides partial mechanical strength and works as a binder for the reinforcing fibrous phase. Due to their good mechanical properties and availability, cellulose fibers are intensively used in bio-composites and natural fiber-reinforced plastics (NFRPs) [124–126]. Several duroplastic or thermoplastic polymers, including poly(vinyl chloride), poly(ethylene), epoxides, and polyurethanes, have been used as matrices. Flax, hemp, cotton, among others are used as reinforcing fibers [124, 127]. Generally, the adhesion at the interface between a hydrophobic polymer matrix and a hydrophilic natural fiber is very weak [77, 127]. As a consequence, mechanical strength, stiffness, and impact strength of the respective composites are low, and the reinforcement potential of the fiber, for example, cellulose is only used to a small extent. To overcome this drawback, all-cellulose composites (ACCs) were first discussed as easily recyclable and biodegradable composites in 2004 [128, 129]. Due to an increased environmental awareness and increasingly restrictive legislative rules, the demand for recyclable and sustainable composites strongly increased over the last few years [130]. Nishino et al. used a 3 wt% solution of cellulose in *N,N*-dimethylacetamide (DMAc) containing 8 wt% LiCl as a matrix solution [128]. Ramie fibers were aligned parallel and impregnated with the cellulose solution under reduced pressure. After 12 h, the fiber-reinforced cellulose gel was washed with methanol to extract LiCl and DMAc to form the final composite. This manufacturing process is a two-step process, as shown in Fig. 10.7. An alternative process [9] in which the fiber surface is partially dissolved by a solvent system to form the matrix in situ is referred to as a one-step process [131].

There are different routes for the preparation of ACCs with derivatizing and non-derivatizing solvent systems described in the literature, such as DMAc/LiCl, NMMO,

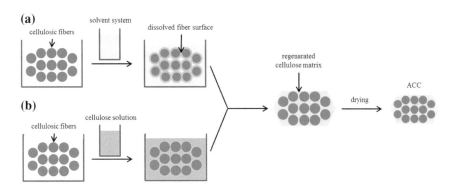

Fig. 10.7 Schematic of **a** one-step and **b** two-step all-cellulose composite preparation

Table 10.4 Overview over unidirectional (UD), bidirectional (BD), and isotropic (ISO) all-cellulose composites prepared by one-step processes

Cellulose source for matrix and reinforcement	Solvent system	Tensile strength (MPa)	Young's modulus (GPa)	Elongation (%)	References
Microcrystalline cellulose (ISO)	DMAc/LiCl	58–105	3.2–6.9	2.5–3.3	[142]
Microcrystalline cellulose (ISO)	[(allyl)C$_1$im][Cl]	34–135	3.6–8.1	0.7–6.2	[143]
Canola straw microfiber sheets (ISO)	[C$_4$C$_1$im][Cl]	208	20	9.8	[144]
Microcrystalline cellulose (ISO)	[C$_4$C$_1$im][Cl]	90–124	7–10.8	1.8–2	[145]
Filter paper (ISO)	[C$_4$C$_1$im][Cl]	15–90	1.8–5.7	2–4	[145]
Rayon (BD)	[C$_4$C$_1$im][Cl]	65–91	1.1–4.1		[146]
Lyocell fabric	[C$_4$C$_1$im][Cl]	32–48	0.25–1.75	17–25	[147]
Jute fabric	[C$_4$C$_1$im][Cl]	22–28	0.24–0.3	24–37	[148]
Cotton cloth	[C$_2$C$_1$im][Ace]/methanol	35–65			[149]
Hemp thread (UD)	[C$_2$C$_1$im][Ace]	28–31	0.45–7.2	11–15.4	[150]

sodium hydroxide/urea (NaOH/urea), and ILs. A summary is provided in Tables 10.4 and 10.5 [132–136]. There are few alternative solvent systems known as well, like various molten salt hydrates, dinitrogen tetroxide/dimethylformamide (N$_2$O$_4$/DMF), and dimethylsulfoxide/tetrabutylammonium fluoride (DMSO/TBAF) [15, 129, 137–139]. However, toxicity, slow dissolution rates, limited dissolution capacity, and non-recyclability prevent most of these solvents from being used on an industrial scale. By contrast, some ILs offer high cellulose dissolution rates. ILs are also easy to reuse and safer to handle [47, 140]. Another benefit of an IL-based process is the easy coagulation of the pre-composite, which can be easily done in water. The water/IL mixture can be separated from the composite, worked up, and the IL can be reused as a matrix solvent again. Like in the spinning process, the use of different ILs is possible. For example, [C$_4$C$_1$im][Cl] was used by Zhao et al. to prepare ACCs based on filter paper and rice husks [141].

Huber et al. used two different types of commercially available, cellulose-based textiles, for example, linen flax and rayon (Cordenka 700) as ACC precursor material [135]. They prepared composites by using a one-step process based on [C$_4$C$_1$im][Ace] as the solvent. In their study, they concluded that man-made cellulosic fibers are preferable to natural fibers, since the rayon fibers dissolved more efficiently in the IL. Indeed, Lindman et al. showed that the higher crystallinity of cellulosic material has a significant influence on the dissolution rate [152]. Since

Table 10.5 Overview over unidirectional (UD), bidirectional (BD), and isotropic (ISO) all-cellulose composites prepared by two-step processes

Cellulose source reinforcement	Cellulose source matrix	Solvent system	Fiber fraction (vol%)	Tensile strength (MPa)	Young's modulus (GPa)	Elongation (%)	References
Ramie fiber (UD)	Wood pulp	DMAc/LiCl	80	480		4	[128]
Hemp fiber (ISO)	Cellulose powder	NMMO	40	28.9	1.8	20.8	[134]
Whiskers (ISO)	Cotton linter pulps	NaOH/Urea	20	117	5.9		[151]
Rice husks (ISO)	Filter paper	$[C_4C_1im][Cl]$	40	58	1.7	5.7	[141]
Rice husks (ISO)	Filter paper	$[C_4C_1im][Cl]$	60	56	2.9	2.7	[141]
Rayon (BD)	Sappi pulp	$[C_2C_1im][Cl]$	73–81	45–75	1.6–2.3	3.8–16.7	[77]
Rayon (UD)	Sappi pulp	$[C_2C_1im][Cl]$	76	144	3.6	14	[77]

the linen fibers had a crystallinity of 81%, the dissolution of those fibers is slower than the dissolution of the man-made fibers having 42% crystallinity. A lower crystallinity of the fiber also leads to a better matrix–fiber interface in the final ACCs and is also reflected in the mechanical properties. Thus, rayon-based ACCs had twice the tensile strength (70 MPa) of the linen-based ones (36 MPa). Beside the properties of the fibers, the mechanical properties of the ACCs can be influenced by using different preparation parameters, like temperature, dissolving time, and cellulose concentration in the matrix solution. Dormanns et al. used a regenerated cellulose fiber (Cordenka 700) in the form of a fabric and $[C_4C_1im][Ace]$ as the solvent [153]. A single-step process was used for the manufacture of their ACCs with a fiber fraction of 93 vol% using a reinforcing fiber for the formation of the matrix by partial dissolution. They reported Young's moduli up to 7.0 GPa.

Despite these most attractive properties in a one-step process, there is the disadvantage that the high-quality reinforcing fiber is often weakened by partial dissolution in the IL during ACC formation. In view of this impediment, an IL-based, two-step method can be applied, in which an IL is used as the pre-matrix solvent, as shown in Fig. 10.7. In such an approach, Spörl et al. used different concentrations of cellulose (3, 6, and 8 wt%) in $[C_2C_1im][Ace]$ [77]. Dormanns et al. used rayon-based tire cord (Cordenka 700) as reinforcement material and reported mechanical properties for the final ACCs fiber fractions of 73–81 vol%. In this approach, the amount of expensive reinforcing material could be significantly reduced by using less expensive pulp material. Generally, the cellulose concentration in the pre-matrix solution has two different effects on the final composite. On one hand, with increasing concentration, the viscosity of the solution increases, which hinders the solution's flow and its ability to embrace the single filaments of the fiber bundle. On the other hand, a

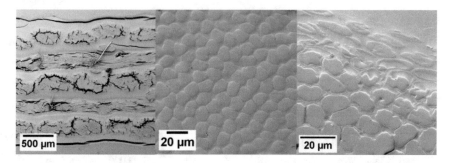

Fig. 10.8 SEM micrographs of an ACC. Adapted with permission from WILEY-VCH [77]

higher cellulose concentration in the IL solution reduces the fiber volume fraction. Thus, the strength and modulus of the composite should decrease with increasing cellulose concentration in the matrix precursor.

The two-step process offers access to well-defined ACCs with the opportunity to modify the fiber content over a large range by variation of the solution concentration or layer thickness of the matrix precursor. Due to the excellent fiber–matrix interaction, ACCs prepared by this approach showed mechanical properties comparable to those of thermoplastic short cut glass fiber-reinforced plastics, as shown in Fig. 10.8. Notably, the partial dissolution of the fiber surfaces is significantly affected by the preparation time, temperature, and the matrix solution.

The dissolution of the fibers depends on the cellulose concentration in the pre-matrix solution, the temperature, and the process time [77, 136–138, 147–150]. Due to the different partial dissolution of the fibers, the mechanical properties of the composites change. For example, a high temperature of the IL of 80 °C during processing decreases the tensile strength, elongation, and Young's modulus with increasing dwell time of the pre-composite. This can be explained by the fact that the IL is able to dissolve the reinforcement fibers, which will ultimately result in decreasing mechanical properties. Successful recycling of ACCs was demonstrated by reusing the ACC for matrix preparation several times [77]. Minor changes were observed in the polydispersity index (*PDI*) and degree of polymerization (*DP*) upon recycling of the material, as shown in Table 10.6.

However, the mechanical properties of the composite remained constant over three recycling steps, as shown in Table 10.7.

10.5   IL-Technology-Derived Cellulosic Fibers as Carbon Fiber Precursors

Carbon fibers (CFs) are described as fibers containing at least 92% carbon obtained by the controlled pyrolysis of polymeric precursor fibers, like poly(acrylonitrile) (PAN), pitch, lignin, or cellulose [154–156]. CFs generally have low densities around

Table 10.6 DP_{visc}, M_n, and polydispersity index (*PDI*) of the recycled composite materials compared to the starting materials

Material	DP_{visc}	M_n (g/mol)	*PDI*
Reinforcing fiber	620 ± 10	48,000	5.7
Pulp for matrix	580 ± 10	16,000	12.1
Pulp regenerated from IL	520 ± 10	19,000	7.2
Composite	470 ± 20	30,000	4.3
Composite, regenerated from IL	490 ± 10	25,000	4.7
Composite 1st recycle	520 ± 15	33,000	4.8
Composite 1st recycle, regenerated from IL	470 ± 10	17,000	6.0
Composite 2nd recycle	500 ± 15	30,000	5.1
Composite 2nd recycle, regenerated from IL	480 ± 10	18,000	5.7
Composite 3rd recycle	500 ± 10	34,000	4.1
Composite 3rd recycle, regenerated from IL	460 ± 10	15,000	5.9

Reproduced with permission from WILEY-VCH [77]

Table 10.7 Mechanical properties of the original and recycled ACCs

	Tensile strength (MPa)	Young's modulus (GPa)	Elongation (%)	Flexural modulus (GPa)	Charpy impact strength (kJ/m^2)
Composite	53 ± 3	2.0 ± 0.04	12.7 ± 0.9	4.1 ± 1.0	70 ± 15
Composite 1st recycle	57 ± 3	1.7 ± 0.10	13.9 ± 2.4	6.3 ± 0.7	65 ± 3
Composite 2nd recycle	41 ± 5	2.0 ± 0.15	14.7 ± 3.7	6.0 ± 0.4	64 ± 2
Composite 3rd recycle	50 ± 2	2.4 ± 0.10	6.9 ± 0.8	7.8 ± 0.1	66 ± 3

Reproduced with permission from WILEY-VCH [77]

1.85 g cm^{-3} paired with excellent tensile properties, good thermal and electrical conductivities, high thermal and chemical stabilities, and excellent creep resistance [157]. The first carbon fibers were made from natural cellulose fibers by Edison as carbon filaments in electric lamps in 1880 [158]. Almost 80 years later, Union Carbide applied a process for CFs made from rayon [159, 160]. Shortly after this, in 1965, carbon fibers with a tensile strength of 1.25 GPa and a Young's modulus of 170 GPa were commercially available from Union Carbide. The production of these high-modulus CFs was based on a post-carbonization hot stretching process at 2800 °C developed by Bacon and Schalamon [161]. Despite major research activity on cellulosic precursors in the 1950–1970s, research in this field was abandoned due to disadvantages, such as high production costs at low carbon yields and the much more promising results with PAN-based CF. However, today, cellulose-based CF precursors have again evoked interest [94, 155, 162–164].

Naturally grown cellulose fibers are not suitable for the production of CFs due to fiber morphology inhomogeneities and impurities, like proteins and lignin. The porous structure leads to high brittleness after carbonization [94]. Well-defined filament fiber dimensions and high-purity cellulosic fibers can be produced by the viscose or Lyocell process and by IL-technology processes. When processed as continuous multi-filaments, they are considered and used as promising precursors. However, during carbonization, the total mass loss might be up to 90% due to degradation reactions and the formation of volatile carbon-containing compounds. The maximum theoretical carbon yield in the carbonization of cellulose is 44.4 wt% corresponding to the formal loss of five molecules of water per anhydroglucose repeat unit (AGU). Strategies have been developed to achieve high-performance CFs with higher carbonization yields of 25–38 wt%. These strategies include low heating rates during pyrolysis up to 400 °C [159, 165, 166], oxidative pretreatment, [167–171], and application of carbonization aids, including catalysts, like ammonium phosphates or sulfates, for dehydration, as shown in Fig. 10.9 [170, 172–182]. An overview of cellulose-based CFs is found in numerous books and reviews [155, 157, 168, 183–186].

So far, IL-technology-based cellulosic fibers and cellulosic blended fibers as precursor materials for CFs have been prepared by several research groups [82, 94, 164]. Byrne et al. showed that the mechanical and morphological properties of the resulting CFs were directly influenced by the physical properties of the precursor [164]. Spörl et al. successfully increased the carbon yield up to 35–38 wt% by using cellulosic derivate (cellulose tosylate/phosphate) precursor fibers spun with IL-technology [82]. Another study used cellulose/lignin blends dissolved in $[C_2C_1im][Ace]$ as spinning dope for blended fibers as CF precursor material. Nonetheless, the CF can be based

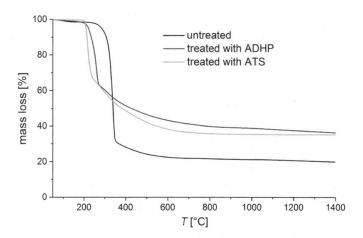

Fig. 10.9 TGA curves of cellulosic fibers treated with ATS (ammonium tosylate) (1 wt% S, green), ammonium dihydrogen phosphate (ADHP) (1 wt% P, blue), and untreated fibers (black) [187]. Adapted with permission from Lenzinger Berichte

on IL-cellulose fibers due to the low space-time yield and mechanical properties with PAN-based CF [82, 155]. However, CF properties still have to be improved to some extent. Once tensile strengths >2.5 GPa and Young's moduli >200 GPa have been reached, PAN-based carbon fibers could be substituted stepwise by this ecologically and economically amended alternative.

10.6 Conclusions

The examples in this chapter have summarized the processing of cellulose using IL-technology, which is a promising research field with great potential for industrial applications. Numerous ILs are available for the dissolution of cellulose matching all requirements for technical conversion. Spinning cellulosic fibers based on the IL-technology is close to industrial conversion; the basics have been developed and ongoing pressure to develop environmentally friendly technology will rapidly increase the interest in this technology.

Based on the flexibility of the IL-based processing of cellulose, there is more interest in developing cellulose-based materials, like ACCs and super-microfibers. Likewise, ACCs can be used in lightweight construction. In addition to this, the exploration of a new class of cellulose-based carbon fibers is made possible by IL-derived cellulosic fibers.

References

1. Moon RJ, Martini A, Nairn J, Simonsen J, Youngblood J (2011) Chem Soc Rev 40:3941–3994. https://doi.org/10.1039/c0cs00108b
2. Tsioptsias C, Stefopoulos A, Kokkinomalis I, Papadopoulou L, Panayiotou C (2008) Green Chem 10:965–971. https://doi.org/10.1039/B803869D
3. Medronho B, Romano A, Miguel MG, Stigsson L, Lindman B (2012) Cellulose 19:581–587. https://doi.org/10.1007/s10570-011-9644-6
4. Fink H-P, Weigel P, Purz HJ, Ganster J (2001) Prog Polym Sci 26:1473–1524. https://doi.org/10.1016/S0079-6700(01)00025-9
5. McCormick CL, Dawsey TR (1990) Macromolecules 23:3606–3610. https://doi.org/10.1021/ma00217a011
6. Ciacco GT, Liebert TF, Frollini E, Heinze TJ (2003) Cellulose 10:125–132. https://doi.org/10.1023/a:1024064018664
7. Swatloski RP, Spear SK, Holbrey JD, Rogers RD (2002) J Am Chem Soc 124:4974–4975. https://doi.org/10.1021/ja025790m
8. Invs.: Swatloski RP, Rogers RD, Holbrey JD (2003) WO03029329
9. Bredereck K, Hermanutz F (2005) Rev Prog Color Relat Top 35:59–75. https://doi.org/10.1111/j.1478-4408.2005.tb00160.x
10. Ingildeev D, Effenberger F, Bredereck K, Hermanutz F (2013) J Appl Polym Sci 128:4141–4150. https://doi.org/10.1002/app.38470
11. Pang F-J, He C-J, Wang Q-R (2003) J Appl Polym Sci 90:3430–3436. https://doi.org/10.1002/app.13063

12. Zhu S, Wu Y, Chen Q, Yu Z, Wang C, Jin S, Ding Y, Wu G (2006) Green Chem 8:325–327. https://doi.org/10.1039/b601395c
13. Liu C, Sun R, Zhang A, Li W (2010) ACS Symp Ser 1033:287–297. https://doi.org/10.1021/bk-2010-1033.ch016
14. Kargl R, Mohan T, Ribitsch V, Saake B, Puls J, Stana Kleinschek K (2015) Nord Pulp Pap Res J 30:6–13. https://doi.org/10.3183/NPPRJ-2015-30-01-p006-013
15. Isik M, Sardon H, Mecerreyes D (2014) Int J Molec Sci 15:11922–11940. https://doi.org/10.3390/ijms150711922
16. Hermanutz F, Vocht MP, Panzier N, Buchmeiser MR (2018) Macromol Mater Eng 304:1800450. https://doi.org/10.1002/mame.201800450
17. Walden P (1914) Bull Imp Acad Sci St Pétersbourg 8:405–422
18. Brennecke JF, Maginn EJ (2001) AIChE J 47:2384–2389. https://doi.org/10.1002/aic.690471102
19. Endres F, Zein El Abedin S (2006) Phys Chem Chem Phys 8:2101–2116. https://doi.org/10.1039/B600519P
20. Pinkert A, Marsh KN, Pang S, Staiger MP (2009) Chem Rev 109:6712–6728. https://doi.org/10.1021/cr9001947
21. Cao Y, Wu J, Zhang J, Li H, Zhang Y, He J (2009) Chem Eng J 147:13–21. https://doi.org/10.1016/j.cej.2008.11.011
22. Tokuda H, Tsuzuki S, Abu Bin Hasan Susan M, Hayamizu K, Watanabe M (2006) J Phys Chem B 110:19593–19600. https://doi.org/10.1021/jp064159v
23. Dupont J (2004) J Braz Chem Soc 15:341–350. https://doi.org/10.1590/S0103-50532004000300002
24. Irge DD (2016) Am J Phys Chem 5:74–79. https://doi.org/10.11648/j.ajpc.20160503.14
25. Ngo HL, LeCompte K, Hargens L, McEwen AB (2000) Thermochim Acta 357:97–102. https://doi.org/10.1016/S0040-6031(00)00373-7
26. Sashina ES, Kashirskii DA, Janowska G, Zaborski M (2013) Thermochim Acta 568:185–188. https://doi.org/10.1016/j.tca.2013.06.022
27. Wendler F, Todi L-N, Meister F (2012) Thermochim Acta 528:76–84. https://doi.org/10.1016/j.tca.2011.11.015
28. Freire MG, Santos LMNBF, Fernandes AM, Coutinho JAP, Marrucho IM (2007) Fluid Phase Equilib 261:449–454. https://doi.org/10.1016/j.fluid.2007.07.033
29. Seddon K, Stark A, Torres M-J (2000) Pure Appl Chem 72:2275–2287. https://doi.org/10.1351/pac200072122275
30. Huddleston JG, Visser AE, Reichert WM, Willauer HD, Broker GA, Rogers RD (2001) Green Chem 3:156–164. https://doi.org/10.1039/B103275P
31. Dupont J, de Souza RF, Suarez PAZ (2002) Chem Rev 102:3667–3692. https://doi.org/10.1021/cr010338r
32. Hayes R, Warr GG, Atkin R (2015) Chem Rev 115:6357–6426. https://doi.org/10.1021/cr500411q
33. Plechkova NV, Seddon KR (2008) Chem Soc Rev 37:123–150. https://doi.org/10.1039/b006677j
34. Wasserscheid P, Keim W (2000) Angew Chem 39:3772–3789. https://doi.org/10.1002/1521-3773(20001103)39:21%3c3772:AID-ANIE3772%3e3.0.CO;2-5
35. Dorn S (2009) Ionische Flüssigkeiten: neuartige Löse- und Reaktionsmedien in der Cellulosechemie. Doctoral thesis. Friedrich-Schiller-Universität Jena, Germany
36. Cvjetko Bubalo M, Radošević K, Radojčić Redovniković I, Halambek J, Gaurina Srček V (2014) Ecotoxicol Environ Saf 99:1–12. https://doi.org/10.1016/j.ecoenv.2013.10.019
37. Zhao D, Liao Y, Zhang Z (2007) Clean 35:42–48. https://doi.org/10.1002/clen.200600015
38. Rogers RD, Seddon KR (2003) Science 302:792–793. https://doi.org/10.1126/science.1090313
39. Pham T, Cho C-W, Yun Y-S (2010) Water Res 44:352–372. https://doi.org/10.1016/j.watres.2009.09.030

40. Wang H, Gurau G, Rogers RD (2012) Chem Soc Rev 41:1519–1537. https://doi.org/10.1039/c2cs15311d
41. Witos J, Russo G, Ruokonen S-K, Wiedmer SK (2017) Langmuir 33:1066–1076. https://doi.org/10.1021/acs.langmuir.6b04359
42. Ruokonen S-K, Sanwald C, Sundvik M, Polnick S, Vyavaharkar K, Duša F, Holding AJ, King AWT, Kilpeläinen I, Lämmerhofer M, Panula P, Wiedmer SK (2016) Environ Sci Technol 50:7116–7125. https://doi.org/10.1021/acs.est.5b06107
43. Warner JC, Anastas PT (1998) Green chemistry: theory and practice. Oxford University Press, Oxford, p 135
44. Anastas PT, Horváth IT (2007) Chem Rev 107:2167–2168. https://doi.org/10.1021/cr0783784
45. Xu A, Wang J, Wang H (2010) Green Chem 12:268–275. https://doi.org/10.1039/B916882F
46. Zhang H, Wu J, Zhang J, He J (2005) Macromol 38:8272–8277. https://doi.org/10.1021/ma0505676
47. Heinze T, Schwikal K, Barthel S (2005) Macromol Biosci 5:520–525. https://doi.org/10.1002/mabi.200500039
48. Kosan B, Michels C, Meister F (2008) Cellulose 15:59–66. https://doi.org/10.1007/s10570-007-9160-x
49. Fukaya Y, Sugimoto A, Ohno H (2006) Biomacromol 7:3295–3297. https://doi.org/10.1021/bm060327d
50. Vitz J, Erdmenger T, Haensch C, Schubert US (2009) Green Chem 11:417–424. https://doi.org/10.1039/B818061J
51. El Seoud OA, Koschella A, Fidale LC, Dorn S, Heinze T (2007) Biomacromol 8:2629–2647. https://doi.org/10.1021/bm070062i
52. Barthel S, Heinze T (2006) Green Chem 8:301–306. https://doi.org/10.1039/B513157J
53. Zavrel M, Bross D, Funke M, Büchs J, Spiess AC (2009) Bioresour Technol 100:2580–2587. https://doi.org/10.1016/j.biortech.2008.11.052
54. Erdmenger T, Haensch C, Hoogenboom R, Schubert US (2007) Macromol Biosci 7:440–445. https://doi.org/10.1002/mabi.200600253
55. Mikkola J-P, Kirilin A, Tuuf J-C, Pranovich A, Holmbom B, Kustov LM, Murzin DYu, Salmi T (2007) Green Chem 9:1229–1237. https://doi.org/10.1039/B708533H
56. Ab Rahim AH, Yunus NM, Man Z, Sarwono A, Hamzah WSW, Wilfred CD (2018) AIP Conf Proc 2016:020010. https://doi.org/10.1063/1.5055412
57. Fukaya Y, Hayashi K, Wada M, Ohno H (2008) Green Chem 10:44–46. https://doi.org/10.1039/b713289a
58. Sashina ES, Novoselov NP (2009) Russ J Gen Chem 79:1057. https://doi.org/10.1134/S1070363209060024
59. Cai T, Yang G, Zhang H, Shao H, Hu X (2012) Polym Eng Sci 52:1708–1714. https://doi.org/10.1002/pen.23069
60. Sashina E, Kashirskii D, Busygin KN (2016) Cellul Chem Technol 50:199–211
61. Sashina ES, Kashirskii DA, Zaborski M, Jankowski S (2012) Russ J Gen Chem 82:1994–1998. https://doi.org/10.1134/S1070363212120158
62. Zhao H, Baker GA, Song Z, Olubajo O, Crittle T, Peters D (2008) Green Chem 10:696–705. https://doi.org/10.1039/B801489B
63. Hermanutz F, Gähr F, Uerdingen E, Meister F, Kosan B (2008) Macromol Symp 262:23–27. https://doi.org/10.1002/masy.200850203
64. BASF SE, Invs.: Abels F, Cwik T, Beyer R, Hermanutz F (2017) WO2017/137284A1
65. Parviainen A, Wahlström R, Liimatainen U, Liitiä T, Rovio S, Helminen JKJ, Hyväkkö U, King AWT, Suurnäkki A, Kilpeläinen I (2015) RSC Adv 5:69728–69737. https://doi.org/10.1039/C5RA12386K
66. Michud A, Tanttu M, Asaadi S, Ma Y, Netti E, Kääriainen P, Persson A, Berntsson A, Hummel M, Sixta H (2015) Text Res J 86:543–552. https://doi.org/10.1177/0040517515591774
67. Kostag M, Jedvert K, Achtel C, Heinze T, El Seoud OA (2018) Molecules 23:511–549. https://doi.org/10.3390/molecules23030511
68. Feng L, Chen Z (2008) J Mol Liq 142:1–5. https://doi.org/10.1016/j.molliq.2008.06.007

69. Sashina ES (2018) Fibre Chem 50:139–143. https://doi.org/10.1007/s10692-018-9949-4
70. Lu B, Xu A, Wang J (2014) Green Chem 16:1326–1335. https://doi.org/10.1039/c3gc41733f
71. Zhang J, Xu L, Yu J, Wu J, Zhang X, He J, Zhang J (2016) Sci China Chem 59:1421–1429. https://doi.org/10.1007/s11426-016-0269-5
72. Zhu C, Koutsomitopoulou AF, Eichhorn SJ, van Duijneveldt JS, Richardson RM, Nigmatullin R, Potter KD (2018) Macromol Mater Eng 303:1800029. https://doi.org/10.1002/mame. 201800029
73. Kilpeläinen I, Xie H, King A, Granstrom M, Heikkinen S, Argyropoulos DS (2007) J Agric Food Chem 55:9142–9148. https://doi.org/10.1021/jf071692e
74. Spörl JM, Hermanutz F, Buchmeiser MR (2017) Nachr Chem 65:998–1003. https://doi.org/ 10.1002/nadc.20174058531
75. Dorn S, Wendler F, Meister F, Heinze T (2008) Macromol Mater Eng 293:907–913. https:// doi.org/10.1002/mame.200800153
76. Chanzy H, Nawrot S, Peguy A, Smith P, Chevalier J (1982) J Polym Sci B Polym Phys 20:1909–1924. https://doi.org/10.1002/pol.1982.180201014
77. Spörl JM, Batti F, Vocht M-P, Raab R, Müller A, Hermanutz F, Buchmeiser MR (2017) Macromol Mater Eng 303:1700335. https://doi.org/10.1002/mame.201700335
78. Olsson C, Westman G (2013) J Appl Polym Sci 127:4542–4548. https://doi.org/10.1002/app. 38064
79. Moosbauer J, Röder T, Kliba G, Schlader S, Zuckerstätter G, Sixta H (2009) Lenzinger Ber 87:98–105
80. Laus G, Bentivoglio G, Schottenberger H, Kahlenberg V, Kopacka H, Röder T, Sixta H (2005) Lenzinger Ber 84:71–86
81. Spörl JM, Beyer R, Abels F, Cwik T, Müller A, Hermanutz F, Buchmeiser MR (2017) Macromol Mater Eng 302:1700195. https://doi.org/10.1002/mame.201700195
82. Spörl JM, Ota A, Son S, Massonne K, Hermanutz F, Buchmeiser MR (2016) Mater Today Commun 7:1–10. https://doi.org/10.1016/j.mtcomm.2016.02.002
83. Luo Z, Wang A, Wang C, Qin W, Zhao N, Song H, Gao J (2014) J Mater Chem A 2:7327–7336. https://doi.org/10.1039/c4ta00225c
84. Rahatekar SS, Rasheed A, Jain R, Zammarano M, Koziol KK, Windle AH, Gilman JW, Kumar S (2009) Polymer 50:4577–4583. https://doi.org/10.1016/j.polymer.2009.07.015
85. Olsson C, Hedlund A, Idström A, Westman G (2014) J Mater Sci 49:3423–3433. https://doi. org/10.1007/s10853-014-8052-3
86. Zhu C, Richardson RM, Potter KD, Koutsomitopoulou AF, van Duijneveldt JS, Vincent SR, Wanasekara ND, Eichhorn SJ, Rahatekar SS (2016) ACS Sustain Chem Eng 4:4545–4553. https://doi.org/10.1021/acssuschemeng.6b00555
87. Hermanutz F, Ingildeev D, Buchmeiser MR (2013) Chem Fibers Int 63:84–87
88. Parviainen A, King AW, Mutikainen I, Hummel M, Selg C, Hauru LK, Sixta H, Kilpelainen I (2013) ChemSusChem 6:2161–2169. https://doi.org/10.1002/cssc.201300143
89. Hauru LKJ, Hummel M, Michud A, Sixta H (2014) Cellulose 21:4471–4481. https://doi.org/ 10.1007/s10570-014-0414-0
90. Michud A, Hummel M, Sixta H (2015) Polymer 75:1–9. https://doi.org/10.1016/j.polymer. 2015.08.017
91. Wanasekara ND, Michud A, Zhu C, Rahatekar S, Sixta H, Eichhorn SJ (2016) Polymer 99:222–230. https://doi.org/10.1016/j.polymer.2016.07.007
92. Michud A, Hummel M, Sixta H (2016) J Appl Polym Sci 133:43718. https://doi.org/10.1002/ app.43718
93. Ma Y, Hummel M, Määttänen M, Särkilahti A, Harlin A, Sixta H (2016) Green Chem 18:858–866. https://doi.org/10.1039/c5gc01679g
94. Spörl JM (2016) Neue Präkursoren für Carbonfasern auf Basis von Cellulose. Doctoral thesis. University of Stuttgart, Cuvillier Verlag, Göttingen
95. Zhang Y, Li H, Li X, Gibril ME, Yu M (2014) Carbohydr Polym 99:126–131. https://doi.org/ 10.1016/j.carbpol.2013.07.084

96. Vo HT, Kim YJ, Jeon EH, Kim CS, Kim HS, Lee H (2012) Chem Eur J 18:9019–9023. https://doi.org/10.1002/chem.201200982
97. Ingildeev D, Hermanutz F, Buchmeiser MR, Onuseit V, Feuer A, Weber R (2013) Stuttgarter Kunststoff-Kolloquium 23:157–161
98. Ingildeev D, Hermanutz F, Bredereck K, Effenberger F (2012) Macromol Mater Eng 297:585–594. https://doi.org/10.1002/mame.201100432
99. Byrne N, Leblais A, Fox B (2014) J Mater Chem A 2:3424–3429. https://doi.org/10.1039/C3TA15227H
100. Mundsinger K, Müller A, Beyer R, Hermanutz F, Buchmeiser MR (2015) Carbohydr Polym 131:34–40. https://doi.org/10.1016/j.carbpol.2015.05.065
101. Ingildeev D (2010) Herstellung und Charakterisierung von Fasern aus Cellulose und Cellulose/Polymer-Blends mittels ionischer Flüssigkeiten. Doctoral thesis. University of Stuttgart, Shaker Verlag, Düren (2011)
102. Wu K, Yao Y, Yu J, Chen S, Wang X, Zhang Y, Wang H (2017) Cellulose 24:3377–3386. https://doi.org/10.1007/s10570-017-1351-5
103. Bengtsson A, Bengtsson J, Olsson C, Sedin M, Jedvert K, Theliander H, Sjöholm E (2018) Holzforschung 72:1007–1016. https://doi.org/10.1515/hf-2018-0028
104. Pillai CKS, Paul W, Sharma CP (2009) Progr Polym Sci 34:641–678. https://doi.org/10.1016/j.progpolymsci.2009.04.001
105. Rinaudo M (2006) Progr Polym Sci 31:603–632. https://doi.org/10.1016/j.progpolymsci.2006.06.001
106. Silva SS, Mano JF, Reis RL (2017) Green Chem 19:1208–1220. https://doi.org/10.1039/C6GC02827F
107. Walther P, Ota A, Müller A, Hermanutz F, Gähr F, Buchmeiser MR (2016) Macromol Mater Eng 301:1337–1344. https://doi.org/10.1002/mame.201600208
108. Qin Y, Lu X, Sun N, Rogers RD (2010) Green Chem 12:968–971. https://doi.org/10.1039/c003583a
109. Barber PS, Kelley SP, Griggs CS, Wallace S, Rogers RD (2014) Green Chem 16:1828–1836. https://doi.org/10.1039/C4GC00092G
110. Shamshina JL, Zavgorodnya O, Bonner JR, Gurau G, Di Nardo T, Rogers RD (2017) ChemSusChem 10:106–111. https://doi.org/10.1002/cssc.201601372
111. Huang Y, Zhong Z, Duan B, Zhang L, Yang Z, Wang Y, Ye Q (2014) J Mater Chem B 2:3427–3432. https://doi.org/10.1039/C4TB00098F
112. Nowotny J, Aibibu D, Farack J, Nimtschke U, Hild M, Gelinsky M, Kasten P, Cherif C (2016) J Biomater Sci Polym Ed 27:917–936. https://doi.org/10.1080/09205063.2016.1155879
113. Toskas G, Brünler R, Hund H, Hund R-D, Hild M, Aibibu D, Cherif C (2013) Autex Res J 13:134–140. https://doi.org/10.2478/v10304-012-0041-5
114. Chen Q, Xu A, Li Z, Wang J, Zhang S (2011) Green Chem 13:3446–3452. https://doi.org/10.1039/C1GC15703E
115. Zhang L, Guo J, Du Y (2002) J Appl Polym Sci 86:2025–2032. https://doi.org/10.1002/app.11156
116. Setoyama M, Kato T, Yamamoto K, Kadokawa J (2013) J Polym Environ 21:795–801. https://doi.org/10.1007/s10924-013-0580-4
117. Kadokawa J, Hirohama K, Mine S, Kato T, Yamamoto K (2012) J Polym Environ 20:37–42. https://doi.org/10.1007/s10924-011-0331-3
118. Dutta PK, Dutta J, Tripathi VS (2004) J Sci Ind Res 63:20–31
119. Holmberg M, Berg J, Stemme S, Ödberg L, Rasmusson J, Claesson P (1997) J Coll Interf Sci 186:369–381. https://doi.org/10.1006/jcis.1996.4657
120. Kuzmina O, Heinze T, Wawro D (2012) ISRN Polym Sci 2012:251950. https://doi.org/10.5402/2012/251950
121. Ma B, Zhang M, He C, Sun J (2012) Carbohydr Polym 88:347–351. https://doi.org/10.1016/j.carbpol.2011.12.020
122. Stefanescu C, Daly WH, Negulescu II (2012) Carbohydr Polym 87:435–443. https://doi.org/10.1016/j.carbpol.2011.08.003

123. Ehrenstein GW (2006) Faserverbund-Kunststoffe: Werkstoffe—Verarbeitung—Eigenschaften, 2nd edn. Carl-Hanser Verlag, München
124. Yazdanbakhsh A, Bank LC (2014) Polymers 6:1810–1826. https://doi.org/10.3390/polym6061810
125. The European Parliament and the Council (2008) Off J Eur Union, L312/313
126. Karus M, Ortmann S, Vogt D, Müssig J (2005) Naturfaserverstärkte Kunststoffe—Pflanzen, Rohstoffe, Produkte, 1st edn. Fachagentur Nachwachsende Rohstoffe e. V. (FNR), Gülzow, pp 1–40
127. Bos HL (2004) The potential of flax fibres as reinforcement for composite materials. Doctoral thesis. Technische Universiteit Eindhoven, Eindhoven, The Netherlands. https://doi.org/10.6100/IR575360
128. Nishino T, Matsuda I, Hirao K (2004) Macromolecules 37:7683–7687. https://doi.org/10.1021/ma049300h
129. Kalka S, Huber T, Steinberg J, Baronian K, Müssig J, Staiger MP (2014) Compos A Appl Sci Manuf 59:37–44. https://doi.org/10.1016/j.compositesa.2013.12.012
130. Beauson J, Lilholt H, Brøndsted P (2014) J Reinf Plast Compos 33:1542–1556. https://doi.org/10.1177/0731684414537131
131. Gindl W, Keckes J (2005) Polymer 46:10221–10225. https://doi.org/10.1016/j.polymer.2005.08.040
132. Piltonen P, Hildebrandt NC, Westerlind B, Valkama J-P, Tervahartiala T, Illikainen M (2016) Compos Sci Technol 135:153–158. https://doi.org/10.1016/j.compscitech.2016.09.022
133. Duchemin B, Le Corre D, Leray N, Dufresne A, Staiger MP (2016) Cellulose 23:593–609. https://doi.org/10.1007/s10570-015-0835-4
134. Qi H, Cai J, Zhang L, Kuga S (2009) Biomacromol 10:1597–1602. https://doi.org/10.1021/bm9001975
135. Huber T, Pang S, Staiger MP (2012) Compos Part A Appl Sci Manuf 43:1738–1745. https://doi.org/10.1016/j.compositesa.2012.04.017
136. Soykeabkaew N, Arimoto N, Nishino T, Peijs T (2008) Compos Sci Technol 68:2201–2207. https://doi.org/10.1016/j.compscitech.2008.03.023
137. Li J, Nawaz H, Wu J, Zhang J, Wan J, Mi Q, Yu J, Zhang J (2018) Compos Commun 9:42–53. https://doi.org/10.1016/j.coco.2018.04.008
138. Adak B, Mukhopadhyay S (2016) J Appl Polym Sci 133:43398. https://doi.org/10.1002/app.43398
139. Huber T, Müssig J, Curnow O, Pang S, Bickerton S, Staiger MP (2011) J Mater Sci 47:1171–1186. https://doi.org/10.1007/s10853-011-5774-3
140. Zhao H, Xia S, Ma P (2005) J Chem Technol Biotechnol 80:1089–1096. https://doi.org/10.1002/jctb.1333
141. Zhao Q, Yam RCM, Zhang B, Yang Y, Cheng X, Li RKY (2009) Cellulose 16:217–226. https://doi.org/10.1007/s10570-008-9251-3
142. Duchemin BJC, Newman RH, Staiger MP (2009) Compos Sci Technol 69:1225–1230. https://doi.org/10.1016/j.compscitech.2009.02.027
143. Zhang J, Luo N, Zhang X, Xu L, Wu J, Yu J, He J, Zhang J (2016) ACS Sustain Chem Eng 4:4417–4423. https://doi.org/10.1021/acssuschemeng.6b01034
144. Yousefi H, Nishino T, Faezipour M, Ebrahimi G, Shakeri A (2011) Biomacromol 12:4080–4085. https://doi.org/10.1021/bm201147a
145. Duchemin BJC, Mathew AP, Oksman K (2009) Compos Part A Appl Sci Manuf 40:2031–2037. https://doi.org/10.1016/j.compositesa.2009.09.013
146. Huber T, Bickerton S, Müssig J, Pang S, Staiger MP (2012) Carbohydr Polym 90:730–733. https://doi.org/10.1016/j.carbpol.2012.05.047
147. Adak B, Mukhopadhyay S (2018) Polymer 141:79–85. https://doi.org/10.1016/j.polymer.2018.02.065
148. Adak B, Mukhopadhyay S (2016) Indian J Fibre Text Res 41:380–384
149. Haverhals LM, Sulpizio HM, Fayos ZA, Trulove MA, Reichert WM, Foley MP, De Long HC, Trulove PC (2012) Cellulose 19:13–22. https://doi.org/10.1007/s10570-011-9605-0

150. Haverhals LM, Reichert WM, De Long HC, Trulove PC (2010) Macromol Mater Eng 295:425–430. https://doi.org/10.1002/mame.201000005
151. Ouajai S, Shanks RA (2009) Compos Sci Technol 69:2119–2126. https://doi.org/10.1016/j.compscitech.2009.05.005
152. Lindman B, Karlström G, Stigsson L (2010) J Mol Liq 156:76–81. https://doi.org/10.1016/j.molliq.2010.04.016
153. Dormanns JW, Weiler F, Schuermann J, Müssig J, Duchemin BJC, Staiger MP (2016) Compos Part A Appl Sci Manuf 85:65–75. https://doi.org/10.1016/j.compositesa.2016.03.010
154. Fitzer E, Edie DD, Johnson DJ (1990) Carbon fibres—present state and future expectations; Pitch and mesophase fibers; Structure and properties of carbon fibers. In: Figueiredo JL, Bernardo CA, Baker RTK, Hüttinger KJ (eds) Carbon fibers filaments and composites. NATO ASI series E, vol 117. Springer, Dordrecht, pp 3–41, pp 43–72, pp 119–146. https://doi.org/10.1007/978-94-015-6847-0_1, https://doi.org/10.1007/978-94-015-6847-0_2, https://doi.org/10.1007/978-94-015-6847-0_5
155. Frank E, Steudle LM, Ingildeev D, Spörl JM, Buchmeiser MR (2014) Angew Chem Int Ed 53:5262–5298. https://doi.org/10.1002/anie.201306129
156. Baker DA, Rials TG (2013) J Appl Polym Sci 130:713–728. https://doi.org/10.1002/app.39273
157. Huang X (2009) Materials 2:2369–2403. https://doi.org/10.3390/ma2042369
158. Inv.: Edison TA (1880) US223898A
159. Union Carbide Corporation, Invs.: Ford CE, Mitchell CV (1963) US3107152
160. Union Carbide Corporation, Invs.: Ford CE, Mitchell CV (1962) DE1130419
161. Union Carbide Corporation, Invs.: Bacon R, Schalamon WA (1973) US3716331A
162. Dumanlı AG, Windle AH (2012) J Mater Sci 47:4236–4250. https://doi.org/10.1007/s10853-011-6081-8
163. Zhang X, Lu Y, Xiao H, Peterlik H (2013) J Mater Sci 49:673–684. https://doi.org/10.1007/s10853-013-7748-0
164. Byrne N, Setty M, Blight S, Tadros R, Ma Y, Sixta H, Hummel M (2016) Macromol Chem Phys 217:2517–2524. https://doi.org/10.1002/macp.201600236
165. Union Carbide Corporation, Invs.: Cross CB, Ecker DR, Stein OL (1964) US3116975
166. Brunner PH, Roberts PV (1980) Carbon 18:217–224. https://doi.org/10.1016/0008-6223(80)90064-0
167. Konkin AA (1985) Production of cellulose based carbon fibrous materials. In: Kelly A, Rabotnov YuN, Watt W, Perov BV (eds) Handbook of composites. Vol 1 Strong fibres, Elsevier, Amsterdam, pp 275–325
168. Bacon R (1973) Carbon fibers from rayon precursors. In: Walker Jr PL, Thrower PA (eds) Chemistry and physics of carbon, vol 9, 1st edn. Marcel Dekker, New York, pp 1–102
169. Ross SE (1968) Text Res J 38:906–913. https://doi.org/10.1177/004051756803800905
170. Karacan I, Soy T (2013) J Mater Sci 48:2009–2021. https://doi.org/10.1007/s10853-012-6970-5
171. Schwenker Jr RF, Pacsu E (1958) Ind Eng Chem 50:91–96. https://doi.org/10.1021/ie50577a043
172. Goodhew PJ, Clarke AJ, Bailey JE (1975) Mater Sci Eng 17:3–30. https://doi.org/10.1016/0025-5416(75)90026-9
173. Tang WK, Neill WK (1964) J Polym Sci C Polym Sympos 6:65–81. https://doi.org/10.1002/polc.5070060109
174. Basch A, Lewin M (1975) Text Res J 45:246–250. https://doi.org/10.1177/004051757504500310
175. Tomlinson JB, Theocharis CR (1992) Carbon 30:907–911. https://doi.org/10.1016/0008-6223(92)90014-N
176. Morozova AA, Brezhneva YuV (1997) Fibre Chem 29:31–35. https://doi.org/10.1007/BF02430683
177. Bhat G, Akato K, Hoffman W (2012) Pretreatment and pyrolysis of rayon-based precursor for carbon fibers. In: Proceedings of the Fiber Society Spring Conference, EMPA, St. Gallen, Switzerland

178. Li H, Yang Y, Wen Y, Liu L (2007) Compos Sci Technol 67:2675–2682. https://doi.org/10.1016/j.compscitech.2007.03.008
179. Wu Q, Pan N, Deng K, Pan D (2008) Carbohydr Polym 72:222–228. https://doi.org/10.1016/j.carbpol.2007.08.005
180. Liu Q, Lv C, Yang Y, He F, Ling L (2005) J Molec Struct 733:193–202. https://doi.org/10.1016/j.molstruc.2004.01.016
181. BASF SE, Invs.: Son S, Massonne K, Hermanutz F, Spörl JM, Buchmeiser MR, Beyer R (2015) WO2015173243
182. Byrne N, Chen J, Fox B (2014) J Mater Chem A 2:15758–15762. https://doi.org/10.1039/C4TA04059G
183. Diefendorf RJ, Tokarsky E (1975) Polym Eng Sci 15:150–159. https://doi.org/10.1002/pen.760150306
184. Morgan P (2005) Carbon fibers and their composits. CRC Press, Taylor and Francis Group, Boca Raton
185. Chand S (2000) J Mater Sci 35:1303–1313. https://doi.org/10.1023/A:1004780301489
186. Donnet J-B, Bansal RC (1984) Carbon fibers, 1st edn. Marcel Dekker, New York
187. Spörl JM, Hermanutz F, Buchmeiser MR (2018) Lenzinger Ber 94:85–94

Chapter 11
Toward Industrialization of Ionic Liquids

Roland S. Kalb

Abstract By the end of 2018, more than 80,000 scientific papers were published, and 17,000 patent applications were filed on ionic liquids. The field experienced an archetypal hype cycle, with inflated expectations followed by a period of disillusionment, and is now on an upward slope of enlightenment. While there are no massive-scale applications of ionic liquids today, and the market remains behind the mid-term forecasts, there are many indications that ionic liquids will become a commercial success story. This chapter summarizes critical statements collected by surveying 25 academic and industrial leaders. Recent market developments are discussed and evaluated, including 57 ionic liquid applications that have been commercialized or are being developed on a pilot scale as of 2019. By comparison, there were only 13 commercial and pilot applications in 2008. In conclusion, the author is confident that the field will soon reach maturity and ionic liquids will enter megatrend mass markets.

Keywords Commercialized ionic liquid applications · Emerging technologies · Future of ionic liquids · Gartner® hype cycle · Implemented ionic liquid applications

11.1 Introduction

The use of high-temperature molten salts dates back to the *Shang Dynasty* (sixteenth to eleventh century B.C.), and in the second century B.C., the ancient Chinese commercialized melts of alkali nitrates and chromates for surface modifications (e.g., hardening and corrosion protection) of bronze handcrafted weapons [1]. In the modern world, high-temperature molten salts have been used for more than two centuries, and generations of chemists and engineers have developed a multitude of industrial processes described elsewhere [2, 3]. With the discovery of the first representative low-temperature molten salt by Paul Walden in 1914 [4], a new class of molten salts would be named ionic liquids by Bockris [5] in 1955. From the late 1950s to

R. S. Kalb (✉)
proionic GmbH, Raaba-Grambach, Austria
e-mail: roland.kalb@proionic.com

© Springer Nature Switzerland AG 2020
M. B. Shiflett (ed.), *Commercial Applications of Ionic Liquids*, Green Chemistry and Sustainable Technology, https://doi.org/10.1007/978-3-030-35245-5_11

Fig. 11.1 Typical phases of a Gartner® hype cycle [7, 8]

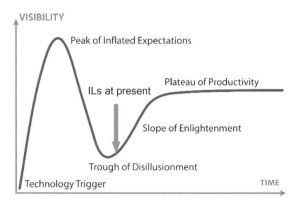

the 1990s, an incrementally increasing annual number of papers and patents were reported. Around the year 2000, there was a rapid, and at times exponential growth that continued until recently. By the end of 2018, more than 80,000 scientific papers were published and 17,000 patent families were filed [6] in a cross-sectional field of highly diverse applications.

As expected with almost any emerging technology, much of the ionic liquid field experienced a typical hype cycle [7] leading to increasingly unrealistic expectations until approximately 2010. The technology "peak" was followed by a phase of frustration and corporate withdrawal from committed strategic investments (Fig. 11.1).

This phase of disillusionment forced both academia and industry to focus on longer-term, mandatory activities involving ionic liquids for future technologies. At present, the field has just passed the *trough of disillusionment*, and a very few percent of applications have reached market adoption, which will steepen the *slope of enlightenment*. It is anticipated that the first massive applications of ionic liquids will appear in the next few years.

11.2 Ionic Liquid Market Growth

> We tend to overestimate the effect of a technology in the short run and underestimate the effect in the long run.
>
> Amara's law [9]

As of 2019, no mass applications of ionic liquids have emerged, despite enthusiastic expectations and cross-sectional applicability in the key industrial sectors including analytics, biotechnology, chemicals and chemical reactions, electrochemistry, energy, engineering liquids, food technology, materials, medical, metal finishing, separations, solvents, performance additives, and pharmaceuticals. An examination of publication statistics and market data for semiconductors and nanoparticles, two

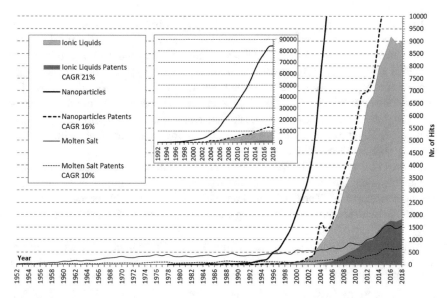

Fig. 11.2 Number of publications and patents found in SciFinder™ database, for the term "ionic liquids" versus other technologies. Compound annual growth rates (CAGR) shown for 2004–2018

technologies that have recently entered mass markets, may help explain the seemingly disjointed situation of ionic liquids.

Figure 11.2 compares total annual journal publications and patents for the concept search terms *ionic liquids*, *molten salts*, and *nanoparticles* [6]. The search results are presented in a non-cumulative plot; for example, *ionic liquids* hits in 2012 did not include *ionic liquids* hits from any previous year(s).

The term *molten salts* resulted in a slow, yet constant growth of publications and patents over the years 1952–2018 and had an initialization period of approximately 25 years between the significant increase in the number of publications and the increase in patents. The compound annual growth rate (CAGR) of patent applications for *molten salts* was approximately 10% over the past 15 years and increased notably over the last 5 years, indicating a type of renaissance for modern applications. Between the years 1952–2004, the CAGR for *molten salts* was only in the range of 5.5%. Analyzing the curves for *nanoparticles* and *ionic liquids*, a completely different picture emerges; both technologies show significantly shorter initialization periods and exponential annual growth rates. The CAGRs over the last 15 years for *nanoparticle* patents are 16% and that of *ionic liquid* patents is 21%. The graph shows a peak for the former in 2017, possibly indicating the beginning of market saturation. Curve shapes for *nanoparticle* and *ionic liquid* patents both indicate rapid research and development within a variety of application fields and are typical initial growth phases for cross-sectional technologies.

Compound annual growth rates of patent applications show strong positive correlations with increased private R&D funding, venture capital, and—most importantly—market growth. In fact, CAGRs are one of the key indicators used by stock exchange analysts and governments to predict market trends [10, 11]. This correlation is entirely plausible considering that patents are a mandatory requirement for the economic exploitation of technologies and can be used to measure the market's ability to transfer scientific results into applied technologies. Technological growth is compounded by the development of applied technologies, which increase intellectual property (IP) developments, creating more patents.

Figures 11.3 and 11.4 illustrate the strong correlation between annual growth rates of patent applications and global revenue for the technology examples of *semiconductor* and *nanoparticles*. The CAGR and revenue curves are roughly congruent to each other. This trend is observed in other high-tech products (e.g., Teflon® [14]and lasers [15]).

A market study published by the Helmut Kaiser Consultancy [16] expected the accessible global annual ionic liquid market to rise from 300 million US dollars in 2012 to 3.4 billion US dollars in 2020. This increase represents a CAGR of 35% and is very close to the CAGR for ionic liquid patents for the years 2002–2012 (34%), as shown in Fig. 11.2. Another recent survey, published by the renowned market research firm Frost & Sullivan, expects the accessible ionic liquid market to rise from 1.3 billion US dollars in 2015 to 2.8 billion US dollars in 2020, with a CAGR of 16.3% [17]. While encouraging, the aforementioned CAGR figures disagree with market surveys published by others (e.g., Grand View Research [18] and Markets and Markets [19]), which forecast the global ionic liquid market to reach only 62.3 million US dollars in 2025 (CAGR 10%) and 39.6 million US dollars in 2021 (CAGR 9.2%).

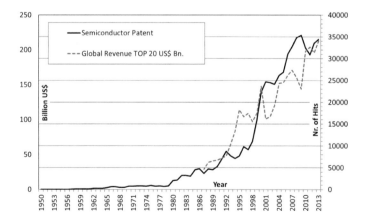

Fig. 11.3 Number of publications and patents (non-cumulative) found in SciFinder™ database, using search term "semiconductor patents" (CAGR 1987–2013 8.9%) versus "annual global semiconductor revenue" of the top 20 companies [12] (CAGR 2004–2013 8.0%)

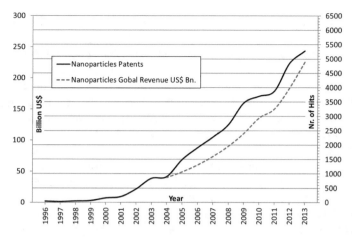

Fig. 11.4 Number of publications and patents (non-cumulative) found in SciFinder™ database, using search term "nanoparticles patents" (CAGR 2004–2013 21.7%) versus "annual global nanoparticle revenue" [13] (CAGR 2004–2013 22.6%)

How can the deviation between the predictions be explained? The analyses and methods used by Helmut Kaiser and Frost & Sullivan predict growth rates based on "accessible" volume for existing ionic liquid technologies and are estimated by approximating 1.5% market penetration of the corresponding global application markets. In contrast, Grand View Research and Markets and Markets reports use the ionic liquid volume "actually sold" globally, which was obtained from data delivered by ionic liquid manufactures and suppliers. The latter method provides a more realistic annual global volume of several hundreds of tons and was certainly less than 1000 tons in 2017. By comparing all market predictions, it can be stated that **the currently served ionic liquid market represents only 1–2% of the potentially accessible market**. The growth of ionic liquid patent applications, as shown in Fig. 11.2, indicates a correspondingly steep market growth because **markets are largely unexploited at the present time**. These findings reflect the commercialization challenges of ionic liquid technologies resulting from diverse commercialization barriers to be overcome in the future. While these challenges pose a seemingly "natural" process for emerging technologies, there is obvious market potential for ionic liquids.

11.3 Commercialized Applications to Date

The large number of ionic liquid patent applications (~17,000 cumulative in 2018) is an optimistic sign toward dissuaded commercial attention, especially in light of the key market indicators as described above. So, when will significant commercial applications come for ionic liquids? The answer is: Commercial applications already exist, they increase every year, and they are here to stay!

Table 11.1 shows a list of carefully selected, industrially implemented ionic liquid applications. By nature of the means by which this information was collected [20], the list cannot be complete, and there are thus additional unpublished and/or confidential applications. It is beyond the scope of this chapter to describe the chemical and technical details of each of these applications, and for more information, the reader is referred to relevant literature [21–32], Web pages, and electronic documents of the cited companies, as well as other chapters of this book. The listed implementations are summarized by the general role the ionic liquid plays in each application and have proven successful within the processing limits of technical effectiveness, robustness, life cycle, safety, and environmental impact. In spite of the rather high prices of ionic liquids, the ionic liquid/application pairs have proven competitive within capital expense (CAPEX) and operating expense (OPEX) budgets and remain clearly attractive to management and investors.

The unusual solvation properties of several ionic liquids are used as **activating solvents**, for example, to highly activate hydrochloric acid and replace phosgene for chlorination of alcohols at BASF ("nucleophilic HCl") and for the dimerization of butenes to octanes, that are ultimately used to manufacture plasticizers at Axens (the "Difasol™" process).

Implemented uses of ionic liquids as **catalysts** include petroleum refinery alkylations with strong Lewis-acidic, chloroaluminate-based ionic liquids, applied by Chevron Honeywell UOP (see Chap. 2) and PetroChina.

Since 2002, BASF has used ionic liquids at the commercial scale as **auxiliaries** to operate the well-known BASIL™ process; the ionic liquids act as acid-scavenging agents during the production of alkoxyphenylphosphines. In contrast to the conventional use of triethylamine to synthesize alkoxyphenylphosphines, which results in the formation of solids, an ionic liquid phase is formed that can be more easily separated. Additionally, BASF is currently using ionic liquids as entrainer components for azeotropic distillation at the pilot scale.

Ionic liquids are also used as **performance additives** in commercial products. For example, polymers made by BASF, Evonik, and 3M exploit the antistatic properties of the ionic liquids. Antistatic fluids are used to clean sensitive and high-value surfaces and also to prevent the formation of solid residues in process spray nozzles (IoLiTec/Wandres). Clariant utilizes a SCILL catalyst (supported catalyst with ionic liquid layer) for the selective hydrogenation of downstream commercial products, such as 1,3-butadiene, propyne, and acetylene [33–35]. Here, the ionic liquid layer modifies the catalytic sites of a heterogeneous solid-supported catalyst in a ligand-like manner. The specific solubility of localized feedstocks and products in the ionic liquid film adjusts their concentrations at the active center causing a crucial modification to the reaction path. Additional performance additives are used for diverse applications at IoLiTec, including ionic liquids for CO_2 separation and protein stabilization at pilot scale. In the latter application, the solvent's unique properties allow for higher temperatures to be tolerated by the proteins, and thus, better crystallization can be achieved. IoLiTec also utilizes ionic liquids as chemical dispersants at

Table 11.1 Implemented ionic liquid applications by March 2019 (pilot or commercial)

No.	Ionic liquid isÖ	Company	Process / Application	Status
1	Activating solvent	Axens	Olefin dimerization (Difasol process)	Pilot
2	Activating solvent	BASF	Chlorination (nucleophilic HCl)	Comm.
3	Auxiliary	BASF	Acid scavenging (BASILô process)	Comm.
4	Auxiliary	BASF	Azeotrope cleavage by extractive distillation	Pilot
5	Catalyst	Chevron / honeywell UOP	Alkylation (ISOALKYô technology)	Pilot**
6	Catalyst	PetroChina	Alkylation (Ionikylation process)	Comm.
7	Electrolyte	C-Tech Innovation	Aluminum electroplating	Pilot
8	Electrolyte	G24 power	Dye sensitized solar cell	Comm.
9	Electrolyte	H.Glass	Dye sensitized solar cell	Pilot
10	Electrolyte	IoLiTec	Aluminum electroplating	Pilot
11	Electrolyte	IoLiTec	Electrochromic windows	Pilot
12	Electrolyte	IoLiTec	Sensor-electrolyte	Comm.
13	Electrolyte	Nante nergy	Zink-air battery	Comm.
14	Electrolyte	Nohms	Lithium ion battery	Comm.
15	Electrolyte	Novasina	Gas sensor	Comm.
16	Electrolyte	Panasonic	Supercapacitor	Comm.
17	Electrolyte	Pionics	Lithium ion battery	Pilot
18	Electrolyte	Scionix	Metal plating and polishing	Pilot
19	Electrolyte	Xtalic	Nanostructured aluminum-alloy plating (XTALIUMô)	Pilot*
20	Electrolyte	Zapgo	Supercapacitor (Carbon-Ionô technology)	Pilot*
21	Intermediate	Proionic	Carbonate based IL synthesis (CBILSÆ process)	Comm.
22	Lubricant	KI, ber lubrication	Electroconductive lubrication	Comm.
23	Lubricant	Proionic	Gas compression	Comm.
24	Operating fluid	BASF / evonik	Absorption cooling	Pilot
25	Operating fluid	Evonik	Dehumidification	Pilot
26	Operating fluid	Linde / flowserve	Gas compression ("Ionic Compressor")	Comm.
27	Operating fluid	Mettop / proionic	High temperature metallurgical cooling (ILTEC)	Comm.
28	Operating fluid	Proionic	Hydraulics for high temperature environment	Pilot
29	Operating fluid	Proionic	Open sorption air condition	Pilot
30	Performance additive	3M	Antistatic polymers	Comm
31	Performance additive	BASF	Antistatic polymers	Comm.
32	Performance additive	Clariant	Selective hydrogenation (SCILL Catalyst)	Comm.
33	Performance additive	Evonik	Antistatic polymers	Comm.
34	Performance additive	IoLiTec	CO2-separation	Pilot
35	Performance additive	IoLiTec	Dispersing agent	Comm.
36	Performance additive	IoLiTec	Optical brightener	Comm.
37	Performance additive	IoLiTec	Protein stabilization	Pilot
38	Performance additive	Wandres / IoLiTec	Cleaning fluid	Comm.
39	Process chemical	IoLiTec	Purification of materials for display technologies	Pilot
40	Reagent	Petronas	Mercury removal from natural gas (HycaPureô)	Comm.
41	Solvent	Arkema	Fluorination	Comm.
42	Solvent	BASF	Cellulose processing and reshaping	Pilot
43	Solvent	Chrysalix technologies	Softwood and agricultural waste deconstruction	Pilot
44	Solvent	Clariant	Water-gas shift reaction (SILP catalyst)	Pilot
45	Solvent	C-Tech Innovation	Recovering high-value polymers from waste	Pilot
46	Solvent	Evonik	Hydroformylation of C3/C4 feed (SILP catalyst)	Pilot
47	Solvent	IoLiTec	Synthesis of alcohols	Pilot
48	Solvent	IoLiTec	Synthesis of inorganic materials	Comm.
49	Solvent	Joint BioEnergy Institute	One pot pretreatment-saccharification switchgrass	Pilot
50	Solvent	Mari Signum	Chitin extraction from shell fish	Pilot
51	Solvent	Metsae spring	Staple fiber from wood-based paper pulp	Pilot*
52	Solvent	Natural fiber welding	Natural composites by fiber welding	Comm.
53	Solvent	Versum materials	Storage of hazardous gases (Gasguardô techn.)	Comm.
54	Solvent & additive	IoLiTec	Redox-flow-electrolyte	Pilot
55	Solvent & reagent	IoLiTec	Gas purification	Pilot
56	Stationary phase	Supelco	Gas chromatography column	Comm.
57	Visualization reagent	Hitachi high-tech	Scanning electron microscopy (HILEMÆIL 1000)	Comm.
* Commercial in 2019; **Commercial in 2020				

*Commercial in 2019; **Commercial in 2020

the commercial scale, as optical brighteners in display applications, and as **process chemicals** for the purification of electronic display materials.

Due to their high electrochemical stability, negligible vapor pressure and flammability, and tunable solvation properties for the dissolution of electrochemically active species, a large number of implemented ionic liquid applications involve **electrolytes**. Several companies use ionic liquids on the pilot scale for electroplating and electropolishing metals, such as aluminum (IoLiTec, Scionix, C-Tech Innovation). Another company, Xtalic, relies on ionic liquids for the production of nanostructured aluminum alloys and plans to commercialize the process in 2019. Many electrolyte applications are energy-related, such as dye-sensitized solar cells (G24 Power, H. Glass), lithium-ion batteries (NOHMs, Pionics), zinc–air batteries (NantEnergy), or supercapacitors (Panasonic, ZapGo). Other applications include electrolytes for gas sensors (IoLiTec, Novasina) and electrochromic windows (IoLiTec). Even though most of the applications are quite expensive due to the very high purity requirements and the fluorine chemistry costs of the anion, ionic liquids electrolytes are competitive due to the significantly increased performance and safety. The higher added value of the final industrial and consumer end products is justified.

Despite the hundreds of papers and patent applications in the field of ionic liquid **lubricants** that can be found in literature—many of them attributing excellent tribological properties—there are only two niche products on the market. Klüber Lubrication offers electro-conductive lubricants that prevent electro-erosion in roller bearings, and proionic produces a lubricant for high-pressure gas compression. Both applications are commercialized. This finding seems to reflect the competitiveness of the commercial lubricant sector, especially when considering state-of-the-art lubricants with specialized functional applications. Moreover, lubricant R&D is sophisticated and expensive, and many critical parameters are met largely by empiricism.

Six engineering applications have been implemented using ionic liquids as **operating fluids**. New working pairs based on ionic liquids and water or alcohols are operated by Evonik (under a BASF license) at the pilot level for absorption chillers or heat pumps. The fluids show less corrosion, lower toxicity, lower flammability, and no crystallization compared to conventional working pairs (e.g., based on LiBr and water). This technology was invented and originally developed by IoLiTec, and in 2014, a patent application was filed by BASF. Ionic liquids for ILDAC applications (ionic liquid desiccant air conditioning), with water being the refrigerant, were developed at proionic and were piloted in the multi-100 kg scale. This process places the ionic liquid in direct contact with breathable air and is, therefore, an open sorption process. In addition to similar advantages over conventional systems, as described above for absorption chillers, the desiccant air-conditioning system requires lowered energy input. A comparable ILDAC system was developed at the pilot stage by Evonik.

The first implementation of an engineering ionic liquid is the well-known "ionic compressor" invented by Linde and proionic in 2005 and commercialized by Linde and Flowserve in a joint venture. This compressor uses an ionic liquid piston and sealing systems for the high compression of, for example, hydrogen gas or natural gas for fueling stations and is one of the guiding examples for disruptive innovations based on the unique properties of ionic liquids. The combined properties of sparingly soluble compressed gases with very low compressibility, high thermal stability, and negligible vapor pressures of ionic liquids were key phenomena to realize this technology.

Another extraordinary industrial application of ionic liquids is their use in cooling high-temperature metallurgical devices, a technology developed jointly by proionic and Mettop and commercialized by Mettop as ILTEC Technology in 2014. Working together with the SMS Group, a world leader in metallurgical plant construction and mechanical engineering, the ILTEC process has been running successfully since 2016 at several industrial sites. Special ionic liquids are used as coolants and have negligible reactivity with molten metals and slags. The ionic liquids are far superior to the previous state-of-the-art cooling medium for metallurgical devices, water, which can cause catastrophic explosions if the cooling device becomes compromised. By developing these ionic liquid cooling media, it is now possible to redesign metallurgical aggregates (e.g., furnaces, lances, tap holes, or off-gas systems) and cross the heretofore forbidden line of cooling underneath the bath level of metal melts. It is also possible to retrofit ionic liquid cooling media into existing plants. In addition to their inherent safety, these ionic liquids operate at much higher temperatures as compared to water, making it possible to recover energy, for example, by Rankine cycle processes. With knowledge gathered during the development of high-temperature metallurgical cooling applications, proionic currently operates a pilot project of non-flammable hydraulic systems for the use in high-temperature environments.

Another major field is the utilization of ionic liquids as **solvents**; no less than 15 implemented applications can be found in Table 11.1. Arkema commercialized a process for the liquid fluorination of saturated or unsaturated compounds bearing C–Cl groups via a catalyzed reaction with hydrogen fluoride in certain ionic liquids. The water-gas shift reaction is active at the pilot stage at Clariant. By the use of a SILP catalyst (supported ionic liquid phase [36]), the reaction can be performed at "ultra-low" temperatures [37, 38], and trace amounts of carbon monoxide can be removed from the water gas. Another SILP-based application with superior performance is operated at Evonik at the pilot scale. The process uses ionic liquids for the hydroformylation of C3/C4 feedstock to yield the corresponding aldehydes [39]. Other chemical reactions with ionic liquids include the commercial synthesis of inorganic materials with defined particle size and a pilot-scale reaction for the synthesis of alcohols, both performed by IoLiTec.

The outstanding solvation properties of ionic liquids are used in the so-called HiPerPol process for the recovery of high-value polymers from waste. This recycling process is underway at the pilot scale at C-Tech Innovation and uses post-industrial composite packaging waste as feedstock; for example, the separation of heavy metal additives from recovered PVC produces high-quality polymer powders [40].

A remarkable property of highly coordinating ionic liquids (e.g., 1,3-dialkylimidazolium acetates, chlorides, or dialkylphosphates) is their ability to dissolve biomass by hydrogen-bond donor–acceptor interactions. Dissolution occurs rapidly, under very mild conditions, while maintaining the structural integrity of the biopolymers [41]. This attractive finding is reflected by approximately 1500 publications and 300 patent applications published to date [42] and a number of implemented applications. The U.S. Department of Energy's Joint BioEnergy Institute (JBEI), a scientific partnership led by the Lawrence Berkeley National Laboratory, currently pilots a one-pot, wash-free process for the ionic liquid pretreatment and saccharification of switchgrass, one of the leading potential biofuel feedstocks [43]. The novel process incorporates amino acid-based ionic liquids and realizes 85–95% glucose yields. This process represents a major step toward an integrated biorefinery in which sugars are extracted from biomass and directly converted into fuels in a single vessel.

A prominent patent by inventors Swatloski, Rogers, and Holbrey, held by The University of Alabama [44] and licensed by BASF, pioneered the technical processing of cellulose for fiber spinning. Metsä Spring, an innovation company of the Finnish Metsä Group, developed an ionic liquid-based process for spinning of staple fiber from wood-based paper pulp, a lower-quality feedstock. The Metsä process will be commercialized in 2019. Both applications described above dissolve biopolymers (e.g., cellulose) in the bulk phase and reshape the biopolymer by coagulation processes using anti-solvents (e.g., water). Due to the large volumes of biomass and ionic liquids used, these processes are very cost-sensitive, and, therefore, the recycling rate of ionic liquid is critical for commercial success. Typically, the concentration of biopolymer does not exceed 10–20 weight %.

Natural fiber welding (NFW) has a completely different approach. By partially dissolving or swelling naturally produced polymers (e.g., cotton, wool, or silk) and subsequently coagulating the fiber, the yarn product exhibits properties of natural and synthetic fibers. This revolutionary process can be used to produce novel composites (e.g., by welding different biopolymers together, modifying surface characteristics, and/or recycling textiles, like denim). The highly improved mechanical properties of the resulting yarn simplify the entire textile production process. Improved properties streamline weaving, knitting, warp knitting, or felting and have the potential to redefine portions of the textile economy. For more details on this process, see Chap. 9.

Another revolutionary process, operated by Mari Signum currently at the pilot scale, is the extraction of chitin from shellfish. Chitin is the planet's second most abundant polysaccharide, after cellulose, and is found in the exoskeletons of arthropods (e.g., crustaceans and insects) as well as in fungi. Though chitin is a valuable raw material, millions of tons of shrimp shell waste are produced globally per year and dumped in landfills. Chitin has unique properties that have the potential to replace traditional petrochemical plastics and to transform filtration, agriculture, animal feed, cosmetic, packaging, textile, and medical industries. Mari Signum has successfully implemented ionic liquids as effective, yet gentle solvents for chitin utilization at the technical scale and has thus become the vanguard for a new, chitin-based economy. For more details, see Chap. 4.

Chrysalix Technologies is a spin-out company from the Imperial College London developing an innovative biomass fractionation process for the supply of raw materials to produce biofuels and sustainable platform chemicals. Softwood and agricultural by-products are used as feedstocks for the replacement of petroleum. Low-cost protic ionic liquids feed the proprietary BioFlex process, which fractionates cellulose, hemicellulose, and lignin. By-products as well as heavy metals present in feedstock (e.g., construction wood waste) can be recovered.

Versum Materials, a spin-off of Air Products, applies ionic liquids for the storage of hazardous gases used in the electronic industry in their GASGUARD® technology. For example, BF_3, PH_3, or AsH_3 are complexed with ionic liquids, such as $[C_n\text{mim}][BF_4]$, $[C_n\text{mim}][Cu_2Cl_3]$, and $[C_n\text{mim}][Cu_2Br_3]$, by Lewis acid–base interactions using sub-atmospheric pressure. The technology circumvents the need for conventional steel cylinders and high-pressure storage conditions. The desired gases are released by applying mild temperatures or vacuum in a safe and controllable manner.

IoLiTec is currently piloting bromide-based ionic liquids as **solvents** and **additives** in a redox flow electrolyte for zinc–bromine batteries. Here, the bromide anion complexes bromine to form tribromide and reduces self-discharge and vapor pressure within the system [23]. IoLiTec is also testing ionic liquids as **solvents** and **reagents** for the purification of biogas in a pilot-scale project.

Supelco commercialized the use of ionic liquids as a **stationary phase** in gas chromatography columns showing high to extremely high polarity, unmatched thermal and chemical stability, low bleed, and unique selectivity (see Chap. 6). PETRONAS operates the first commercial application of the SILP technology [36] to remove mercury from natural gas. The chlorocuprate(II) ionic liquid **reagent** is immobilized on a solid silica support and effectively removes elemental, organic, and inorganic mercury with an expected service life up to three times greater than conventional sorbents. The technology was developed in cooperation with Clariant and QUILL and commercialized in 2014 [45] (see also Chap. 3). Hitachi High-Tech uses ionic liquids as **visualization reagents** for scanning electron microscopy (SEM). By coating samples—including biological specimens—with a thin layer of a hydrophilic ionic liquid, observation in the SEM proceeds efficiently without the need for prior preparation or fixation. Sample visualization is facilitated by the negligible vapor pressure, electrical conductivity, and polarity of the ionic liquids. The product was commercialized under the label HILEM® IL 1000.

Finally, proionic, one of the leading manufacturers of ionic liquids worldwide, commercialized its proprietary Carbonate Based Ionic Liquid Synthesis (CBILS®) in 2003. CBILS® is one of the most advanced commercialized processes for a "green" industrial production of high-purity ionic liquids, and proionic has reached an annual production capacity in the 100 metric ton range. This entirely halide-free, and virtually waste-free, production route [46, 47] uses carbonic acid esters as quaternization reagents forming ionic liquid key-**intermediates** with ammonium-, phosphonium-, imidazolium-, pyrrolidinium-, or morpholinium-type cations and methylcarbonate anions. The ionic liquid methylcarbonates are true platform chemicals, which can easily be converted into final products upon a facile hydrolytic reaction with Brønsted acids or ammonium salts. The conjugate base of the Brønsted acid or the ammonium salt anion becomes the anion of the ionic liquid final product. CBILS® chemistry overcomes most drawbacks associated with conventional ionic liquid syntheses and minimizes the use of noxious chemicals.

In 2008, M. Maase reported 13 implemented applications of ionic liquids [25]. Today's 57 implemented applications show a strong increase in commercialization efforts and offer optimism for future development of ionic liquid markets.

11.4 A Survey on Some Common Statements About Ionic Liquids

Despite the encouraging increase in commercial ionic liquid applications, there are still no mass applications, and producers of ionic liquids agree that today's market is considerably smaller than anticipated 15 years ago. In this context, statements like "ionic liquids are too expensive" or "at the end of the day, ionic liquids have disappointing performance" might be heard in discussions behind closed doors. Are these and other statements true, and do they explain the slow growth of the ionic liquid market? To shed light on public opinion, the author performed a survey among 25 academic and industrial leaders in the field of ionic liquids, asking them to rate the following statements on a Likert scale.

1. Today, commercially available ionic liquids fully meet the technical requirements of the industry.
2. Ionic liquids are much too expensive in general.
3. Regulatory requirements are getting more and more strict and expensive.
4. Today, there is no reliable supply of ionic liquids in larger ton scale.
5. The intellectual property rights landscape got very complicated in the last decades, and the large number of patents is a roadblock for the breakthrough of ionic liquids.
6. There is low willingness of enterprises to take risks, especially of large, stock-listed ones.
7. Material compatibility with ionic liquids is challenging.
8. "Hardware is struggling, software is surging": Until today, there was no sufficiently substantial investment in ionic liquid technology in order to "think big."
9. Ionic liquids have passed a hype and are getting mature now; as with any cross-technological breakthrough, this process takes a long time until substantial adoption.
10. There is only a low number of convincing, in particular disruptive value propositions.
11. Ionic liquids are too viscous for most applications.
12. Ionic liquids disappoint in their performance, and they were just overestimated.
13. Most ionic liquids are safer and greener than their molecular counterparts.
14. Though there is a lot of research in ionic liquids, the fraction of strategic, applied research still is way too small.

15. There is a growing number of successful implementations in industry today, but most are not visible to the public.
16. Ionic liquids are too difficult in handling, operation, and recycling.
17. Ionic liquids have the potential to revolutionize the world.

Figure 11.5 summarizes the results of the survey. Though more than 60% of the survey participants agreed that ionic liquids "meet the technical requirements of the industry" (statement 1), only 26% of participants totally or mostly agree with this statement. Industry has not yet fully defined appropriate technical specifications and standards (e.g., purity, electrochemical stability, rheology, and corrosion) to implement ionic liquids [48]. During discussions regarding technical requirements, several participants stated that the overall quality of commercial ionic liquids has increased over the years, but individual batch quality remains a fluctuating problem. Some participants believe that today's ionic liquids are kind of "old fashioned," and new structures are needed.

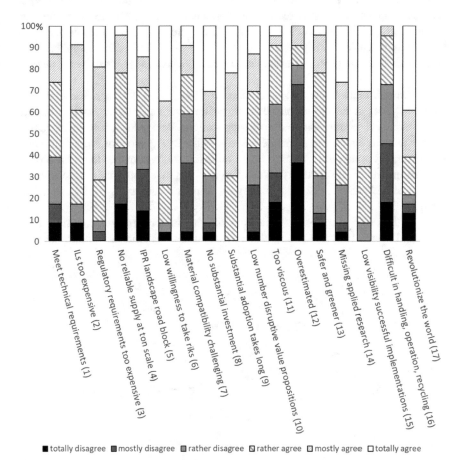

Fig. 11.5 Survey on some common ionic liquid statements

More than 80% of the participants believe that "ionic liquids are too expensive in general" (statement 2), but only 40% totally or mostly agree, and 9% totally disagree. This result reflects a broad distribution of opinions and is focused on the term "too expensive." Because chemical value-chain competition always depends on the intended application of the final product, pricing of intermediates or components frequently invites prejudice. However, it is important to recognize the broad range of both well-established and other commercially available specialty chemicals were once at a higher price point than ionic liquids are today. Currently, ionic liquids cannot compete with low-price commodity chemicals, and most probably never will, but economies of scale can dramatically lower price points. Figure 11.6 shows the price decrease with scaled production of two standard aprotic ionic liquids, both composed of the 1,3-dialkylimidazolium cation. Ionic liquid production at low costs and large volumes frequently causes a chicken-and-egg problem, which makes high-risk investments so important. In fact, the total cost of ownership should be considered when discussing chemical price points. The ionic liquid may be more expensive per kilogram, but a process could require less volume if it is more effective than the state of the art. Ionic liquids may have a longer life cycle, may lead to lower operation costs, may be easier to recycle, and so on. Or the ionic liquid can be applied as an additive or in a thin layer, for example, in a SILP design [36]—to show the desired effect.

In general, there is certainly a demand for lower cost, hydrophobic ionic liquids. However, there are current examples of hydrophobic ionic liquids that are economically competitive for particular applications. Persuading conservative industries requires a fundamental change made possible by (1) superior ionic liquid performance, (2) competitive pricing based on calculations showing the total cost of ownership, and (3) instinctively visualizing new business opportunities. In order to argue against long-established technologies, being too conservative contains risk as well!

More than 90% of the participants think that "regulatory requirements are getting more and more strict and expensive" (statement 3). Without any doubt, stronger

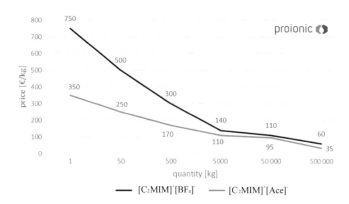

Fig. 11.6 Price trend of some ionic liquids produced with proionic's CBILS® process [46, 47]

chemical regulations, like REACH in Europe and TSCA in the USA, make our world safer [49, 50]; however, this safety comes at a price. For example, the typical REACH direct costs required to register new chemicals in the 1–10 metric tons per year (tpy) is at least 66,000 US dollars, increases to 330,000 US dollars when stepping up to the 10–100 tpy range, and at least another 1,100,000 US dollars for 100–1000 tpy. There are other significant costs that can arise from higher safety standards (e.g., packaging, transport, and storage), not only for ionic liquids, but for all new chemicals. These costs directly affect profit margins, especially for smaller companies.

Slightly more than the half of the participants agree that "there is no reliable supply of ionic liquids on the ton (or greater) scale" (statement 4). Opinions on this topic are well-distributed: 22% mostly or totally agree and 35% totally or mostly disagree. In reality, a reliable supply of ionic liquids at any commercial scale is available today. As with any other chemical on the market, when a quote is requested, lead times and possible regulatory affairs are considered, and delivery is guaranteed by a couple of established medium-sized manufacturers, like IoLiTec or proionic. The latter produce at two- to three-digit ton scale per year, and big players, like BASF, Evonik, or Merck, will increase their activities again as soon as demand emerges for significantly larger volumes.

The statement "The IPR landscape got very complicated in the last decades" (statement 5) resulted in an equal distribution of responses. Undoubtedly, there is a steadily increasing number of patent applications in the world [51] with average annual growth rates between 4 and 5% and reaching 25% in China. In the early days of the "ionic liquid gold rush," a large number of patent applications with broad claims and questionable quality emerged. This led to decreased activity in the field. Many of these patent applications took a long time before they were dropped, but some are still active. This problem will soon solve itself due to patent expiration. In the meantime, it seems to the author that patent office reviewers have become more critical in their evaluation of ionic liquid patents, thus forcing applicants to be more specific in their claims.

"Is there low willingness of large, especially stock-listed enterprises to take risks" (statement 6)? More than 90% of the participants agree with this statement and almost 75% mostly to totally agree, which aligns with the personal experience of the author. This is clearly the result of a shareholder-value economy driven by large corporations, whereby it is difficult to enforce risky strategic decisions against the will of the shareholders. Therefore, it is expected that major future breakthroughs in the field of ionic liquids will come from small companies.

Almost 60% of the participants disagree that "material compatibility with ionic liquids is challenging" (statement 7), but 40% are not completely convinced. Material compatibility of new compounds has always been a challenge for the chemical industry; for example, general costs of corrosion are 3–4% of the GDP of industrialized countries per year, which equals about 2.5 trillion US dollars globally [52]. Ionic liquids represent a special case; they are a large class of new liquid salts, with very different chemistries than molecular solvents and salt solutions. In this respect, all direct corrosion comparisons between ionic liquids and inorganic salts fall short and are not meaningful at all. There exists a practically endless number of combinations of cations and anions to make different ionic liquids for a variety of product and process parameters. The published knowledge we have today is insufficient, and there remains a lot of work to be done in the future for applied researchers. However, to the best knowledge of the author including unpublished results from our laboratory, there are already a large number of standard ionic liquids and materials that show excellent general compatibility. While all applications will always be evaluated on a case-by-case basis, solutions have been determined for the majority of material compatibility problems to date.

"Hardware is struggling; software is surging": Until today, there was no sufficiently substantial investment in ionic liquid technology in order to "think big" (statement 8). Only 8.7% of the participants totally or mostly disagree with this statement, about 70% agree, and more than 50% totally or mostly agree. This supports the opinion of the author; not nearly enough investment has been made in ionic liquids. In order to fully exploit the potential of ionic liquids (e.g., in giving the world access to biomass in a way that it has never been accessed before, committing to total substitution of fossil oil, and solving the pressing problem of replacing non-degradable plastics from the planet [53, 54]), more strategic investment must be made in promising ionic liquid technologies.

Of course, this is not meant to undervalue the millions of dollars that have been invested in ionic liquids to date. However, these investments have supported single applications and not mass applications of globally strategic importance. To achieve the latter, investors and governments might have to invest hundreds of millions of dollars to really move forward. While 100 million U.S. dollars seems to be a lot of money, it is on par with the amount earned by companies like Apple, Microsoft, Shell, or Volkswagen within a few hours [55] and equals only 6 ppm (0.0006%) of the GDP of the European Union [56]. Instead of smartphone apps, this planet needs more investment in hardware!

"Ionic liquids have passed a hype and are getting mature now; as with any cross-technological breakthrough, this process takes a long time until substantial adoption" (statement 9). As to be expected, 100% of the participants agreed with this statement and almost 70% agreed mostly to totally. This indicates that the change in expectations from unrealistic to realistic has taken place over the last decade. The party is over; let's do the real work now!

The evaluation of statement 10: "There are only a low number of convincing, in particular disruptive value propositions" was surprisingly distributed among all possible ratings with 56% agreement and 44% disagreement. On the one hand, it is the nature of disruptive technologies to be rare. On the other hand, the ratings show that almost half of the participants think that there is still a lot of remaining potential in the field of ionic liquids.

It frequently can be heard that "ionic liquids are too viscous" (statement 11). While 64% of the participants disagree with this, only 32% disagree totally to mostly, and 60% of participants rather disagree or rather agree. It seems the statement is understood by participants as being oversimplified, but there is a grain of truth in it. It is a fact that compared to classical organic solvents, viscosities of ionic liquids are 1–3 orders of magnitude higher. As such, applications which depend on these low viscosities will prevent the successful use of ionic liquids. In practice, these cases are rather rare, and as with material compatibility, there are numerous options to find a solution or workaround. For example, viscosity can be most often decreased by the addition of sometimes small amounts of co-solvents, increasing temperature, making eutectic mixtures of ionic liquids, changing process parameters, or process design in order to increase mass transport. The latter can be achieved by increasing interfacial surface area by high-shear mixing, ultrasonication, use of continuous-flow reactors, coating of porous supports with the ionic liquid, and so on.

Only 18% of the participants mostly or rather agree, and none totally agree that "ionic liquids disappoint in their performance and were just overestimated" (statement 12). This result again indicates that the overhyped expectations have been overcome and are understood nowadays in a historical context and from a much more mature perspective. It is well-accepted today that ionic liquids are a key-enabling technology for a broad range of application fields.

Are "most ionic liquids safer and greener than their molecular counterparts" (statement 13)? Although 70% of the participants agreed with this statement, approximately 50% rather agreed. There is a trend, but not a strong consent. The marketing of ionic liquids in the early days as being "green solvents" was another overgeneralization, which probably has done more harm than good to the field [57]. Meghna Dilip writes in a paper [58] that: "claims of 'green-ness'…rest mainly on [ionic liquids] ability to serve as non-volatile and non-flammable alternatives to traditional molecular solvents … it is increasingly apparent that no single property other than melting point (inherent in the definition of this class of compounds) can be used to describe the entire class of compounds …. one can no longer classify them unequivocally as green but one can definitely not dismiss ionic liquids as un-green." Based on, for example, life-cycle analysis and application, a particular ionic liquid could be called "green" in one application but determined to be "un-green" in another. Clearly, there is a number of sustainable energy applications (e.g., energy storage, CO_2 capture, and biomass utilization) where ionic liquids play a key-enabling role. Many molecular fluids used every day, like gasoline fuel or engine oil, are classified as substantially more dangerous and environmentally harmful than most ionic liquids, for example, after GHS-CLP (United Nations' Globally Harmonized System of Classification and Labelling of Chemicals).

"The fraction of strategic, applied research still is way too small" (statement 14) is agreed upon by 74% of the participants. Here, 52% mostly to totally agree, and only 9% mostly or totally disagree. The point emerging from this response aligns with the impression one gathers from visiting diverse international conferences year after year. Most of the presentations have been, and still are, very focused on basic research. While basic research is critically valuable and important for the field, scientists can miss the chance to apply their work, sometimes even when the application is clearly evident. Understandably, dialkylimidazolium cations and bistriflamide anions are often the best candidates for studying physicochemical properties of ionic liquids, but it is of great strategic importance to work on new ionic liquid structures (e.g., those that are cheaper and greener).

There is a lot of potential in simple quaternary ammonium-based, protic ionic liquids and ionic liquid mixtures. Any of which could be an economical game changer for a multitude of applications. For all structures, there is need for much more fundamental data on material compatibility, thermal and chemical long-term stability under application conditions, toxicity of degradation products, recycling and disposal, product and by-product separation, safety, health, environmental impact, and so on. In order to move forward to mass applications of ionic liquids, a substantial increase in strategic applied research funding would better enable the industry to succeed, and the money earned could be reinvested in third-party funds needed by academia. To address those specific problems, the industry could reciprocate through a more open approach to academia.

More than 90% of the participants agree that "there is a growing number of successful implementations in industry today, but most are not visible to the public" (statement 15). Everyone working in the field of commercial ionic liquids knows this to be true to some degree, yet very difficult to overcome. Regardless of the IPR situation, public visibility for the field of ionic liquids would be of great value and could provide indirect returns to the industry.

"Ionic liquids are too difficult in handling, operation, and recycling" (statement 16). Only 4.5% of the participants mostly or totally agree with this statement, while 73% disagree. This statement is understood by the participants to be in line with the overgeneralizations, like ionic liquids being too viscous or incompatible with materials. Again, if problems arise there are many options to find a solution or workaround to succeed.

"Do ionic liquids have the potential to revolutionize the world" (statement 17)? This rather pointed statement was surprisingly agreed upon by almost 80% of the participants, with 22% mostly and 39% totally agreeing. Only 17% mostly or totally disagreed. Obviously, any opinion about this statement will be based upon more personal belief than hard facts, but there are some strong indications. How ionic liquids may have significant influence on future technologies is of global importance. For example, the unique solvation properties of ionic liquids when applied to biomass have never before seen results when compared to traditional solvents. These solvation phenomena have the potential to impact megatrends in neo-ecology (e.g., green tech, zero emission, reuse–reduce–recycle, urban mining, post-carbon society, and smart buildings) and health (e.g., drug delivery, drug formulation, drug extraction, and artificial skin). The supreme electrochemical properties combined with safety aspects, like non-flammability and impact on neo-ecology (e.g., sustainable energy, e-mobility, energy grids, and smart buildings). In addition to this, the negligible volatility of ionic liquids and the degrees of freedom allowing the introduction of several functionalities in parallel via modification and combination of cations and anions should not be underestimated.

11.5 The Future

Though the market for ionic liquids was well below the expectations in the last decade, there is an increasing number of encouraging trends. Most impressive is the 57 implemented ionic liquid applications known to date, which is a clear improvement compared to 13 in 2008. Implemented applications with a potential for mass applications are now a reality, and convincing examples can be found in this book. Market studies by renowned consultancies estimate the market penetration in the range of 1–2% of the accessible market; this highlights the huge growth potential for future industrial applications—including mass applications. The continuously growing number of annual patent applications, with a compound annual growth rate of 21% over the last 15 years, is another strong key indicator. This is especially impressive when considering that these patent applications cover a broad range of cross-sectional fields. A recent paper was published by patent attorneys of Mathys & Squire LLP, London, [59] entitled "Ionic liquids—The beginning of the end or the end of the beginning? A look at the life of ionic liquids through patent claims." The authors reached the same conclusion as this chapter's author; there will be an optimistic future for ionic liquids, and "this is very much the end of the beginning regarding ionic liquid technologies, and not the beginning of the end."

In total, the publication data and survey data clearly reveal that ionic liquids will continue to cross the frontier from laboratory experiments to implemented technologies. Without doubt, ambitious developments will lead to mass applications, resulting in annual value creation of multibillion US dollars. The toolbox of ionic liquids is established and ready, what remains lies in the power of imagination, inventive genius, committed investors, public awareness, and open-minded industries to show that the best is yet to come for the commercialization of ionic liquids.

Acknowledgements The author is grateful to the participants of the survey for their support and to Dr. Aaron Socha (Queens University of Charlotte) for his proofreading. He is also grateful to colleagues from industry and academia for helpful comments during the extensive research for ionic liquid implementations: Dr. Uwe Vagt (BASF SE), Dr. Thomas Schubert (IoLiTec GmbH), Dr. Hye Kyung Timken (Chevron), PD Dr. Marco Haumann (Friedrich-Alexander University Erlangen-Nürnberg), and Prof. Dr. James H. Davis, Jr. (University of South Alabama).

References

1. Qiu Z (1993) Molten salt and slag in ancient China. Trans Nonferrous Met Soc China 3(1):91–94. ISSN: 1003-6326
2. Hatt BW, Kerridge DH (1979) Industrial applications of molten salts. Chem Brit 15(2):78–79. ISSN: 0009-3106
3. Gaune-Escard M, Haarberg GM (eds) (2014) Molten salts chemistry and technology, 1st ed. Wiley, New York. https://doi.org/10.1002/9781118448847
4. Walden P (1914) Molecular weights and electrical conductivity of several fused salts. Bull Imp Acad Sci St-Pétersbourg 8:405–442. ISSN: 0366-3914

5. Bockris JO'M, McKenzie JD, Kitchener JA (1955) Viscous flow in silica and binary liquid silicates, Trans Faraday Soc 51:1734–1748. https://doi.org/10.1039/tf9555101734
6. SciFinder™ Database, concept-search (accessed 2019-01-08)
7. Gartner, Inc., Stamford CT, USA. https://www.gartner.com/en/research/methodologies/gartner-hype-cycle. Accessed 2019-01-08
8. ©Jeremy Kemp at English Wikipedia, CC BY-SA 3.0, https://commons.wikimedia.org/w/index.php?curid=10547051. Accessed 2019-01-08
9. Ratcliffe S (ed) (2016) Roy Amara 1925–2007, American futurologist. In: Oxford essential quotations, 4th ed. Oxford University Press. https://doi.org/10.1093/acref/9780191826719.001.0001
10. Frietsch R, Kladroba A, Markianidon P, Neuhäusler P, Peter V, Ravet J, Rothengatter O, Schneider J (2017) Final report on the collection of patents and business indicators by economic sector: societal grand challenges and key enabling technologies. Publications Office of the European Union, Luxembourg. https://doi.org/10.2760/39818
11. Hullmann A (2006) The economic development of nanotechnology—an indicator based analysis. European Commission, DG Research, Unit "Nano S&T—convergent science and technologies. http://cordis.europa.eu/nanotechnology
12. Gartner Inc., Stamford CT, USA, IHS Inc., Englewood CO, USA. Annual market trends semiconductor industry. https://www.gartner.com, https://www.ihs.com. Accessed 2014-04-19
13. Roco MC (2011) The long view of nanotechnology development: the national nanotechnology initiative at 10 years. Nanopart Res 13:427–445. https://doi.org/10.1007/s11051-010-0192-z
14. Grand View Research, Inc., San Francisco CA, USA. http://www.grandviewresearch.com/press-release/PTFE-Market. Accessed 2014-05-14
15. BCC Research LLC, Wellesley MA, USA (2012) Global markets for laser systems, components and materials. Report PHO002A. https://www.bccresearch.com/market-research/photonics/laser-markets-systems-components-materials-pho002a.html. Accessed 2014-04-19
16. Helmut Kaiser Consultancy (2012) Ionic liquids 2030—ionic liquid technologies, markets, applications, companies and developments worldwide 2008 to 2020 and 2030. Tübingen, Germany. http://www.hkc22.com/ionicliquids.html. Accessed 2015-05-02
17. Frost & Sullivan, Mountain View CA, USA (2016) Technology advancements in ionic liquids. Demand across industries being driven by sustainability initiatives, industrial solvents to be major application area, Tech vision report D6E-01-00-00-00
18. Grand View Research, San Francisco CA, USA (2016) Ionic liquids market size and forecast by application (solvents & catalysts, extractions & separations, bio-refineries, energy storage), by region (North America, Europe, Asia Pacific, Latin America and Middle East & Africa) and trend analysis from 2018 to 2025. report ID: GVR-1-68038-283-9. https://www.grandviewresearch.com/press-release/global-ionic-liquids-market. Accessed 2019-01-02
19. Markets and Markets Research, Hadapsar, Pune, Maharashtra, India (2016) Ionic liquids market by application (solvents & catalyst, process & operating fluids, plastics, batteries & electrochemistry, bio-refineries) and by region (North America, Europe, Asia-Pacific and rest of world)—global forecast to 2021. Report Code: CH 4297. https://www.marketsandmarkets.com/Market-Reports/ionic-liquid-market-163716481.html. Accessed 2019-01-02
20. Sources: Personal communication with authorized managers and validated, publicly accessible documents of listed companies. Dec 2018–Mar 2019
21. Bailey MP (2015) Ionic liquids create more sustainable processes. Chem Eng, October 1, 2015, pp 18–24. ISSN: 0009-2460
22. Plechkova NV, Seddon KR (2008) Applications of ionic liquids in the chemical industry. Chem Soc Rev 37:123–150. https://doi.org/10.1039/B006677J
23. Schubert TJS (2017) Current and future ionic liquid markets. In: Shiflett MB, Scurto AM (eds) Ionic liquids: current state and future directions, ACS Symposium Series 1250. American Chemical Society, Washington DC, pp 35–65. https://doi.org/10.1021/bk-2017-1250.ch003
24. Durga G, Mishra A (2016) Ionic liquids: industrial applications. Encycl Inorg Bioinorg Chem, 1–13. https://doi.org/10.1002/9781119951438.eibc2434

25. Maase M (2008) Industrial applications of ionic liquids. In: Wasserscheid P, Welton T (eds) Ionic liquids in synthesis, 2nd edn. Wiley-VCH, Weinheim, pp 663–687. https://doi.org/10. 1002/9783527621194.ch9
26. Rogers RD, Seddon KR, Volkov S (eds) (2002) Green industrial applications of ionic liquids. NATO science series, vol 92. Kluwer Academic Publishers, Dordrecht, Boston, London. ISBN 978-1-4020-1137-5
27. Holbrey JD (2004) Industrial applications of ionic liquids. Chim Oggi 22(6):35–37. ISSN: 0392-839X
28. Meksi N, Moussa A (2017) A review of progress in the ecological application of ionic liquids in textile processes. J Clean Prod 161:105–126. https://doi.org/10.1016/j.jclepro.2017.05.066
29. Toledo Hijo AAC, Maximo GJ, Costa MC, Batista EAC, Meirelles AJA (2016) Applications of ionic liquids in the food and bioproducts industries. ACS Sustain Chem Eng 4(10):5347–5369. https://doi.org/10.1021/acssuschemeng.6b00560
30. Singh NR, Speight JG (2011) Applications of ionic liquids in industry. Chem Technol Indian J 6(2):114–122. ISSN: 0974-7443
31. Han S, Li J, Zhu S, Chen R, Wu Y, Zhang X, Yu Z (2009) Potential applications of ionic liquids in wood related industries. BioResources 4(2):825–834. ISSN: 1930-2126
32. Vallée C, Olivier-Bourbigou H (2005) Industrial production of ionic liquids. In: Cornils B, Herrmann WA, Horváth IT, Leitner W, Mecking S, Olivier-Bourbigou H, Vogt D (eds) Multiphase homogeneous catalysis. Wiley-VCH, Weinheim, pp 581–587. ISBN: 978-3-527-30721-0
33. Barth T, Korth W, Jess A (2017) Selectivity-enhancing effect of a SCILL catalyst in butadiene hydrogenation. Chem Eng Technol 40(2):395–404. https://doi.org/10.1002/ceat.201600140
34. Barth T (2016) Selektivhydrierung von 1,3-Butadien an mit ionischen Fluiden beschichteten heterogenen Katalysatoren, Doctoral thesis, University of Bayreuth, http://nbn-resolving.org/ urn:nbn:de:bvb:703-epub-2927-6
35. Szesni N, Fischer R, Hagemeyer A, Großmann F, Hou HC, Boyer J, Sun M, Urbancic M, Lugmair C, Lowe DM (2013) Catalyst composition for selective hydrogenation with improved characteristics, WO2013057244
36. Supported Ionic Liquid Phase (SILP) Technology. http://www.silp-technology.de. Accessed 2019-01-17
37. Werner S, Haumann M (2014) Ultralow temperature water-gas shift reaction enabled by supported ionic liquid phase catalysts. In: Fehrmann R, Riisager A, Haumann M (eds) Supported ionic liquids. Wiley-VCH, Weinheim, pp 327–350. https://doi.org/10.1002/9783527654789. ch16
38. Werner S (2011), Ultra-low temperature water-gas shift reaction with supported ionic liquid phase (SILP) catalysts. Doctoral thesis, Friedrich-Alexander University Erlangen-Nürnberg (FAU). http://nbn-resolving.de/urn:nbn:de:bvb:29-opus-26028
39. Franke R, Hahn H (2014) A catalyst that goes to its limits. PharmaChem 14(11/12):2–5. ISSN: 1720-4003
40. C-Tech Innovation Ltd., Chester, United Kingdom; personal communication. See also HiPerPol project 2011, UK Research and Innovation, https://gtr.ukri.org/projects?ref=100009. Accessed 2019-01-17
41. Wang H, Gurau G, Rogers RD (2014) Dissolution of biomass using ionic liquids. In: Zhang S, Wang J, Lu X, Zhou Q (eds) Structures and interactions of ionic liquids. Structure and bonding 151. Springer, Berlin, Heidelberg, pp 79–105. https://doi.org/10.1007/978-3-642-38619-0_3
42. SciFinder® concept search using term "ionic liquids biomass"
43. Shi J, Gladden JM, Sathitsuksanoh N, Kambam P, Sandoval L, Mitra D, Zhang S, George A, Singer SW, Simmons BA, Singh S (2013) One-pot ionic liquid pretreatment and saccharification of switchgrass. Green Chem 15(9):2579–2589. https://doi.org/10.1039/c3gc40545a
44. Swatloski RP, Rogers RD, Holbrey JD (2003) Dissolution and processing of cellulose using ionic liquids, cellulose solution, and regenerating cellulose. WO2003029329
45. Abai M, Atkins MP, Hassan A, Holbrey JD, Kuah Y, Nockemann P, Oliferenko AA, Plechkova NV, Rafeen S, Rahman AA, Ramli R, Shariff SM, Seddon KR, Srinivasan G, Zoub Y (2015) An ionic liquid process for mercury removal from natural gas. Dalton Trans 44(18):8617–8624. https://doi.org/10.1039/C4DT03273J

46. Kalb RS, Stepurko EN, Emel'yanenko VN, Verevkin SP (2016) Carbonate based ionic liquid synthesis (CBILS®): thermodynamic analysis. Phys Chem Chem Phys 18(46):31904–31913. https://doi.org/10.1039/C6CP06594E
47. Kalb RS, Damm M, Verevkin SP (2017) Carbonate Based Ionic Liquid Synthesis (CBILS®): Development of the continuous flow method for preparation of ultra-pure ionic liquids. React Chem Eng 2(4):432–436. https://doi.org/10.1039/C7RE00028F
48. Kalb RS (2018) What is purity of ionic liquids? A definition from a real-world perspective. In: Santos F, Lourenço MJ, Nieto de Castro C (eds) Abstracts of the 27th EuChemS conference on molten salts and ionic liquids, Lisbon, Portugal 7–12 Oct 2018, p 19. ISBN: 978-972-96653-7-0
49. https://echa.europa.eu/regulations/reach/understanding-reach. Accessed 2019-01-22
50. https://www.epa.gov/tsca-inventory. Accessed 2019-01-22
51. http://www.fiveipoffices.org/statistics.html. Accessed 2019-01-22
52. http://corrosion.org/. Accessed 2019-01-22
53. Unlocking the Chitin Economy, Mari Signum, Mid Atlantic LLC, Richmond VA, USA. https://www.marisignum.com. Accessed 2019-01-24
54. U.S. Department of Energy, Bioenergy Technologies Office. Replacing the whole barrel. To reduce U.S. dependence on oil. Report DOE/EE-0920, July 2013
55. https://money.cnn.com/magazines/fortune/global500/2012/full_list/. Accessed 2019-01-22
56. https://www.imf.org. Accessed 2019-01-22
57. Cvjetko Bubalo M, Radošević K, Radojčić Redovniković I, Halambek J, Gaurina Srček V (2014) A brief overview of the potential environmental hazards of ionic liquids. Ecotoxicol Environ Saf 99:1–12. https://doi.org/10.1016/j.ecoenv.2013.10.019
58. Dilip M (2012) Cradle to grave: how green are ionic liquids? Nanomater Energy 1(4):193–206. https://doi.org/10.1680/nme.12.00009
59. Morton MD, Hamer CK (2018) Ionic liquids—the beginning of the end or the end of the beginning? A look at the life of ionic liquids through patent claims. Sep Purif Technol 196:3–9. https://doi.org/10.1016/j.seppur.2017.11.023

Author Index

© Springer Nature Switzerland AG 2020
M. B. Shiflett (ed.), *Commercial Applications of Ionic Liquids*, Green Chemistry
and Sustainable Technology, https://doi.org/10.1007/978-3-030-35245-5

283

Subject Index

© Springer Nature Switzerland AG 2020
M. B. Shiflett (ed.), *Commercial Applications of Ionic Liquids*, Green Chemistry and Sustainable Technology, https://doi.org/10.1007/978-3-030-35245-5

Printed in the United States
By Bookmasters